D0605290

Photoelastic Stress Analysis

Photoelastic Stress Analysis

Albrecht Kuske

Professor and Director
Institut für Maschinenwesen
Technische Universität, Clausthal

and

George Robertson

Principal Lecturer in Mechanical Engineering
Dundee College of Technology

A Wiley—Interscience Publication

JOHN WILEY & SONS

London · New York · Sydney · Toronto

Library of Congress Catalog card No. 73-2788

ISBN 0 471 51101 3

Printed in Great Britain by
J. W. Arrowsmith Ltd.
Winterstoke Road, Bristol BS3 2NT

PREFACE

The aim of this book is to present the photoelastic method of stress analysis—in which field the authors have a combined experience of some 60 years—in its contemporary state as comprehensively as possible within the limitations imposed by practical considerations.

It is true that, where computer facilities are available, certain problems of stress analysis which might formerly have been investigated by photoelasticity may now be more expeditiously solved numerically by finite element procedures. On the other hand the availability of instruments such as the laser, the photomultiplier and the computer itself, together with the development of new methods and more sophisticated techniques have greatly increased the range of problems for which photoelasticity is the most viable means of investigation. The net result is that, instead of diminishing, the importance of the photoelastic method of stress analysis as an aid to design has continued to increase.

The authors have endeavoured to present the subject in such a manner that the book will meet the requirements of the beginner as well as being of assistance to the experienced research worker.

For this reason, the earlier chapters are of an introductory character. The first two chapters provide an introduction to the theoretical analysis of stress and strain. The treatment of more specialized theory such as that relating to the bending of plates and dynamic problems, which may not be of immediate interest to all readers, is deferred till the chapters devoted to these topics. The following two chapters provide an introduction to elementary optics, including the transmission of light through crystals. It is the opinion of the authors that a sound knowledge of basic optics is an invaluable aid towards the correct application of the principles involved and the recognition and avoidance of errors which may arise from faulty technique or equipment. For this reason this subject is dealt with in rather more detail than is customary in books on photoelasticity. The remaining sixteen chapters deal with the theory and practice of modern photoelasticity.

A special feature of the book is its presentation of the *j*-circle method of analysis. This concept, the general theoretical basis of which is developed in Chapter 13, simplifies the derivation of optical equations even in complicated problems and at the same time provides a model by means of which the various optical phenomena can readily be visualized. The method may be applied to the solution of certain general three-dimensional problems, such as the analysis of shells and folded plates described in detail in Chapter 17. The large amount of optical data required for this particular application was obtained by computer and is recorded in eighty charts. It was hoped by the authors that these charts would be reproduced in this book but this was found to be impracticable because of production difficulties. They have been published separately, however, and are available from Kommisionsverlag Hubert Hövelborn, 5216 Niederkassel-Mondorf, Provinzialallee 25, West Germany. Applications of the *j*-circle method to the frozen stress and scattered light techniques are contained in Chapters 15 and 16 respectively.

Other chapters of the book which it is hoped will be of special interest are those on Interferometric Methods (Chapter 10), Plates under Transverse Bending (Chapter 12), Scattered Light Method (Chapter 16), Photothermoelasticity (Chapter 18) and Dynamic Photoelasticity (Chapter 19), in each of which the topic is treated in a more comprehensive manner than in previous books on photoelasticity.

The authors take this opportunity of expressing their appreciation of the laboratory assistance of Dr.-Ing. R. Harlfinger, Dr.-Ing. R. Kayser, Dr.-Ing. B. Nill, Dr.-Ing. G. Sembritzki and Dr.-Ing. G. Steinhart.

The authors also wish to express their gratitude to those investigators in various countries who willingly provided details or examples of special interest of their work and to whom individual acknowledgement has been made in the text.

Finally, the authors would like to acknowledge their indebtedness to Deutsche Forschungsgemeinschaft who sponsored part of the research work described and to the following bodies for their kind permission to reproduce material which previously appeared in their publications: American Institute of Aeronautics and Astronautics, Butterworths & Co. (Publishers) Limited, Eidgenossische Technische Hochschule (Zurich), Institution of Mechanical Engineers, Institute of Physical and Chemical Research (Tokyo), Society for Experimental Stress Analysis, and Verein Deutscher Ingenieure.

ALBRECHT KUSKE
GEORGE ROBERTSON

CONTENTS

1

THE ANALYSIS OF STRESS

1.1 Force and stress. Differences in character

External forces which may act on a body can be classified into two groups, namely, surface forces and body forces. Surface forces are produced by physical contact of the body with other bodies, including fluids, and act on the external surfaces only. Body forces act directly on every particle within the body. Common examples of body forces are those produced by gravitational and inertia effects.

When a body is subjected to the actions of external forces, the effects are transmitted through the material. Internal forces are thus induced with the result that the material on one side of any section of the body will in general exert a force on the material on the other side. The mean or average stress σ_m due to a resultant force F acting over a section of area A is defined $\sigma_m = F/A$. The intensity of the internal forces usually varies from point to point over the section and to define the stress at a point we consider any small elementary area δA of the section and let δF be the force transmitted across it (Figure 1.1). The limiting value of the ratio $\delta F/\delta A$ as δA approaches zero about a point defines the resultant stress at the point. In general, δF is inclined to δA and can be resolved into components δN normal to the section and δS in the plane of the section. The limiting values of $\delta N/\delta A$ and $\delta S/\delta A$ define the normal and shear stress components which will be denoted by σ and τ respectively. The direction of the shear force δS in the plane of the section depends on the direction of δF and it is usually convenient to replace δS by its components δS_1 and δS_2 parallel to any two perpendicular axes ox, oy in the plane. The limiting values of $\delta S_1/\delta A$ and $\delta S_2/\delta A$ as δA tends towards zero are the component shear stresses in the directions of these axes and will be denoted by τ_{zx} and τ_{zy} respectively. In this notation the first subscript indicates by the direction on its normal the plane on which the shear stress acts while the second indicates the direction of the shear stress in that plane. The normal stress requires only a single subscript which indicates the direction in which it acts. In Figure 1.1 the axis oz is normal to the plane of δA and parallel to the normal stress which is accordingly represented by σ_z.

1

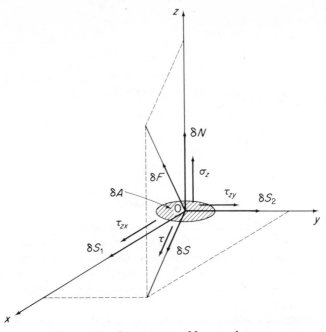

Figure 1.1.   Components of force and stress

The three values σ_z, τ_{zx} and τ_{zy} do not completely define the state of stress at the point considered since they obviously depend on the direction of the plane of the section which can be chosen arbitrarily. Thus, while several forces acting at a point can be added vectorially and represented by a single vector, the state of stress requires for its complete specification the stress components with reference to three different planes passing through the point. If these planes are mutually perpendicular and normal to the x, y, z axes we thus obtain three values of the normal stresses, $\sigma_x, \sigma_y, \sigma_z$ and six values of the shear stresses, i.e. $\tau_{xy}, \tau_{xz}, \tau_{yz}, \tau_{yx}, \tau_{zx}, \tau_{zy}$. These are known as the nine components of stress and are illustrated with reference to an infinitesimal rectangular element about a point in Figure 1.2. The shear stress components are not all independent however, as we can show by taking moments of the forces acting on the element about each of the co-ordinate axes in turn.

The effects of body forces or of variation in the stresses across the faces of the element can here be neglected since the moments due to these are small quantities of higher order than those due to the shear stresses and therefore ultimately vanish. In this way we obtain

$$\tau_{xy} = \tau_{yx}, \qquad \tau_{yz} = \tau_{zy}, \qquad \tau_{zx} = \tau_{xz},$$

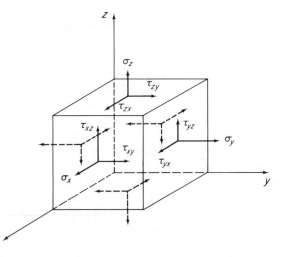

Figure 1.2.   Stress components in rectangular co-ordinates
on an infinitesimal element

i.e. the component shear stresses normal to the line of intersection of two mutually perpendicular planes through the point are equal. Thus we see that while a force can be completely specified by three values such as its magnitude and two angles indicating its direction, a state of stress requires six independent values.

The character of a state of stress becomes more evident when we consider the deformations or strains produced. A simple normal or direct stress produces a certain elongation or contraction of the material. This is accompanied by a lateral deformation which we need not take into account in our present consideration. We begin by considering a spherical element in an unstressed body. If the body is now subjected to a uniaxial stress, say tensile, all chords of the sphere will be elongated by amounts proportional to their original length. In this way the sphere is transformed into an ellipsoid. If we now imagine a further stress to be applied in any other direction, the chords of the ellipsoid are deformed in a similar way and the element will be transformed into another ellipsoid of different shape and this will apply no matter what the state of stress on the original elementary sphere may be. The shape of the ellipsoid characterizes the state of stress on the element and can be specified by the independent lengths of the three mutually perpendicular principal axes and their directions.

In the directions of the axes of the ellipsoid, the resulting displacements of points on the surface of the sphere are purely radial while those of all other points are partly radial and partly tangential. Since radial displacements result from normal stresses and tangential displacements from shear stresses, it follows that at points in the directions of the principal axes the stresses are purely normal while at all other points both normal and shear stresses are produced. If we now imagine the sphere to be indefinitely reduced in dimension and concentrated about a point, these arguments will apply for different directions through the point. Thus we see that at every point in a stressed material no matter what the state of stress may be, there exist three mutually perpendicular directions in which the stress is purely normal. These normal stresses are called principal stresses and will be denoted by $\sigma_1, \sigma_2, \sigma_3$. One of the principal stresses is algebraically the greatest and another the smallest in any direction at the point. The remaining principal stress is a minimum in a plane containing it and the greatest principal stress and a maximum in a plane containing it and the smallest principal stress and thus corresponds to what is known as a saddle point. The three principal stresses together with three angles indicating their directions completely specify the state of stress at a point.

The planes normal to the principal stresses are called principal planes and may be defined as planes on which the shear stress is zero.

The stresses and deformations produced within any chosen plane, e.g. the xy plane, passing through a point can also be studied using the ellipsoid. The deformations within this plane and the corresponding stresses are represented by the central elliptical section in which the plane cuts the ellipsoid. If we consider only the displacements in the plane of the section, those in the directions of the axes of the ellipse will be purely radial. Thus, in planes perpendicular to the axes, the shear stresses acting parallel to the section must be zero and only shear stresses normal to the section can exist. The normal stresses acting in the xy plane assume extreme values in the directions of the axes of the ellipse. These extreme stresses represent a form of principal stress relative to the plane considered and are known as secondary or pseudo principal stresses. We shall denote them by σ'_1, σ'_2 with additional subscripts to indicate the plane parallel to which they act. In photoelastic experiments their difference is determined and this for the plane considered would be represented by $(\sigma'_1 - \sigma'_2)_{xy}$.

1.2 Plane stress

A body is said to be in a state of plane stress or alternatively in a two-dimensional or biaxial state of stress when it can be assumed that all the stresses act parallel to one plane. This condition is approximately realized in plates loaded in their own plane by forces applied at the edges, the distribution of these forces being more or less uniform over the thickness. If the plane of the plate is parallel to the xy plane, then since the components of stress normal to the two surfaces of the plate must be zero we have for these surfaces $\sigma_z = \tau_{xz} = \tau_{yz} = 0$. Provided the thickness is not too great, these components can be assumed to be zero throughout the thickness while the other components $\sigma_x, \sigma_y, \tau_{xy}$, can be assumed constant throughout the thickness of the plate, i.e. independent of z. All planes parallel to the plane of the plate are therefore principal planes so that, at any point, two of the principal axes of the ellipsoid described in Section 1.1 lie within this plane.

If within the plane of the plate the stresses are all parallel to one direction, the state of stress is said to be uniaxial. If the axis of stress coincides with the x axis, then of the six stress components only σ_x has a finite value while all the others are zero. In this case the ellipsoid is one of revolution, the single axis of which coincides with the axis of stress. The condition of uniaxial stress is closely approximated in prismatic bars subjected to pure axial tensile or compressive loading.

Strictly speaking, a state of stress should always be described as triaxial since any body is three-dimensional; finite stresses exist on all planes except that normal to the vanishing principal stress in a biaxial state of stress or the two such planes if the stress is uniaxial.

The assumption that the stress components normal to the plane of the plate are all zero is only valid when the material can contract or expand laterally without restriction. Thus, a state of plane stress implies a three-dimensional state of strain.

1.3 Stresses at a point in a system of plane stress

We now consider the stresses at a point in a plate which is in a state of plane stress and assume that the plane of the plate is parallel to the xy plane. Let us imagine a small triangular element ABC as shown in Figure 1.3 cut from the plate about the point by planes parallel to the xz and yz planes and a plane inclined at an angle φ to the yz plane. The normal and shear stresses on the inclined face AB we denote by σ and τ respectively.

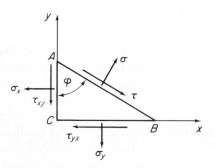

Figure 1.3. Normal and shear stresses on an infinitesimal triangular element

By considering the equilibrium of the forces acting on the element in the direction normal to AB we obtain

$$\sigma = \sigma_x \cos^2 \varphi + \sigma_y \sin^2 \varphi + 2\tau_{xy} \sin \varphi \cos \varphi,$$

or   (1.1)

$$\sigma = \frac{\sigma_x + \sigma_y}{2} + \frac{\sigma_x - \sigma_y}{2} \cos 2\varphi + \tau_{xy} \sin 2\varphi.$$

From the conditions of equilibrium in the direction parallel to AB, we likewise obtain

$$\tau = \frac{\sigma_x - \sigma_y}{2} \sin 2\varphi - \tau_{xy} \cos 2\varphi. \tag{1.2}$$

Thus, knowing the stresses on any two perpendicular planes, the stresses on any other plane can be determined by means of equations (1.1) and (1.2). To determine the planes on which σ is a maximum and a minimum we differentiate equation (1.1) with respect to φ and equate to zero. This yields

$$\tan 2\varphi = \frac{2\tau_{xy}}{\sigma_x - \sigma_y}, \tag{1.3}$$

and hence

$$\varphi = \tfrac{1}{2} \tan^{-1} \frac{2\tau_{xy}}{\sigma_x - \sigma_y} \quad \text{or} \quad \tfrac{1}{2} \tan^{-1} \frac{2\tau_{xy}}{\sigma_x - \sigma_y} + \frac{\pi}{2}.$$

We thus obtain the directions of the two principal planes normal to the plate on one of which σ is a maximum while on the other it is a minimum.

To determine the principal stresses, we insert the values of $\sin 2\varphi$ and $\cos 2\varphi$ obtained from equation (1.3) into equation (1.1). This produces

$$\sigma_1, \sigma_2 = \frac{\sigma_x + \sigma_y}{2} \pm \sqrt{\left(\left(\frac{\sigma_x - \sigma_y}{2} \right)^2 + \tau_{xy}^2 \right)}. \tag{1.4}$$

The same substitutions in equation (1.2) give $\tau = 0$, i.e. the principal planes are free from shear stress. From equation (1.4) we see that

$$\sigma_1 + \sigma_2 = \sigma_x + \sigma_y. \tag{1.5}$$

This equation is independent of φ, showing that the sum of the normal stresses on any two mutually perpendicular planes is constant and equal to the sum of the principal stresses.

The considerations involved in the development of equations (1.1) and (1.2) are unaffected by any stress components which may act in the z-direction.

Further, since we are considering the stresses at a point, we imagine the element to be indefinitely reduced in dimensions so that variation over the thickness of the element of any shear stresses τ_{zx} and τ_{zy} which may act on the faces parallel to the xy plane can be neglected. It follows that if the stress components in any two mutually perpendicular directions at a point in a three-dimensional stress system are known, the stress components in any other direction in the same plane can be determined from equations (1.1) and (1.2). Equations (1.3) and (1.4) also apply but σ_1 and σ_2 are then in general the secondary principal stresses in the plane considered; the planes defined by equation (1.3) are those on which the secondary principal stresses act. These are not in this case planes of vanishing shear stress but planes on which there is no component of shear stress acting parallel to the xy plane. If the plane considered contains one of the principal axes of the ellipsoid, then one true principal stress and one secondary principal stress are obtained. Only when the plane contains two principal axes are both σ_1 and σ_2 true principal stresses.

Equations (1.1) and (1.2) simplify in particular two-dimensional cases:

(a) $\tau_{xy} = \tau_{yx} = 0$. In this case AC and BC are principal planes and σ_x, σ_y are identical with the principal stresses σ_1, σ_2. Thus for a plane inclined at an angle φ to the principal plane on which σ_1 acts (Figure 1.4), we have

$$\sigma = \frac{\sigma_1 + \sigma_2}{2} + \frac{(\sigma_1 - \sigma_2)}{2} \cos 2\varphi = \tfrac{1}{2}[(\sigma_1 + \sigma_2) + (\sigma_1 - \sigma_2) \cos 2\varphi], \quad (1.6)$$

$$\tau = \tfrac{1}{2}(\sigma_1 - \sigma_2) \sin 2\varphi. \quad (1.7)$$

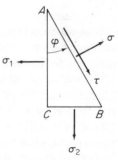

Figure 1.4.   Stresses on an element bounded by two principal planes and an oblique plane

From equation (1.7) it can be seen that the maximum shear stress is given by

$$\tau_{max} = \frac{\sigma_1 - \sigma_2}{2} = \sqrt{\left(\left(\frac{\sigma_x - \sigma_y}{2}\right)^2 + \tau_{xy}^2\right)}, \quad (1.8)$$

and occurs on planes inclined at 45° to the principal planes of stress.

(*b*) $\tau_{xy} = \tau_{yx} = \sigma_y = 0$. This represents a simple uniaxial stress system in which the axis of stress is parallel to the *x* axis. Equations (1.1) and (1.2) now reduce to

$$\sigma = \sigma_1 \cos^2 \varphi = \frac{\sigma_1}{2}(1 + \cos 2\varphi), \tag{1.9}$$

$$\tau = \frac{\sigma_1}{2} \sin 2\varphi. \tag{1.10}$$

From these equations, the normal and shear stresses acting on any inclined section of a prismatic bar in simple tension or compression can be calculated.

1.4 Mohr's circle for stress

If we eliminate φ from equations (1.6) and (1.7) we obtain

$$\sigma = \frac{\sigma_1 + \sigma_2}{2} + \sqrt{\left(\left(\frac{\sigma_1 - \sigma_2}{2}\right)^2 - \tau^2\right)},$$

i.e.

$$\left\{\sigma - \frac{\sigma_1 + \sigma_2}{2}\right\}^2 = \left(\frac{\sigma_1 - \sigma_2}{2}\right)^2 - \tau^2$$

or

$$\left\{\sigma - \frac{\sigma_1 + \sigma_2}{2}\right\}^2 + \tau^2 = \left(\frac{\sigma_1 - \sigma_2}{2}\right)^2$$

It is obvious from this equation that if we plot τ as ordinates against σ as abscissae we obtain a circle with its centre at $((\sigma_1 + \sigma_2)/2, 0)$ and of radius $(\sigma_1 - \sigma_2)/2$. This is known as Mohr's circle for stress and is shown in Figure 1.5. Each point on the circumference of this circle represents a particular plane through the point where the principal stresses are σ_1 and σ_2. The point corresponding to any particular plane is obtained by drawing a chord through *B* parallel to the direction of the plane. Thus, the point *C* corresponds to the plane *AB* in Figure 1.4. The co-ordinates of *C* then represent the normal and shear stresses on the plane *AB*. The stresses on a plane perpendicular to *AB* are similarly represented by the co-ordinates of the point *C'* lying diametrically opposite *C*. Obviously $C'E' = CE$ so that the shear stresses on these perpendicular planes are numerically equal in agreement with the result which we obtained previously from the conditions of equilibrium. *CE* lies above the σ axis however while *C'E'* lies below it and this has a special meaning in connection with Mohr's circle. The sign of the shear stresses will be consistent with the direction shown in Figure 1.4 if we regard positive

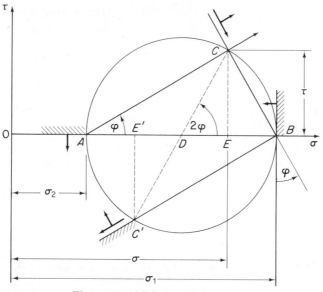

Figure 1.5.   Mohr's circle for stress

shear stresses to be those which produce clockwise moments and the reverse. Thus, in Figure 1.4, τ produces a clockwise moment on AB and is positive in the Mohr circle.

Since chords drawn from A and B which intersect on the circumference of the circle are mutually perpendicular, it follows that if we choose A as pole the direction of any chord now represents the direction of the normal stress instead of the direction of the plane on which it acts. This is illustrated with reference to the point C in Figure 1.5.

Mohr's circle can also be used when one or both of the principal stresses is compressive. In such cases, the compressive stresses are set off in the negative direction of the σ axis.

Mohr's circle in more general form is shown in Figure 1.6. Let us assume that the stress components $\sigma_x, \sigma_y, \tau_{xy}$ acting on any two mutually perpendicular planes AC, BC (Figure 1.3) are known. Mohr's circle is drawn by setting off $OF_1, OF_2 = \sigma_x, \sigma_y$ and $F_1B = \tau_{xy}$. Since τ_{xy} acts in a counterclockwise direction about the plane AC, it is set off in the negative direction of the τ axis. The centre D of the circle bisects F_1F_2 and its radius is equal to DB. Choosing the plane AC as our reference plane, we now produce the perpendicular BF_1 to intersect the circle in the pole point P. The normal and shear stresses acting on any plane passing through the point considered in the body are then given by the co-ordinates of the point in which a chord drawn through the pole parallel to the plane intersects the circle. Thus, the stresses on the planes AB, BC and AC in Figure 1.3 are represented by the co-ordinates of the

Photoelastic Stress Analysis

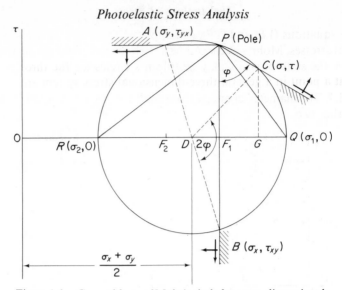

Figure 1.6.   General form of Mohr's circle for a two-dimensional stress system

points C, A and B respectively in Figure 1.6. The direction of the shear stress on each plane is indicated using the sign convention previously adopted.

The principal planes of stress are parallel to the chords PQ, PR drawn through the extremities of the diameter QR and the principal stresses σ_1, σ_2 are represented by OQ, OR respectively.

To show the validity of the above construction, for example for point C, we note that $C\hat{D}B = 2C\hat{P}B = 2\varphi$. Then

$$OG = OD + DG = OD + DC \cos CDF_1$$

$$= OD + DC \cos (2\varphi - BDF_1)$$

$$= OD + DC (\cos BDF_1 \cos 2\varphi + \sin BDF_1 \sin 2\varphi)$$

$$= OD + DF_1 \cos 2\varphi + BF_1 \sin 2\varphi$$

$$= \frac{\sigma_x + \sigma_y}{2} + \frac{\sigma_x - \sigma_y}{2} \cos 2\varphi + \tau_{xy} \sin 2\varphi,$$

which agrees with equation (1.1).

Also,

$$CG = DC \sin (2\varphi - BDF_1)$$

$$= DF_1 \sin 2\varphi - BF_1 \cos 2\varphi$$

$$= \frac{\sigma_x - \sigma_y}{2} \sin 2\varphi - \tau_{xy} \cos 2\phi,$$

which is in agreement with equation (1.2).

Since equations (1.6) and (1.7) also apply when σ_1 and σ_2 are secondary principal stresses, Mohr's circle can be drawn for any plane through a point whatever the state of stress may be. Mohr's circles for the three principal planes at a point in a general three-dimensional stress system are shown in Figure 1.7. The diameter of one of these circles is obviously equal to the sum of the other two.

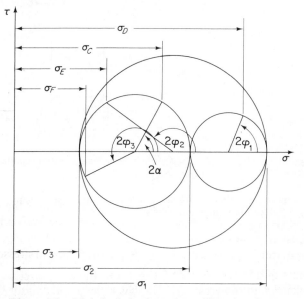

Figure 1.7.   Mohr's circles for a three-dimensional stress system

The normal stress acting in any random direction can be obtained in the following way. Let us assume that the plane on which the stress is required is parallel to the tangent plane to the point A on a sphere which has its centre at the point considered (Figure 1.8). The principal axes of stress (which coincide with the principal axes of the ellipsoid) intersect the sphere in three points. The great circle passing through A and the point B in which the σ_1 axis cuts the sphere intersects the plane of σ_2 and σ_3 in the point C. The normal stress σ_C acting in the direction OC can be read from Figure 1.7, using the appropriate circle which in this case is the $\sigma_2\sigma_3$ circle. Mohr's circle for the plane through A and B can now be drawn (Figure 1.9) using the values of the principal stress σ_1 and the secondary principal stress σ_c. From this circle the required normal stress σ_A can be determined in the manner indicated.

If Mohr's circle is required for any oblique plane passing through the point O, such as that through A normal to the plane of AOB, this can be drawn after the normal stresses in three different directions in the required plane

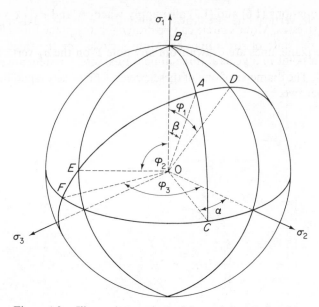

Figure 1.8.   Illustrating method of determining stresses in a
random direction or plane from Mohr's circles

have been found in the above manner. For this purpose it is convenient to
use the points of intersect *D*, *E* and *F* between the great circles lying in the
principal planes and the great circle in which the plane considered cuts the
sphere. The normal stresses σ_D, σ_E and σ_F are then measured using the
appropriate Mohr circles as shown in Figure 1.7. To obtain the required

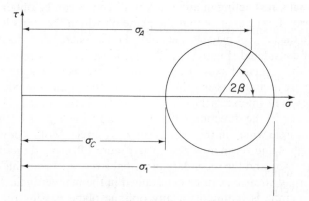

Figure 1.9.   Mohr's circle for the plane containing the
points *A* and *B* in Figure 1.8

Mohr circle, vertical lines are drawn at distances from the τ-axis representing σ_D, σ_E and σ_F (Figure 1.10). Choosing the vertical plane on which σ_F acts as the reference plane, lines are drawn from any pole F on the σ_F vertical parallel

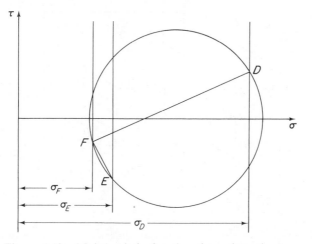

Figure 1.10. Mohr's circle for the plane through A normal to plane AOB in Figure 1.8

to the other two planes to intersect the appropriate verticals in D and E. The circle described through the points D, E and F is Mohr's circle for the plane through these points.

1.5 Equations of equilibrium

We consider a small rectangular element or parallelepiped of dimensions $dx\, dy\, dz$ with its edges parallel to the co-ordinate axes x, y, z as shown in Figure 1.11. Here we wish to determine the manner in which the stresses vary from point to point through the body and we must therefore allow for the small differences which exist between the stresses on opposite faces of the element. If, for example, the normal stress on one of the faces normal to the x axis is σ_x, then that on the opposite face can be expressed as $\sigma_x + \partial\sigma_x/\partial x\, dx$ and so on. Body forces which may act on the element must also be taken into consideration. These we shall represent by their components X, Y, Z per unit volume which we assume to act in the positive directions of the x, y, z axes respectively.

If we equate the sum of all the forces acting on the element in the direction of each axis in turn to zero and divide by $dx\, dy\, dz$ we obtain the following

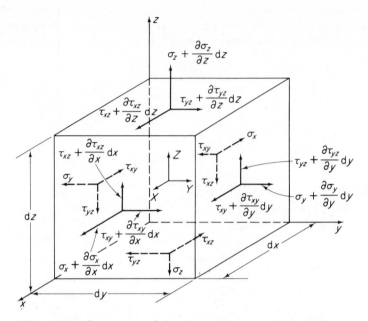

Figure 1.11. Stresses on a finite element in rectangular co-ordinates

three equations:

$$\frac{\partial \sigma_x}{\partial x} + \frac{\partial \tau_{xy}}{\partial y} + \frac{\partial \tau_{xz}}{\partial z} + X = 0,$$

$$\frac{\partial \sigma_y}{\partial y} + \frac{\partial \tau_{yz}}{\partial z} + \frac{\partial \tau_{xy}}{\partial x} + Y = 0, \qquad (1.11)$$

$$\frac{\partial \sigma_z}{\partial z} + \frac{\partial \tau_{zx}}{\partial x} + \frac{\partial \tau_{zy}}{\partial y} + Z = 0.$$

Equations (1.11) are the differential equations of equilibrium in rectangular co-ordinates and must be satisfied by the stress components at every point in a body.

When dealing with bodies which have rotational symmetry it is usually more convenient to use cylindrical co-ordinates. Figure 1.12 shows the stresses in cylindrical notation acting on the surfaces of a small element having inner radius r, radial width dr, axial width dz and subtending an angle $d\theta$ at the z axis.

If we equate the sums of the forces acting on the element in the radial, tangential and axial directions in turn to zero, neglect products of small

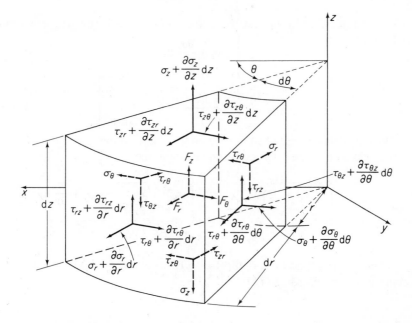

Figure 1.12. Stresses on a finite element in cylindrical co-ordinates

quantities of order higher than the third and divide by $r \, dr \, d\theta \, dz$, we obtain

$$\frac{\partial \sigma_r}{\partial r} + \frac{1}{r} \frac{\partial \tau_{r\theta}}{\partial \theta} + \frac{\partial \tau_{rz}}{\partial z} + \frac{\sigma_r - \sigma_\theta}{r} + F_r = 0,$$

$$\frac{\partial \tau_{r\theta}}{\partial r} + \frac{1}{r} \frac{\partial \sigma_\theta}{\partial \theta} + \frac{\partial \tau_{\theta z}}{\partial z} + \frac{2\tau_{r\theta}}{r} + F_\theta = 0, \qquad (1.12)$$

$$\frac{\partial \tau_{rz}}{\partial r} + \frac{1}{r} \frac{\partial \tau_{\theta z}}{\partial \theta} + \frac{\partial \sigma_z}{\partial z} + \frac{\tau_{rz}}{r} + F_z = 0,$$

in which F_r, F_θ and F_z are the components of the body forces per unit volume.

1.6 Equations of equilibrium in two dimensions

In the case of plane stress, the equations of equilibrium are considerably simplified. If, as before, the xy plane is chosen parallel to the plane of the plate, then from the preceding section we have

$$\sigma_z = \tau_{xz} = \tau_{yz} = 0.$$

If we assume in addition that there are no body forces present, equation (1.11) reduces to

$$\frac{\partial \sigma_x}{\partial x} + \frac{\partial \tau_{xy}}{\partial y} = 0, \qquad (1.13a)$$

$$\frac{\partial \sigma_y}{\partial y} + \frac{\partial \tau_{xy}}{\partial x} = 0. \qquad (1.13b)$$

The equations of equilibrium in cylindrical co-ordinates, equation (1.12), reduce likewise to the polar co-ordinate equations:

$$\frac{\partial \sigma_r}{\partial r} + \frac{1}{r}\frac{\partial \tau_{r\theta}}{\partial \theta} + \frac{\sigma_r - \sigma_\theta}{r} = 0, \qquad (1.14a)$$

$$\frac{\partial \tau_{r\theta}}{\partial r} + \frac{1}{r}\frac{\partial \sigma_\theta}{\partial \theta} + \frac{2\tau_{r\theta}}{r} = 0. \qquad (1.14b)$$

Equations (1.13) and (1.14) apply also to the case of a body in a state of plane strain as described in Section 2.5.

1.7   Equations of equilibrium in terms of the principal stresses

From equations (1.6) and (1.7) we have

$$\sigma_x = \tfrac{1}{2}[(\sigma_1 + \sigma_2) + (\sigma_1 - \sigma_2)\cos 2\varphi],$$

$$\sigma_y = \tfrac{1}{2}[(\sigma_1 + \sigma_2) - (\sigma_1 - \sigma_2)\cos 2\varphi],$$

$$\tau_{xy} = \tfrac{1}{2}(\sigma_1 - \sigma_2)\sin 2\varphi,$$

where φ is the angle between the x axis and the direction of σ_1.

Inserting these values in equations (1.13), we obtain

$$\frac{\partial(\sigma_1 + \sigma_2)}{\partial x} + \frac{\partial(\sigma_1 - \sigma_2)}{\partial x}\cos 2\varphi - 2(\sigma_1 - \sigma_2)\sin 2\varphi\frac{\partial \varphi}{\partial x}$$

$$+ \frac{\partial(\sigma_1 - \sigma_2)}{\partial y}\sin 2\varphi + 2(\sigma_1 - \sigma_2)\cos 2\varphi\frac{\partial \varphi}{\partial y} = 0,$$

$$\frac{\partial(\sigma_1 + \sigma_2)}{\partial y} - \frac{\partial(\sigma_1 - \sigma_2)}{\partial y}\cos 2\varphi + 2(\sigma_1 - \sigma_2)\sin 2\varphi\frac{\partial \varphi}{\partial y}$$

$$+ \frac{\partial(\sigma_1 - \sigma_2)}{\partial x}\sin 2\varphi + 2(\sigma_1 - \sigma_2)\cos 2\varphi\frac{\partial \varphi}{\partial x} = 0.$$

$$(1.15)$$

Equations (1.15) are the equations of equilibrium for the two mutually perpendicular directions inclined at an angle φ to the principal stresses.

1.8 Equations of equilibrium along stress trajectories

A stress trajectory, line of principal stress or isostatic is a line the tangents to which at every point coincide with the directions of one of the principal stresses. Since the two principal stresses at every point are mutually perpendicular, it follows that, in a field of continuously varying stress, two families of orthogonal stress trajectories can be drawn, one representing the directions of the σ_1 stresses and the other those of the σ_2 stresses.

Let us consider a small element of dimensions ds_1, ds_2 about a point P in a two-dimensional state of stress bounded by stress trajectories as shown in Figure 1.13. Let ρ_1, ρ_2 be the radii of curvature of the stress trajectories in

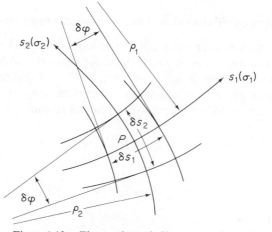

Figure 1.13.   Element bounded by stress trajectories

the directions of σ_1, σ_2, respectively at the point P. To obtain the equations of equilibrium in these directions, we substitute $\varphi = 0$, $\partial x = \partial s_1$, $\partial y = \partial s_2$ in equation (1.15) which then reduces to

$$\frac{\partial(\sigma_1 + \sigma_2)}{\partial s_1} + \frac{\partial(\sigma_1 - \sigma_2)}{\partial s_1} + 2(\sigma_1 - \sigma_2)\frac{\partial\varphi}{\partial s_2} = 0,$$

$$\frac{\partial(\sigma_1 + \sigma_2)}{\partial s_2} - \frac{\partial(\sigma_1 - \sigma_2)}{\partial s_2} + 2(\sigma_1 - \sigma_2)\frac{\partial\varphi}{\partial s_1} = 0,$$

or

$$\frac{\partial\sigma_1}{\partial s_1} + (\sigma_1 - \sigma_2)\frac{\partial\varphi}{\partial s_2} = 0,$$

$$\frac{\partial\sigma_2}{\partial s_2} + (\sigma_1 - \sigma_2)\frac{\partial\varphi}{\partial s_1} = 0.$$

Then, since

$$\frac{\partial \varphi}{\partial s_1} = \frac{1}{\rho_1}, \quad \frac{\partial \varphi}{\partial s_2} = \frac{1}{\rho_2},$$

we obtain

$$\frac{\partial \sigma_1}{\partial s_1} + \frac{(\sigma_1 - \sigma_2)}{\rho_2} = 0, \quad \frac{\partial \sigma_2}{\partial s_2} + \frac{(\sigma_1 - \sigma_2)}{\rho_1} = 0. \tag{1.16}$$

Equations (1.16) are known as the Lamé–Maxwell equations of equilibrium and have important applications in photoelasticity.

1.9 Equations of equilibrium along lines of maximum shear stress

A line of maximum shear stress is a line which is tangential at every point to one of the two directions in which the shear stress is a maximum. Since the maximum shear stress at any point acts at 45° to the directions of the principal stresses, the lines of maximum shear stress form two orthogonal families which are at 45° to the stress trajectories as shown in Figure 1.14.

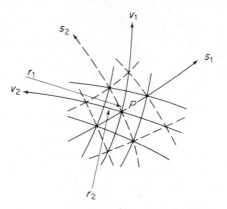

Figure 1.14. Element bounded by lines
of maximum shear stress

Let r_1, r_2 denote the radii of curvature of the lines of maximum shear stress v_1, v_2 at the point P and let ψ be the angle between v_1 and the x axis, i.e. $\psi = \varphi + 45°$. We then have

$$\frac{\partial \varphi}{\partial x} = \frac{\partial \psi}{\partial x}, \quad \frac{\partial \varphi}{\partial y} = \frac{\partial \psi}{\partial y},$$

$$\cos 2\varphi = \sin 2\psi, \quad \sin 2\varphi = -\cos 2\psi.$$

If we now take the x axis to be parallel to v_1, then $\psi = 0$. With the above substitutions, equation (1.15) reduces to

$$\frac{\partial(\sigma_1 + \sigma_2)}{\partial v_1} + 2(\sigma_1 - \sigma_2)\frac{\partial\psi}{\partial v_1} - \frac{\partial(\sigma_1 - \sigma_2)}{\partial v_2} = 0,$$

$$\frac{\partial(\sigma_1 + \sigma_2)}{\partial v_2} - 2(\sigma_1 - \sigma_2)\frac{\partial\psi}{\partial v_2} - \frac{\partial(\sigma_1 - \sigma_2)}{\partial v_1} = 0,$$

or since

$$\frac{\partial\psi}{\partial v_1} = \frac{1}{r_1}, \qquad \frac{\partial\psi}{\partial v_2} = \frac{1}{r_2},$$

$$\frac{\partial(\sigma_1 + \sigma_2)}{\partial v_1} - \frac{\partial(\sigma_1 - \sigma_2)}{\partial v_2} + \frac{2(\sigma_1 - \sigma_2)}{r_1} = 0,$$

$$\frac{\partial(\sigma_1 + \sigma_2)}{\partial v_2} - \frac{\partial(\sigma_1 - \sigma_2)}{\partial v_1} - \frac{2(\sigma_1 - \sigma_2)}{r_2} = 0.$$

(1.17)

1.10 Boundary conditions

The equations of equilibrium, equations (1.11), must be satisfied by the components of stress at every point in a body including all points on the boundary. Let us consider a small triangular element ABC at the boundary of a two-dimensional system (Figure 1.15) and let $\overline{X}$, $\overline{Y}$ denote the components

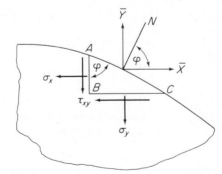

Figure 1.15.   Stresses on an element at a
boundary of a two-dimensional system

parallel to the x, y axes respectively of the surface forces per unit area acting on the boundary. If we assume the side AC to be of unit area, then the areas of the sides AB and BC are $\cos\varphi$ and $\sin\varphi$ respectively.

From the conditions of equilibrium of the forces acting on the element in the directions of the axes, we obtain

$$\sigma_x \cos \varphi + \tau_{xy} \sin \varphi = \overline{X},$$
$$\sigma_y \sin \varphi + \tau_{xy} \cos \varphi = \overline{Y},$$

(1.18)

or, as they are commonly expressed,

$$l\sigma_x + m\tau_{xy} = \overline{X},$$
$$m\sigma_y + l\tau_{xy} = \overline{Y},$$

where l, m, are the direction cosines of the outward normal N to the boundary. Equations (1.18) express the boundary conditions of the body. At points on a straight boundary it is usually convenient to take one of the axes, say the x axis, parallel to the boundary. We then have $\varphi = 90°$ and equations (1.18) reduce to

$$\tau_{xy} = \overline{X}; \qquad \sigma_y = \overline{Y}.$$

2

THE ANALYSIS OF STRAIN

2.1 Components of strain in rectangular co-ordinates

The displacement of any point in a body may be conveniently expressed in terms of its components u, v, w parallel to the co-ordinate axes x, y, z respectively. To determine the strains at a point O parallel to one of the co-ordinate planes, e.g. the xy plane, we consider two infinitesimal line elements OA, OB (Figure 2.1) originally of lengths dx, dy and parallel to the x, y axes

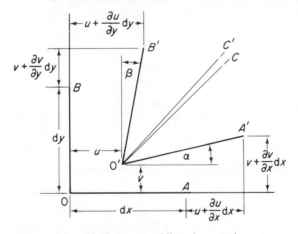

Figure 2.1. Displacement of line elements in rectangular co-ordinates

intersecting at the point O. Let these elements be displaced in the strained body into the positions $O'A'$, $O'B'$. If we denote the displacements of the point O' by u, v, the corresponding displacements of A', B' are

$$u + \frac{\partial u}{\partial x}\mathrm{d}x, v + \frac{\partial v}{\partial x}\mathrm{d}x \quad \text{and} \quad u + \frac{\partial u}{\partial y}\mathrm{d}y, v + \frac{\partial v}{\partial y}\mathrm{d}y$$

respectively.

Normal or linear strain is defined as the change in length per unit length. The change in length of the element OA in the x-direction is $(\partial u/\partial x)\,dx$ so that the normal strain ϵ_x in this direction is equal to $\partial u/\partial x$.

The shear strain γ_{xy} of the element between the xz and yz planes is defined as the difference between the angle $A'O'B'$ and a right angle, i.e. by the sum of the angles α and β. We shall assume that these angles are very small so that their sum may be taken to equal $(\partial v/\partial x) + (\partial u/\partial y)$.

The normal and shear strains relative to the other co-ordinate axes and planes can be determined in the same manner. We thus obtain the following equations relating the six components of strain at a point to the displacements:

$$\epsilon_x = \frac{\partial u}{\partial x}; \qquad \epsilon_y = \frac{\partial v}{\partial y}; \qquad \epsilon_z = \frac{\partial w}{\partial z},$$

$$\gamma_{xy} = \frac{\partial v}{\partial x} + \frac{\partial u}{\partial y}; \qquad \gamma_{yz} = \frac{\partial w}{\partial y} + \frac{\partial v}{\partial z}; \qquad \gamma_{zx} = \frac{\partial u}{\partial z} + \frac{\partial w}{\partial x}.$$

(2.1)

In this notation, the subscripts for the strains ϵ and γ have the same meaning as assigned previously to those for the stresses σ and τ.

The angle $\omega_z = \frac{1}{2}(\alpha - \beta)$ through which the bisector $O'C'$ of the angle $A'O'B'$ has rotated from its original position OC is a measure of the rotation of an element at the point as a rigid body about the z axis. This rotation does not affect the strains produced. Similar rigid body rotations in general occur about the x and y axes. Hence

$$\omega_x = \frac{1}{2}\left(\frac{\partial w}{\partial y} - \frac{\partial v}{\partial z}\right), \qquad \omega_y = \frac{1}{2}\left(\frac{\partial u}{\partial z} - \frac{\partial w}{\partial x}\right), \qquad \omega_z = \frac{1}{2}\left(\frac{\partial v}{\partial x} - \frac{\partial u}{\partial y}\right). \quad (2.2)$$

2.2 Components of strain in cylindrical co-ordinates

When dealing with problems involving bodies having rotational symmetry, it is usually advantageous to use cylindrical co-ordinates.

We first determine the strains at a point A due to the displacements in the plane normal to the axis of symmetry which we take to be the z axis. In Figure 2.2, AB represents a radial line element of length dr and AC an elementary arc subtending an angle $d\theta$ at the z axis. These elements originally intersect orthogonally at A and are displaced into the positions $A'B'$, $A'C'$ when the body is strained. Let the components of displacement of the point A be u, v and w in the radial, tangential and axial directions respectively. The radial displacement of the point B is then $u + (\partial u/\partial r)\,dr$ so that for the radial strain we obtain $\epsilon_r = \partial u/\partial r$.

The tangential strain is partly due to the radial displacement since this obviously changes the length of the arc and partly to the differential tangen-

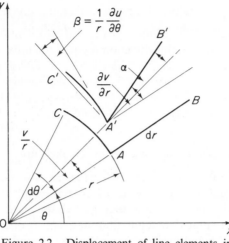

Figure 2.2. Displacement of line elements in
cylindrical co-ordinates

tial displacements. The tangential strain due to the radial displacement is

$$\frac{(r + u)\,\mathrm{d}\theta - r\,\mathrm{d}\theta}{r\,\mathrm{d}\theta} = \frac{u}{r}.$$

The tangential displacement of the point C is $v + (\partial v/\partial \theta)\,\mathrm{d}\theta$ and therefore the tangential strain due to the displacements in this direction is

$$\frac{[v + (\partial v/\partial \theta)\,\mathrm{d}\theta] - v}{r\,\mathrm{d}\theta} = \frac{1}{r}\frac{\partial v}{\partial \theta}.$$

The total tangential strain ϵ_θ is the sum of these two quantities, i.e.

$$\epsilon_\theta = \frac{u}{r} + \frac{1}{r}\frac{\partial v}{\partial \theta}.$$

The shear strain $\gamma_{r\theta}$ is equal to the difference between the angle $B'A'C'$ and a right angle. This difference is equal to the sum of the angle α between $A'B'$ and the radius through A' and the angle β between the tangent to $A'C'$ at A' and the normal to the radius through A'. From Figure 2.2 we see that

$$\gamma_{r\theta} = \alpha + \beta = \frac{1}{r}\frac{\partial u}{\partial \theta} + \frac{\partial v}{\partial r} - \frac{v}{r}.$$

In this equation the term v/r is equal to the angle through which the radial line through the point considered rotates as it is displaced from A to A' and represents the rigid body rotation of an element at A about the z axis.

By considering the displacements in the rz and θz planes, the strain components ϵ_z, γ_{rz} and $\gamma_{z\theta}$ can be obtained. These are derived in exactly the same manner as the corresponding strains in rectangular co-ordinates. We then have the following equations for the six components of strain:

$$\epsilon_r = \frac{\partial u}{\partial r}; \qquad \epsilon_\theta = \frac{u}{r} + \frac{1}{r}\frac{\partial v}{\partial \theta}; \qquad \epsilon_z = \frac{\partial w}{\partial z};$$

$$\gamma_{r\theta} = \frac{1}{r}\frac{\partial u}{\partial \theta} + \frac{\partial v}{\partial r} - \frac{v}{r}; \qquad \gamma_{rz} = \frac{\partial u}{\partial z} + \frac{\partial w}{\partial r}; \qquad \gamma_{z\theta} = \frac{\partial v}{\partial z} + \frac{1}{r}\frac{\partial w}{\partial \theta}.$$

(2.3)

2.3 Relations between stresses and strains. Hooke's law

All bodies are deformed or strained when loaded. An elastic material is one in which all deformations vanish when the load is removed. In the theory of elasticity it is usual to postulate also that in an elastic material the strains are proportional to the applied load and that the elastic properties are the same in every direction. Such a material is said to be isotropic. These conditions are more or less realized by many engineering materials over a considerable range in stress up to a certain limiting value known as the elastic limit or limit of proportionality.

Consider a long prismatic bar in a uniaxial state of stress and let σ_x be the applied stress in the direction of the x axis taken coincident with the axis of the bar. Within the elastic range, the axial or longitudinal strain ϵ_x is given by

$$\epsilon_x = \frac{\sigma_x}{E},$$

(2.4)

in which E is a constant known as the modulus of elasticity. Equation (2.4) expresses the linear relationship between stress and strain known as Hooke's law. The axial strain ϵ_x is accompanied by a simultaneous lateral strain of opposite sign which is also proportional to the applied stress and is the same in every direction perpendicular to the axis of stress. The lateral strain in the directions of any two mutually perpendicular axes y and z in a plane normal to the axis of stress is given by

$$\epsilon_y = \epsilon_z = -\frac{v\sigma_x}{E} = -v\epsilon_x,$$

(2.5)

in which v is a constant known as Poisson's ratio and represents the ratio of the lateral to the longitudinal strain. For constant volume deformation, v has a value of 0·5 and this value is approached in certain soft photoelastic materials such as urethane rubbers and in hard plastics when at their softening temperatures (see Chapter 7). For steel and most other metals, v has a value of about 0·3.

In an isotropic body in a biaxial or triaxial state of stress, the resultant strains in the direction of the axes can be obtained by superposition.

Thus, for a body in a biaxial or plane state of stress in the xy plane we obtain

$$\epsilon_x = \frac{1}{E}(\sigma_x - v\sigma_y),$$

$$\epsilon_y = \frac{1}{E}(\sigma_y - v\sigma_x). \tag{2.6}$$

In the same manner, for the three-dimensional case represented by the parallelepiped in Figure 1.2 we have

$$\epsilon_x = \frac{1}{E}[\sigma_x - v(\sigma_y + \sigma_z)], \tag{2.7a}$$

$$\epsilon_y = \frac{1}{E}[\sigma_y - v(\sigma_z + \sigma_x)], \tag{2.7b}$$

$$\epsilon_z = \frac{1}{E}[\sigma_z - v(\sigma_x + \sigma_y)]. \tag{2.7c}$$

These equations express Hooke's law in generalized form for isotropic materials. When applied to shear stresses and strains, Hooke's law can be expressed by

$$\gamma = \frac{\tau}{G}, \tag{2.8}$$

where G is a constant known as the modulus of rigidity or shear modulus. The shear strains produced by the shear stresses acting on the parallelepiped are therefore

$$\gamma_{xy} = \frac{\tau_{xy}}{G}; \qquad \gamma_{yz} = \frac{\tau_{yz}}{G}; \qquad \gamma_{zx} = \frac{\tau_{zx}}{G}. \tag{2.9}$$

If equations (2.7) are inverted to give the stresses in terms of the strains we obtain

$$\sigma_x = \frac{E}{(1 + v)(1 - 2v)}[(1 - v)\epsilon_x + v(\epsilon_y + \epsilon_z)],$$

$$\sigma_y = \frac{E}{(1 + v)(1 - 2v)}[(1 - v)\epsilon_y + v(\epsilon_z + \epsilon_x)],$$

$$\sigma_z = \frac{E}{(1 + v)(1 - 2v)}[(1 - v)\epsilon_z + v(\epsilon_x + \epsilon_y)].$$

With the substitutions

$$\lambda = \frac{vE}{(1 + v)(1 - 2v)}; \qquad \mu = \frac{E}{2(1 + v)}, \tag{2.10}$$

these equations abbreviate to

$$\sigma_x = \lambda e + 2\mu\epsilon_x,$$
$$\sigma_y = \lambda e + 2\mu\epsilon_y, \qquad (2.11)$$
$$\sigma_z = \lambda e + 2\mu\epsilon_z,$$

where $e = \epsilon_x + \epsilon_y + \epsilon_z$ represents the dilatation or change in volume per unit volume. The constants λ and μ are known as Lamé's constants. It is shown in the following section that μ is identical with the modulus of rigidity G so that equations (2.9) can be written as

$$\tau_{xy} = \mu\gamma_{xy}; \qquad \tau_{yz} = \mu\gamma_{yz}; \qquad \tau_{zx} = \mu\gamma_{zx}. \qquad (2.12)$$

2.4 Relations between elastic constants. Bulk modulus

Lamé's constants λ and μ are independent constants which completely specify the elastic behaviour of a material. All other elastic constants can therefore be expressed in terms of these two.

Equations (2.11) when applied to an element in a uniaxial stress system in which the axis of stress coincides with the x axis become

$$\sigma_x = \lambda e + 2\mu\epsilon_x, \qquad (a)$$
$$0 = \lambda e + 2\mu\epsilon_y, \qquad (b)$$
$$0 = \lambda e + 2\mu\epsilon_z. \qquad (c)$$

By addition, we obtain

$$\sigma_x = (3\lambda + 2\mu)e,$$

or

$$e = \frac{\sigma_x}{3\lambda + 2\mu} = \frac{E\epsilon_x}{3\lambda + 2\mu}, \qquad (2.13)$$

and substituting in (a),

$$\sigma_x = \frac{\lambda\sigma_x}{3\lambda + 2\mu} + 2\mu\epsilon_x,$$

from which

$$\frac{\sigma_x}{\epsilon_x} = E = \frac{\mu(3\lambda + 2\mu)}{\lambda + \mu}.$$

From (b) and (c) we have

$$\epsilon_y = \epsilon_z = -\frac{\lambda e}{2\mu},$$

and from (2.13) and (2.14),

$$e = \frac{\mu \epsilon_x}{\lambda + \mu},$$

so that

$$-\frac{\epsilon_y}{\epsilon_x} = -\frac{\epsilon_z}{\epsilon_x} = v = \frac{\lambda}{2(\lambda + \mu)}. \tag{2.14}$$

In the case of a material subjected to pure hydrostatic compression, the normal stresses in every direction are equal to the applied pressure p and the shear stresses are zero. Substituting in equations (2.11) and adding we obtain

$$-3p = (3\lambda + 2\mu)e$$

i.e.,

$$p = -\tfrac{1}{3}(3\lambda + 2\mu)e.$$

The ratio between the applied pressure p and the unit change in volume e is known as the bulk modulus or modulus of volume expansion and usually denoted by K. Thus

$$K = \tfrac{1}{3}(3\lambda + 2\mu). \tag{2.15}$$

In order to determine the relation between the modulus of rigidity G and the other elastic constants let us consider an originally square element $OABC$ of unit edge subjected to pure or irrotational shear by the application of shear stresses τ_{xy} along its edges as shown in Figure 2.3. Since the normal stresses σ_x, σ_y are zero we find from equation (1.3) that $\tan 2\varphi$ is infinite so that $2\varphi = 90°$, i.e. $\varphi = 45°$. The principal planes of stress therefore coincide

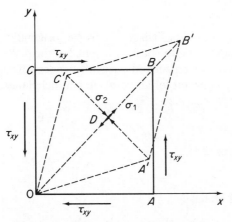

Figure 2.3.   Deformation of a square element
subjected to pure shear

with the diagonals OB', $A'C'$. From equation (1.4), the principal stresses are $\sigma_1, \sigma_2 = \pm\tau_{xy}$, being tensile along OB' and compressive along $A'C'$.

By definition, the shear strain γ_{xy} is equal to $90° - C'OA'$ and since γ can be assumed to be a very small angle, we have approximately

$$\gamma_{xy} = \sin \gamma_{xy} = \cos C'OA'$$

$$= 2\cos^2 C'OD - 1 = \frac{2OD^2}{OC'^2} - 1.$$

Noting that $OD = (1/\sqrt{2})(1 + \epsilon)$, where ϵ is the normal strain in the direction OB' and $OC' = 1$, we then have

$$\gamma_{xy} = (1 + \epsilon)^2 - 1,$$

which to the first order of small quantities gives

$$\gamma_{xy} = 2\epsilon.$$

From equation (2.6) we have with $\sigma_1 = -\sigma_2 = \sigma$,

$$\epsilon = \frac{\sigma}{E}(1 + v),$$

and therefore

$$\gamma_{xy} = \frac{2\sigma}{E}(1 + v) = \frac{2\tau_{xy}}{E}(1 + v),$$

from which, by comparison with equations (2.8) and (2.9), we find

$$G = \frac{E}{2(1 + v)} = \mu. \tag{2.16}$$

2.5 Plane strain

A state of plane strain exists when all particles originally in one plane in a body remain coplanar after the body is strained. If the plane is normal to the z axis, then the displacements u and v are independent of z and the displacement w is zero or a constant for all points in parallel planes. We then have

$$\frac{\partial u}{\partial z} = \frac{\partial v}{\partial z} = \frac{\partial w}{\partial x} = \frac{\partial w}{\partial y} = \frac{\partial w}{\partial z} = 0,$$

and, from equations (2.1)

$$\epsilon_z = \gamma_{yz} = \gamma_{zx} = 0.$$

From equation (2.7c),

$$\epsilon_z = \frac{1}{E}[\sigma_z - v(\sigma_x + \sigma_y)] = 0,$$

and therefore

$$\sigma_z = v(\sigma_x + \sigma_y). \tag{2.17}$$

From equations (2.9),

$$\tau_{yz} = \tau_{zx} = 0.$$

Substituting for σ_z from equation (2.17) in equations (2.7a) and (2.7b) we obtain for the normal strains

$$\epsilon_x = \frac{1 + v}{E}[(1 - v)\sigma_x - v\sigma_y],$$

$$\epsilon_y = \frac{1 + v}{E}[(1 - v)\sigma_y - v\sigma_x]. \tag{2.18}$$

We observe that while a plane or two-dimensional state of strain involves a three-dimensional state of stress, the solution of such problems requires the determination of the same three stress components σ_x, σ_y and τ_{xy} as plane stress problems.

In cylindrical co-ordinates we have

$$\frac{\partial u}{\partial z} = \frac{\partial v}{\partial z} = \frac{\partial w}{\partial r} = \frac{\partial w}{\partial \theta} = \frac{\partial w}{\partial z} = 0,$$

so that, from equations (2.3)

$$\epsilon_z = \gamma_{rz} = \gamma_{z\theta} = 0.$$

Plane strain is encountered in a prismatic body having one dimension (in the case considered this is in the z-direction) much greater than the perpendicular dimensions and loaded by uniform transverse forces along its length. Typical examples are dams, retaining walls and long cylinders loaded radially by fluid pressure or centrifugal forces. In such cases the plane strain state is induced in that part of the body remote from the ends by the restraint imposed on the lateral straining of any transverse section by the material on each side of the section. Plane strain can also be induced in bodies of other shapes such as plates by preventing lateral expansion by external means.

2.6 Strains at a point in a plane. Mohr's circle for strain

If the normal and shear strains at a point relative to any two mutually perpendicular axes are known, the strains in any other direction in the plane formed by these axes can be determined.

In Figure 2.4, $OABC$ represents a rectangular element of dimensions $dx\,dy$ lying in the xy plane. After deformation, the element occupies the position $OA'B'C'$. Let u and v be the displacements of point B in the x and

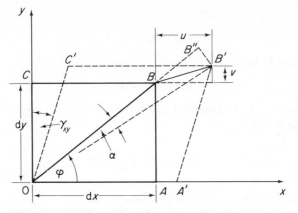

Figure 2.4. Deformation of a rectangular element subjected to normal and shear stresses—determination of strains in an arbitrary direction

y-directions respectively. Neglecting higher order terms, we then have

$$u = \epsilon_x\,dx + \gamma_{xy}\,dy,$$

$$v = \epsilon_y\,dy.$$

The displacement of B in the direction of the diagonal OB is

$$BB'' = u \cos \varphi + v \sin \varphi$$

$$= (\sigma_x\,dx + \gamma_{xy}\,dy) \cos \varphi + \epsilon_y\,dy \sin \varphi.$$

The strain ϵ in this direction is approximately equal to BB''/OB and since $OB = dx/\cos \varphi = dy/\sin \varphi$ we obtain

$$\epsilon = (\epsilon_x \cos \varphi + \gamma_{xy} \sin \varphi) \cos \varphi + \epsilon_y \sin^2 \varphi$$

$$= \epsilon_x \cos^2 \varphi + \epsilon_y \sin^2 \varphi + \tfrac{1}{2}\gamma_{xy} \sin 2\varphi,$$

or

$$\epsilon = \frac{\epsilon_x + \epsilon_y}{2} + \frac{\epsilon_x - \epsilon_y}{2} \cos 2\varphi + \frac{\gamma_{xy}}{2} \sin 2\varphi. \tag{2.19}$$

The displacement of B perpendicular to the direction OB is $B''B' = u \sin \varphi - v \cos \varphi$. The clockwise rotation α of OB is therefore approximately given by

$$\alpha = (u \sin \varphi - v \cos \varphi)/OB$$

$$= [(\epsilon_x \, dx + \gamma_{xy} \, dy) \sin \varphi - \epsilon_y \, dy \cos \varphi]/OB$$

$$= \epsilon_x \sin \varphi \cos \varphi + \gamma_{xy} \sin^2 \varphi - \epsilon_y \sin \varphi \cos \varphi$$

$$= \frac{(\epsilon_x - \epsilon_y) \sin 2\varphi}{2} + \gamma_{xy} \sin^2 \varphi. \tag{2.20}$$

The shear strain γ is given by the sum of the clockwise rotation α of OB and the corresponding counterclockwise rotation β of a line originally perpendicular to OB. We obtain β from equation (2.20) by reversing the sign and replacing φ by $90 + \varphi$. Hence

$$\beta = \frac{(\epsilon_x - \epsilon_y) \sin 2\varphi}{2} - \gamma_{xy} \cos^2 \varphi,$$

and

$$\gamma = \alpha + \beta = (\epsilon_x - \epsilon_y) \sin 2\varphi - \gamma_{xy} \cos 2\varphi,$$

or

$$\frac{\gamma}{2} = \frac{(\epsilon_x - \epsilon_y)}{2} \sin 2\varphi - \frac{\gamma_{xy}}{2} \cos 2\varphi. \tag{2.21}$$

Comparing these equations with those for the corresponding stresses, we see that equations (2.19) and (2.21) are identical in form with equations (1.1) and (1.2) respectively. The equations for the stresses can be transformed into their equivalents for strains by replacing σ by ϵ and τ by $\gamma/2$ in each term. This transformation can be applied to all the stress equations developed subsequently in Chapter 1. If, for example, we transform equation (1.4) in this manner we obtain for the limiting strains in the xy plane:

$$\epsilon_1, \epsilon_2 = \frac{\epsilon_x + \epsilon_y}{2} \pm \sqrt{\left(\left(\frac{\epsilon_x - \epsilon_y}{2}\right)^2 + \left(\frac{\gamma_{xy}}{2}\right)^2\right)}. \tag{2.22}$$

If the system is two dimensional in the xy plane, ϵ_1 and ϵ_2 are the principal strains at the point. In a general three-dimensional system, ϵ_1 and ϵ_2 are the secondary principal strains in the plane through the point formed by the x, y axes; these are analogous to the secondary principal stresses discussed in Section 1.1.

It also follows that strains can be obtained graphically in the same manner as stresses from Mohr's circle with the abscissa now representing ϵ and the ordinates representing $\gamma/2$ as shown in Figure 2.5. From this diagram we can obtain the strains in any required direction if the normal and shear strains with reference to any two mutually perpendicular directions are known.

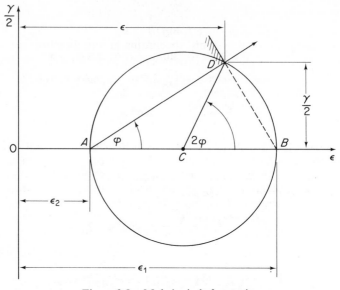

Figure 2.5.   Mohr's circle for strain

Figure 2.6 shows the construction for determining the principal strains and their directions in a two-dimensional system when the normal strains in three arbitrary directions are known. The method of construction is exactly the same as that for drawing Mohr's circle for stress from the values

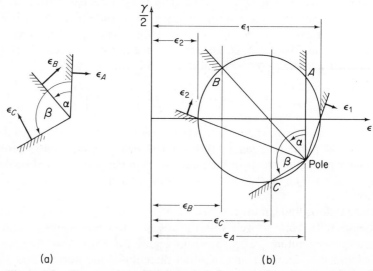

(a)   (b)

Figure 2.6.   Construction of Mohr's circle for strain from known values of the normal strain in three directions

of the normal stresses in three different directions as described in Section 1.4.

In three-dimensional systems, the strains in any direction can be determined from Mohr's circles if the components of strain with reference to any three mutually perpendicular axes are known. Here again, the method is identical with that described for three-dimensional stress systems in Section 1.4.

2.7 Equations of compatibility

The six components of strain can be expressed, as in equations (2.1), in terms of the three components of displacement u, v, w. The individual components of strain cannot therefore be independent functions of x, y and z but must be related in a way which is compatible with the existence of single-valued continuous functions for the displacements, i.e. with the conditions of continuity of the material. The necessary relations can be derived from equations (2.1) which give

$$\gamma_{xy} = \frac{\partial v}{\partial x} + \frac{\partial u}{\partial y}; \qquad \gamma_{yz} = \frac{\partial w}{\partial y} + \frac{\partial v}{\partial z}; \qquad \gamma_{zx} = \frac{\partial u}{\partial z} + \frac{\partial w}{\partial x}.$$

Differentiating the equation for γ_{xy} with respect to x and y we obtain

$$\frac{\partial^2 \gamma_{xy}}{\partial x\, \partial y} = \frac{\partial^2}{\partial x\, \partial y} \cdot \frac{\partial v}{\partial x} + \frac{\partial^2}{\partial x\, \partial y} \cdot \frac{\partial u}{\partial y}.$$

Assuming that u and v are single-valued continuous functions in x and y, this equation can be written

$$\frac{\partial^2 \gamma_{xy}}{\partial x\, \partial y} = \frac{\partial^2}{\partial x^2} \cdot \frac{\partial v}{\partial y} + \frac{\partial^2}{\partial y^2} \cdot \frac{\partial u}{\partial x},$$

or, from equations (2.1),

$$\frac{\partial^2 \epsilon_x}{\partial y^2} + \frac{\partial^2 \epsilon_y}{\partial x^2} = \frac{\partial^2 \gamma_{xy}}{\partial x\, \partial y}. \tag{2.23}$$

Differentiating γ_{xy} with respect to x and z and γ_{zx} with respect to x and y and adding we obtain

$$\frac{\partial^2 \gamma_{xy}}{\partial x\, \partial z} + \frac{\partial^2 \gamma_{zx}}{\partial x\, \partial y} = \frac{\partial^2}{\partial x\, \partial z} \frac{\partial v}{\partial x} + \frac{\partial^2}{\partial x\, \partial z} \frac{\partial u}{\partial y} + \frac{\partial^2}{\partial x\, \partial y} \frac{\partial u}{\partial z} + \frac{\partial^2}{\partial x\, \partial y} \frac{\partial w}{\partial x}$$

$$= \frac{2\partial^2}{\partial y\, \partial z} \frac{\partial u}{\partial x} + \frac{\partial^2}{\partial x^2} \left(\frac{\partial w}{\partial y} + \frac{\partial v}{\partial z} \right)$$

$$= \frac{2\partial^2 \epsilon_x}{\partial y\, \partial z} + \frac{\partial^2 \gamma_{yz}}{\partial x^2}.$$

Hence

$$\frac{2\partial^2 \epsilon_x}{\partial y\, \partial z} = \frac{\partial}{\partial x}\left[\frac{\partial \gamma_{xy}}{\partial z} - \frac{\partial \gamma_{yz}}{\partial x} + \frac{\partial \gamma_{zx}}{\partial y}\right]. \tag{2.24}$$

Two other equations similar to each of equations (2.23) and (2.24) can be written down by cyclic permutation. These six equations are known as the equations of compatibility and must be satisfied, in addition to the equations of equilibrium and the boundary conditions, by the solution to any general three-dimensional problem.

2.8 Equations of compatibility in two dimensions

In plane stress problems the system of six equations of compatibility reduces to one. Taking the xy plane as the plane of strain this equation is

$$\frac{\partial^2 \epsilon_x}{\partial y^2} + \frac{\partial^2 \epsilon_y}{\partial x^2} = \frac{\partial^2 \gamma_{xy}}{\partial x\, \partial y}. \tag{2.25}$$

Since this equation is independent of strains which may exist in the z-direction it can also be applied to a body in a state of plane stress in the xy plane.

Equations of compatibility in terms of stresses

Equation (2.25) may be expressed in terms of stresses instead of strains for both plane stress and plane strain systems using the appropriate stress-strain relations derived from Hooke's equations. In the case of plane stress, the normal strains are given by equations (2.6):

$$\epsilon_x = \frac{1}{E}(\sigma_x - v\sigma_y); \qquad \epsilon_y = \frac{1}{E}(\sigma_y - v\sigma_x),$$

and the shear strain, from equations (2.9) and (2.16), by

$$\gamma_{xy} = \frac{\tau_{xy}}{G} = \frac{2(1+v)\tau_{xy}}{E}.$$

Inserting these values in equation (2.25) we obtain

$$\frac{\partial^2}{\partial y^2}(\sigma_x - v\sigma_y) + \frac{\partial^2}{\partial x^2}(\sigma_y - v\sigma_x) = 2(1+v)\frac{\partial^2 \tau_{xy}}{\partial x\, \partial y}. \tag{2.26}$$

From equations (1.11) the equations of equilibrium in two dimensions when body forces are present are

$$\frac{\partial \sigma_x}{\partial x} + \frac{\partial \tau_{xy}}{\partial y} + X = 0,$$

$$\frac{\partial \sigma_y}{\partial y} + \frac{\partial \tau_{xy}}{\partial x} + Y = 0.$$

Differentiating the first of these with respect to x and the second with respect to y and adding we obtain

$$\frac{2\partial^2 \tau_{xy}}{\partial x\,\partial y} = -\left(\frac{\partial^2 \sigma_x}{\partial x^2} + \frac{\partial^2 \sigma_y}{\partial y^2} + \frac{\partial X}{\partial x} + \frac{\partial Y}{\partial y}\right). \tag{2.27}$$

Substituting this in equation (2.26) gives

$$\frac{\partial^2 \sigma_x}{\partial y^2} + \frac{\partial^2 \sigma_y}{\partial x^2} + \frac{\partial^2 \sigma_x}{\partial x^2} + \frac{\partial^2 \sigma_y}{\partial y^2} = -(1 + \nu)\left(\frac{\partial X}{\partial x} + \frac{\partial Y}{\partial y}\right),$$

or (2.28)

$$\left(\frac{\partial^2}{\partial x^2} + \frac{\partial^2}{\partial y^2}\right)(\sigma_x + \sigma_y) = -(1 + \nu)\left(\frac{\partial X}{\partial x} + \frac{\partial Y}{\partial y}\right).$$

If we follow the same procedure for plane strain, substituting in this case the values for ϵ_x and ϵ_y given by equation (2.18) in equation (2.25) we obtain the compatibility equation in the form

$$\left(\frac{\partial^2}{\partial x^2} + \frac{\partial^2}{\partial y^2}\right)(\sigma_x + \sigma_y) = -\frac{1}{1 - \nu}\left(\frac{\partial X}{\partial x} + \frac{\partial Y}{\partial y}\right). \tag{2.29}$$

If the body forces are zero or constant, both equations (2.28) and (2.29) reduce to

$$\left(\frac{\partial^2}{\partial x^2} + \frac{\partial^2}{\partial y^2}\right)(\sigma_x + \sigma_y) = 0, \tag{2.30}$$

showing that in this case the equation of compatibility is the same for plane stress and plane strain. It was shown in Section 1.3 that the sum of the normal stresses σ_x and σ_y at any given point is a constant for all directions of the axes. Denoting this sum by Σ we can write equation (2.30) in the form

$$\frac{\partial^2 \Sigma}{\partial x^2} + \frac{\partial^2 \Sigma}{\partial y^2} = 0. \tag{2.31}$$

Equation (2.31) is known as Laplace's equation. It is also called a harmonic equation and any function which satisfies it is called a harmonic function. The operator $(\partial^2/\partial x^2 + \partial^2/\partial y^2)$ is termed the Laplace operator and denoted by ∇^2. Thus, equation (2.31) for instance can be written in the form

$$\nabla^2 \Sigma = 0. \tag{2.32}$$

A point of great practical importance to be observed is that none of the elastic constants appear in equation (2.30). This implies that, when the body forces are zero or constant, the stress distribution in geometrically similar bodies made of different elastic materials will be the same provided they are subjected to the same system of forces. This is the mechanical basis of the photoelastic method. Equations (2.28) and (2.29) show that with variable body forces the results depend on the value of Poisson's ratio v.

2.9   The stress function

As mentioned previously, the problem in theory of elasticity is to obtain a solution to the equations of equilibrium which will also satisfy the equations of compatibility and the boundary conditions. It was shown by Airy that the stress components which satisfy the equations of equilibrium in two dimensions can be expressed in terms of a single function in x, y known as the stress function or Airy's stress function. If the only body force is gravity and we take the y axis vertically upwards, the equations of equilibrium, equations (1.11) become

$$\frac{\partial \sigma_x}{\partial x} + \frac{\partial \tau_{xy}}{\partial y} = 0,$$

$$\frac{\partial \sigma_y}{\partial y} + \frac{\partial \tau_{xy}}{\partial x} - \rho g = 0,$$

where ρ is the mass density of the material. As we can verify by substitution, these equations are satisfied by the stress components

$$\sigma_x = \frac{\partial^2 \phi}{\partial y^2}; \qquad \sigma_y = \frac{\partial^2 \phi}{\partial x^2}; \qquad \tau_{xy} = -\frac{\partial^2 \phi}{\partial x\, \partial y} + \rho g x, \qquad (2.33)$$

where ϕ is an arbitrary function in x and y. If we substitute from these relations in the equation of compatibility, equation (2.30), we obtain

$$\frac{\partial^4 \phi}{\partial x^4} + 2\frac{\partial^4 \phi}{\partial x^2\, \partial y^2} + \frac{\partial^4 \phi}{\partial y^4} = 0, \qquad (2.34)$$

which may be written as

$$\nabla^4 \phi = 0,$$

where

$$\nabla^4 = \left(\frac{\partial^2}{\partial x^2} + \frac{\partial^2}{\partial y^2} \right)^2.$$

Equation (2.34) is called the biharmonic equation. Any solution of this equation satisfies the equations of equilibrium and the equation of compatibility; if the boundary conditions are also satisfied, the solution is the correct one and is unique.

If polar co-ordinates are used, the equations of equilibrium when the body forces are zero or can be neglected are expressed by equations (1.14). It can be verified that these equations are satisfied by the stress components

$$\sigma_r = \frac{1}{r}\frac{\partial\phi}{\partial r} + \frac{1}{r^2}\frac{\partial^2\phi}{\partial\theta^2},$$

$$\sigma_\theta = \frac{\partial^2\phi}{\partial r^2}, \tag{2.35}$$

$$\tau_{r\theta} = \frac{1}{r^2}\frac{\partial\phi}{\partial\theta} - \frac{1}{r}\frac{\partial^2\phi}{\partial r\,\partial\theta} = -\frac{\partial}{\partial r}\left(\frac{1}{r}\frac{\partial\phi}{\partial\theta}\right),$$

where ϕ is a function of r and θ.

To express the biharmonic equation in polar co-ordinates we replace x and y by r and θ using the relations

$$r^2 = x^2 + y^2; \qquad \theta = \tan^{-1}\frac{y}{x}.$$

These give

$$\frac{\partial\phi}{\partial x} = \frac{\partial\phi}{\partial r}\cos\theta - \frac{1}{r}\frac{\partial\phi}{\partial\theta}\sin\theta,$$

$$\frac{\partial\phi}{\partial y} = \frac{\partial\phi}{\partial r}\sin\theta + \frac{1}{r}\frac{\partial\phi}{\partial\theta}\cos\theta.$$

Substituting in equation (2.34) we obtain

$$\left(\frac{\partial^2}{\partial r^2} + \frac{1}{r}\frac{\partial}{\partial r} + \frac{1}{r^2}\frac{\partial^2}{\partial\theta^2}\right)\left(\frac{\partial^2\phi}{\partial r^2} + \frac{1}{r}\frac{\partial\phi}{\partial r} + \frac{1}{r^2}\frac{\partial^2\phi}{\partial\theta^2}\right) = 0. \tag{2.36}$$

3

GENERAL OPTICS

3.1 Nature of light

Light is a visible form of the radiation of energy. Theories which have been advanced in an attempt to explain its nature have been based in general terms on one or other of two different concepts, namely, that the transmission of light energy was accomplished by some form of wave motion or by moving corpuscles or particles.

The transverse wave theory of light was established by Fresnel in his well known elastic-solid theory. This theory, although superseded by the more modern electromagnetic theory, still provides the simplest explanation of many optical phenomena.

In the electromagnetic theory, light is assumed to consist of electromagnetic disturbances and as such to follow the fundamental laws of electromagnetism formulated by James Clerk Maxwell in 1865. In this context, light represents the visible portion only of a much wider spectrum which extends from the very high frequency cosmic radiations to the low frequency radio waves.

The particle theory in its modern state exists in the various formulations of the quantum theory. In this theory, developed largely from the work of Planck and Einstein, it is assumed that the emission and absorption of light energy by an atom or molecule is not a continuous process but occurs in steps in which the energy transferred for radiation of a given frequency is always of the same amount. This energy is transmitted through space in the form of bundles or concentrations of energy known as quanta or photons.

Both the wave and particle theories can be used to explain certain, but not all, of the known optical phenomena. The wave theory adequately explains the transmission of light through space and transparent media and all the associated phenomena such as reflection, refraction, interference and polarization. It does not, however, offer any explanation of the interaction of light and matter such as the photoelectric and Compton effects. The particle theory provides an explanation for these but not for such phenomena as interference and diffraction.

Since 1927, the modern theory of quantum mechanics has been developed from the work of de Broglie, Heisenberg, Schrödinger and others. This is a unified theory including the properties of both light and matter. It is essentially mathematical and cannot properly be represented in physical terms. It shows, however, that under certain conditions particles have associated wave characteristics so that the wave and particles theories can be regarded as complementary rather than in opposition to one another.

On this basis, it is justifiable to utilize the concepts of either wave or particles, and to choose whichever will most directly yield a solution to the particular problem involved.

All the phenomena encountered in photoelasticity can be adequately represented in terms of the transverse wave theory and this concept has been adopted in its simplest form in this book.

3.2 Simple harmonic motion

The simplest form of wave motion consists of the combination of a transverse simple harmonic vibration and a uniform translation. A particle moving in a straight line is said to have simple harmonic motion if its acceleration at any instant is proportional to its displacement from a fixed point in that line and is directed towards that point. This can be expressed by the equation

$$\frac{d^2y}{dt^2} = -\omega^2 y \tag{3.1}$$

in which y is the displacement at time t and ω is a constant.

One instance of simple harmonic motion is that of the projection on any diameter of a point which describes a circular path with uniform angular velocity. Consider a point Q which describes a circle about centre O with uniform angular velocity ω, Figure 3.1. P is the projection of Q on any diameter $A'A$. The acceleration of Q is $\omega^2 QO$ along QO. The acceleration of P is equal to the component of the acceleration of Q in the direction of AA', i.e. $\omega^2 PO$. The acceleration of P is therefore proportional to its distance from O and is directed towards O and therefore the motion of P is simple harmonic.

Let the time t be measured from the instant when Q occupies the position Q_0. Then $Q_0\hat{O}Q = \omega t$. Denoting the angle BOQ_0 by α and the radius OQ by a, then the position of P is given by the ordinate

$$y = a \sin(\omega t + \alpha),$$

which represents a general solution of equation (3.1)

The angle $(\omega t + \alpha)$ is known as the phase and α is the initial phase. If the phase is measured anticlockwise from OA instead of from OB, the solution is obtained as

$$y = a \cos(\omega t + \alpha).$$

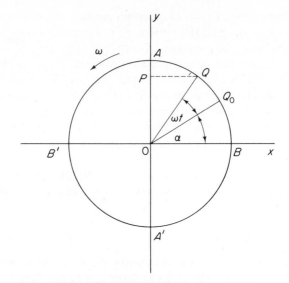

Figure 3.1. Circular representation of simple
harmonic motion

Each of the above two equations can be written in the form

$$y = b \cos \omega t + c \sin \omega t,$$

where

$$b^2 + c^2 = a^2 \quad \text{and} \quad c/b = \tan \alpha.$$

The general solution of the simple harmonic equation (3.1) can therefore be expressed in the alternative forms

$$y = a \sin (\omega t + \alpha)$$

$$y = a \cos (\omega t + \alpha)$$

$$y = b \cos \omega t + c \sin \omega t.$$

The maximum displacement a of P is known as the amplitude of the vibration. The time required for the point P to perform one complete oscillation is called the periodic time and is denoted by T. This is obviously the same as the time required for the point Q to make one complete revolution so that

$$T = 2\pi/\omega. \tag{3.2}$$

The reciprocal of T, i.e. the number of vibrations per second, is called the frequency and will be denoted by f. Then

$$f = \omega/2\pi.$$

ω is called the angular or circular frequency.

3.3 Fundamental wave equation

Consider a simple harmonic vibration represented by $y = a \sin(\omega t + \alpha)$, which is propagated in the direction of the x axis perpendicular to the direction of vibration with uniform velocity v, Figure 3.2. The displacement at the

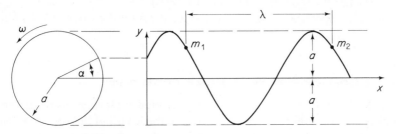

Figure 3.2.   Generation of a sinusoidal wave

origin at any instant will reach a point at distance x along the x axis after an interval of time $t = x/v$. The phase of the vibration at any point x therefore lags behind that of the vibration at the origin by the amount $\omega x/v$. The vibration at any point on the line of propagation is therefore represented by

$$y = a \sin\left[\omega(t - x/v) + \alpha\right]. \tag{3.3}$$

The distance between any two corresponding points m_1 and m_2 on successive waves is called the wavelength and is denoted by λ. Obviously the disturbance at any point will again have the same value after a time λ/v, which is equal to the periodic time T. Hence, using equation (3.2) we obtain

$$T = \lambda/v = 2\pi/\omega,$$

and

$$\omega = 2\pi v/\lambda. \tag{3.4}$$

Inserting the value of ω in equation (3.3) gives

$$y = a \sin\left[\frac{2\pi v}{\lambda}\left(t - \frac{x}{v}\right) + \alpha\right],$$

i.e.

$$y = a \sin\left[\frac{2\pi}{\lambda}(vt - x) + \alpha\right], \tag{3.5}$$

which is the fundamental equation to the wave. The constant $2\pi/\lambda$ is known as the wavelength constant.

If x is treated as constant, equation (3.5) represents the disturbance at a distance x from the origin. If t is treated as constant, it represents the wave profile at that instant. Since the profile follows a sine curve, waves of the form given by equation (3.5) are called sinusoidal waves.

Since wavelengths within the visible spectrum are very small, it is convenient to measure them in terms of units much smaller than the centimetre or millimetre. Those most frequently used are the Ångström unit (A.U. or Å = 10^{-8} cm) and the millimicron ($m\mu = 10^{-7}$ cm).

The visible spectrum ranges from about 3900 Å for the violet to 7700 Å for the red.

3.4   Vectorial representation of simple harmonic motion

Simple harmonic motion may be represented vectorially using either a rotating or a stationary vector. In each case the length of the vector denotes the amplitude of the motion. The rotating vector corresponds to the radius OQ in Figure 3.1. It rotates with uniform angular velocity ω about the centre and its inclination to the x axis at any instant is equal to the phase $(\omega t + \alpha)$. If a stationary vector is used it represents the initial conditions of the motion and is accordingly inclined to the x axis at an angle equal to the initial phase α.

In photoelasticity we are frequently concerned with the combined effects of two simple harmonic vibrations of the same period. The resultant effect can be obtained by superposition using either algebraic or vector addition.

3.5   Combination of two collinear simple harmonic vibrations of the same frequency

Let the component vibrations be represented by

$$y_1 = a_1 \sin(\omega t + \alpha_1) \quad \text{and} \quad y_2 = a_2 \sin(\omega t + \alpha_2).$$

Then the resultant disturbance $y = y_1 + y_2$.

In Figure 3.3, OQ_1 and OQ_2 are the stationary vectors representing y_1 and y_2, respectively. Their vector sum is OQ which therefore represents their resultant y.

Then

$$y = A \sin(\omega t + \gamma). \tag{3.6}$$

The values of the initial phase γ and the amplitude A can be read by inspection from the diagram:

$$\tan \gamma = \frac{a_1 \sin \alpha_1 + a_2 \sin \alpha_2}{a_1 \cos \alpha_1 + a_2 \cos \alpha_2},$$

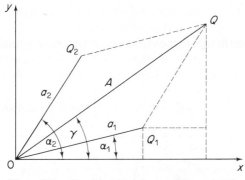

Figure 3.3. Combination of two collinear simple
harmonic vibrations of the same frequency

and

$$A = \sqrt{((a_1 \cos \alpha_1 + a_2 \cos \alpha_2)^2 + (a_1 \sin \alpha_1 + a_2 \sin \alpha_2)^2)}$$

$$\dot{=} \sqrt{(a_1^2 + a_2^2 + 2a_1 a_2 \cos(\alpha_1 - \alpha_2))}$$

$$= \sqrt{(a_1^2 + a_2^2 + 2a_1 a_2 \cos \alpha)}, \tag{3.7}$$

where α = phase difference = $(\alpha_1 - \alpha_2)$.

3.6 Combination of two mutually perpendicular simple harmonic vibrations of the same frequency

We now consider the resultant vibration of a particle which is subjected simultaneously to two simple harmonic vibrations of the same frequency acting in directions at right angles to each other. Let time be measured from the instant when the rotating vector which represents the y vibration crosses the x axis so that the initial phase of this component is zero and let the corresponding phase of the x vibration be α.

In Figure 3.4, let the rotating vectors which represent the x and y component vibrations occupy the positions OP and OQ at any instant. The point S has the same projections as P on the x axis and Q on the y axis respectively and is therefore identical with the tip of the resultant vector at that instant. The resultant vibration is represented by the trajectory described by S as the vibration proceeds.

The component vibrations are expressed by

$$x = a_1 \sin(\omega t + \alpha); \qquad y = a_2 \sin \omega t. \tag{3.8}$$

From equations (3.8),

$$\frac{y}{a_2} = \sin \omega t; \qquad \frac{x}{a_1} = \sin \omega t \cos \alpha + \cos \omega t \sin \alpha,$$

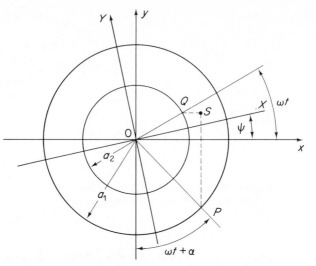

Figure 3.4. Combination of two mutually perpendicular simple harmonic vibrations of the same frequency

which may be written

$$\frac{y}{a_2}\sin\alpha = \sin\omega t\sin\alpha; \qquad \frac{x}{a_1} - \frac{y}{a_2}\cos\alpha = \cos\omega t\sin\alpha.$$

Squaring and adding, we obtain

$$\frac{y^2}{a_2^2}\sin^2\alpha + \frac{x^2}{a_1^2} - \frac{2xy\cos\alpha}{a_1 a_2} + \frac{y^2}{a_2^2}\cos^2\alpha = \sin^2\omega t\sin^2\alpha + \cos^2\omega t\sin^2\alpha,$$

from which

$$\frac{x^2}{a_1^2} - \frac{2xy\cos\alpha}{a_1 a_2} + \frac{y^2}{a_2^2} = \sin^2\alpha. \tag{3.9a}$$

This is the equation to an ellipse since $(-\cos\alpha/a_1 a_2)^2 < (1/a_1^2)(1/a_2^2)$. To determine the directions of the axes of this ellipse, let the axes of reference be rotated counterclockwise through an angle ψ so that the new co-ordinates (X, Y) are related to (x, y) by

$$x = X\cos\psi - Y\sin\psi; \qquad y = X\sin\psi + Y\cos\psi.$$

Replacing (x, y) in equation (3.9a) by X, Y we obtain

$$\frac{X^2 \cos^2 \psi - 2XY \sin \psi \cos \psi + Y^2 \sin^2 \psi}{a_1^2}$$

$$- \frac{2 \cos \alpha (X \cos \psi - Y \sin \psi)(X \sin \psi + Y \cos \psi)}{a_1 a_2}$$

$$+ \frac{X^2 \sin^2 \psi + 2XY \sin \psi \cos \psi + Y^2 \cos^2 \psi}{a_2^2} = \sin^2 \alpha. \quad (3.9b)$$

The coefficient of the term in XY is

$$-\frac{\sin 2\psi}{a_1^2} - \frac{2 \cos \alpha}{a_1 a_2}(\cos^2 \psi - \sin^2 \psi) + \frac{\sin 2\psi}{a_2^2},$$

i.e.

$$-\frac{\sin 2\psi}{a_1^2} - \frac{2 \cos \alpha}{a_1 a_2} \cos 2\psi + \frac{\sin 2\psi}{a_2^2}.$$

The axes of reference coincide with the axes of the ellipse when the XY term vanishes. Then

$$\sin 2\psi \left(\frac{1}{a_2^2} - \frac{1}{a_1^2} \right) = \frac{2 \cos \alpha}{a_1 a_2} \cos 2\psi,$$

and

$$\tan 2\psi = \frac{2a_1 a_2 \cos \alpha}{a_1^2 - a_2^2}$$

$$= \tan 2\theta \cos \alpha, \quad (3.10)$$

where $\theta = \tan^{-1} a_2/a_1$.

The two values of ψ ($< 2\pi$) differing by $\pi/2$ which are identified by equation (3.10) define the directions of the major and minor axes of the ellipse. Inserting these values in equation (3.9b) and equating Y and X in turn to zero, we obtain the lengths of the semi-axes OA, OB of the ellipse:

$$OA, OB = \frac{1}{\sqrt{2}} \sqrt{\{(a_1^2 + a_2^2) \pm \sqrt{[(a_1^2 - a_2^2)^2 + 4a_1 a_2 \cos^2 \alpha]}\}}$$

$$\quad (3.11)$$

$$= \frac{\alpha}{\sqrt{2}} \sqrt{[1 \pm \sqrt{(\cos^2 2\theta + \sin^2 2\theta \cos^2 \alpha)}]},$$

where $a = \sqrt{(a_1^2 + a_2^2)}$.

To determine the sense in which the ellipse $x = a_1 \sin(\omega t + a)$, $y = a_2 \sin \omega t$ is described we shall consider the instant when $\omega t = \pi/2$.

Then

$$x = a_1 \cos \alpha, \qquad y = a_2.$$

The corresponding position of S is shown in Figure 3.5.

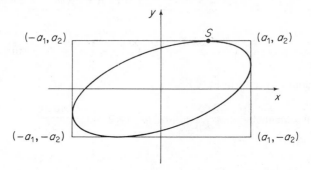

Figure 3.5. Elliptical vibration produced by the combination of two mutually perpendicular simple harmonic vibrations

The component velocities dx/dt and dy/dt at any given instant are given by

$$dx/dt = a_1 \omega \cos(\omega t + \alpha); \qquad dy/dt = a_2 \omega \cos \omega t.$$

For the instant under consideration,

$$dx/dt = -a_1 \omega \sin \alpha; \qquad dy/dt = 0.$$

As would be expected, dy/dt is zero since the y-component is at maximum displacement. If $0 < \alpha < \pi$, dx/dt is negative so that S moves to the left and the ellipse is described in an anticlockwise sense. If however, $\pi < \alpha < 2\pi$, dx/dt is positive and the ellipse is described in a clockwise sense.

We shall now consider some special cases.

(i) $\alpha = 2m\pi$, where m is zero or any integer.
Equations (3.8) reduce to

$$x = a \sin \omega t; \qquad y = \alpha \sin \omega t,$$

from which

$$y = \frac{a_2}{a_1} x.$$

This is the equation to a straight line lying in the first and third quadrants and passing through the origin and the corners $(a_1, a_2), (-a_1, -a_2)$ of the rectangle $x = \pm a_1, y = \pm a_2$.

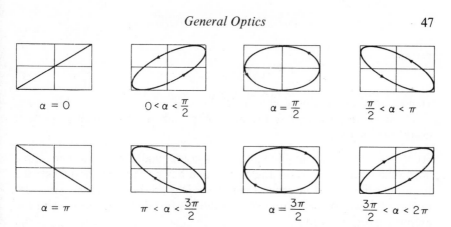

Figure 3.6   Resultant vibrations produced with various phase differences between the components

(ii) $\alpha = (2m + 1)\pi$.
We now obtain

$$x = -a_1 \sin \omega t; \qquad y = a_2 \sin \omega t,$$

giving

$$y = -\frac{a_2}{a_1}x,$$

which again represents a straight line. In this case the line lies in the second and fourth quadrants and passes through the points $(-a_1, a_2)$ and $(a_1, -a_2)$

(iii) $\alpha = (2m + 1)\dfrac{\pi}{4}$.

In this case,

$$x = a_1 \cos \omega t; \qquad y = a_2 \sin \omega t,$$

giving

$$\frac{x^2}{a_1^2} + \frac{y^2}{a_2^2} = 1.$$

This is the equation to an ellipse with principal axes coinciding with the co-ordinate axes.

(iv) $\alpha = (2m + 1)\dfrac{\pi}{2}$ and $a_1 = a_2$.

We now have

$$x = a_1 \cos \omega t; \qquad y = a_1 \sin \omega t,$$

giving

$$x^2 + y^2 = a_1^2.$$

Hence the resultant vibration is circular.

We shall now summarize the more important results which have been established above.

(1) When a simple harmonic vibration is compounded with another simple harmonic vibration at right angles to it and having the same frequency, the resultant vibration is in general elliptical. This vibration has the same frequency as each of the component vibrations but the initial phase and the amplitude are changed.

(2) If the component vibrations have a constant phase difference between them of zero or any multiple of π, the ellipse degenerates into a straight line, i.e. the resultant vibration is linear.

(3) If the component vibrations have equal amplitudes and a constant phase difference between them of $(2m + 1)\pi/2$, the resultant motion is circular.

The variations in the character of the resultant vibration for progressively increasing values of the phase difference α between zero and 2π are shown in Figure 3.5. It is assumed that the component vibrations have different amplitudes and that $a_1 > a_2$. The major axis of the ellipse rotates between limiting angles of $\pm \tan^{-1} a_2/a_1$ with respect to the x axis. If $a_2 > a_1$, the major axis rotates between $\tan^{-1} a_2/a_1$ and $\pi - \tan^{-1} a_2/a_1$. If the amplitudes of the component vibrations are equal, i.e. $a_1 = a_2$, then, as can be seen from equation (3.10), $\tan 2\psi$ is infinite for all values of α except zero or $(2m + 1)(\pi/2)$ when it is indeterminate. In this case, therefore, the angular positions of the principal axes do not rotate but maintain fixed directions at 45° to the directions of the component vibrations except when the resultant vibration is circular.

3.7 Natural light

The characteristic feature of natural or ordinary light is that the transverse vibration pattern exhibits no directional or rotational properties. A beam of natural light may be regarded as consisting of waves vibrating with different amplitudes in all possible azimuths. This is illustrated in Figure 3.7a in which the individual vibrations are represented by radii vectors known as light vectors, the length of each being proportional to the corresponding amplitude. It can be assumed that the vibration pattern changes form in a random manner so rapidly that the mean effect over any finite interval of time is that of a ray which is perfectly symmetrical about the axis of propagation.

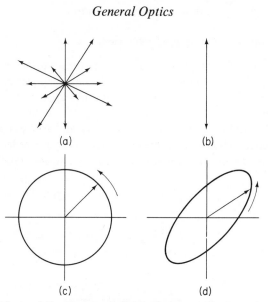

Figure 3.7. Forms of light vibration. (*a*) Natural.
(*b*) Plane polarized. (*c*) Circularly polarized. (*d*) Elliptically polarized

3.8 Polarized light

When the waves in a beam of light are constrained to vibrate in a systematic manner in planes normal to the direction of propagation, the beam is said to be polarized. The form of polarization is described with reference to the behaviour of the representative light vector. When the waves in a beam of light vibrate in parallel planes so that the orientation of the light vector is constant, the beam is said to be plane polarized.

If the vibration is such that the amplitude remains constant while the orientation of the light vector changes uniformly so that the tip of the vector traces out a circle, the light is said to be circularly polarized. If both the amplitude and the orientation vary in a related way so that the vector tip traces out an ellipse, the light is said to be elliptically polarized. These different forms of polarization are illustrated in Figures 3.7(*b*), (*c*) and (*d*), respectively.

3.9 White and monochromatic light

Colour depends on the frequency or wavelength of the light waves. Thus waves of any given frequency correspond to light of a definite colour and are said to be monochromatic.

The light emitted from a source appears white when it includes waves of all frequencies within the visible spectrum in roughly the same amount. Such light is called white light.

3.10   Reflection and refraction at plane surfaces

When a parallel beam of light is incident on a surface separating two transparent media, it is found that, in general, part of the light is reflected back into the medium of the incident beam and part is transmitted into the second medium.

In Figure 3.8, *AO* represents an incident ray and *NN'* the normal at the point of incidence. *OB* and *OC* represent respectively the reflected and transmitted rays. The direction of the transmitted ray does not coincide with that

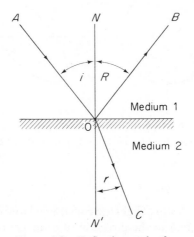

Figure 3.8.   Reflection and refraction at a plane surface

of the incident ray and the transmitted ray is said to be refracted. The angles which the incident, reflected and refracted rays make with the normal to the surface at the point of incidence are known as the angles of incidence, reflection and refraction and are denoted by *i*, *R* and *r* respectively.

Reflection and refraction in isotropic media obey the following laws which can readily be verified by observation:

1. The incident, reflected and refracted rays and the normal to the surface are coplanar.

2. The angle of reflection is equal to the angle of incidence.

3. The ratio of the sine of the angle of incidence to the sine of the angle of refraction is a constant depending on the nature of the media and the wavelength of the light. This constant is called the relative refractive index of the two media for light travelling from medium 1 into medium 2 and is denoted by n_{12}.

Thus,

$$\frac{\sin i}{\sin r} = n_{12}. \tag{3.12}$$

This is usually known as Snell's law after its discoverer.

If light enters a medium from a vacuum, the above ratio is called the absolute refractive index of the medium and is denoted simply by n, a single subscript being added when it is necessary to differentiate between different media.

For light travelling in the reverse direction, i.e. from medium 2 into medium 1, the relative refractive indices are related by

$$n_{21} = 1/n_{12}. \tag{3.13}$$

Refraction is explained in terms of the wave theory by assuming that the velocity of light depends on the medium through which it passes and is a maximum in a vacuum. If v_1 and v_2 are the velocities of light in the two media then an alternative definition of the relative refractive index n_{12} is

$$n_{12} = v_1/v_2.$$

If c is the velocity of light in a vacuum (about 3×10^{10} cm/sec), the absolute refractive indices are

$$n_1 = c/v_1 \quad \text{and} \quad n_2 = c/v_2,$$

so that

$$n_{12} = v_1/v_2 = n_2/n_1,$$

i.e.

$$n_2 = n_1 n_{12}. \tag{3.14}$$

If the first medium is air at standard conditions of temperature and pressure, $n_1 = 1 \cdot 000278$. Equation (3.14) shows, therefore, that for most practical purposes, the absolute refraction index of a solid material and its relative refractive index from air may be regarded as the same.

When light travels from one isotropic medium to another, the number of waves passing in any given time interval must be the same at every point in both media, i.e. the frequency is unchanged. It follows that the wavelength will vary according to the medium. The frequency f is given by

$$f = v/\lambda.$$

Hence, if λ_1 and λ_2 denote the wavelengths in the two media, we have

$$\lambda_1/\lambda_2 = v_1/v_2 = n_{12}. \tag{3.15}$$

If λ_0 is the wavelength in a vacuum, the absolute refractive indices are

$$n_1 = \lambda_0/\lambda_1 \quad \text{and} \quad n_2 = \lambda_0/\lambda_2. \tag{3.16}$$

When light travels in succession through m different media we obtain

$$n_{1m} = n_{12} \cdot n_{23} \cdot n_{34} \ldots n_{(m-1)m}.$$

3.11 Total internal reflection

When light travels from a denser to a rarer medium, in general $n_{12} < 1$, i.e. $\sin i/\sin r < 1$. Thus, $\sin r = 1$ when i is still less than $90°$ so that a further increase in i would imply that $\sin r > 1$. This is impossible and experiment shows that this imaginary case represents a transition from refraction to total reflection. The value of i when this first occurs is known as the critical angle. From the above, the critical angle i_c is obviously given by

$$\sin i_c = n_{12}.$$

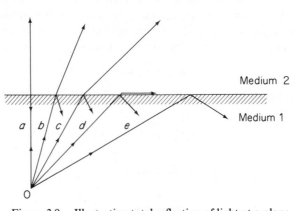

Figure 3.9.   Illustrating total reflection of light at a plane surface

This phenomenon is illustrated in Figure 3.9 which shows various rays originating at a point source O in the denser medium. Ray a is reflected and transmitted without deviation. Rays b and c give rise to reflected and refracted rays. Ray d is at the critical angle and the refracted ray is parallel to

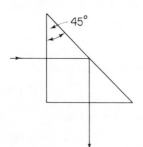

Figure 3.10.   Total reflection of light in a glass prism

the surface. Ray *e* is totally reflected. For glass *n* is not usually less than 1·5 so that the critical angle is less than 45°. Hence, if light enters a 45° glass prism as shown in Figure 3.10 it will be totally reflected, the reflected ray being perpendicular to the incident ray. Prisms of this form are employed in many optical instruments.

3.12 Refraction through a parallel plate

Let $(i_1, r_1), (i_2, r_2)$ be the angles of incidence and refraction at the upper and lower surfaces respectively of the parallel plate of isotropic material shown in Figure 3.11. From equation (3.12) and (3.13) we have

$$n_{12} = \frac{\sin i_1}{\sin r_1} = \frac{1}{n_{21}} = \frac{\sin r_2}{\sin i_2}.$$

Also, since the surfaces are parallel, $r_1 = i_2$ so that $i_1 = r_2$. Thus, the emergent rays are parallel to the incident rays but suffer a lateral displacement.

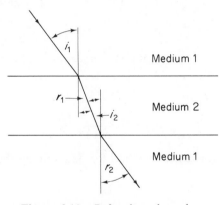

Figure 3.11. Refraction through a parallel plate

3.13 Refraction through a prism

When a ray of light passes through a prism as shown in Figure 3.12(*a*) it is refracted (and reflected) at both the incident and emergent surfaces. The angle γ between the incident and emergent rays is the angular deviation produced by the prism. It can be shown that the angular deviation is a minimum when the ray passes symmetrically through the prism so that $i_1 = r_2$. Then also, from Figure 3.12(*b*), $r_1 = i_2 = \delta/2$, and since

$$n_{12} = \sin i_1 / \sin r_1, \qquad \sin i_1 = n_{12} \sin \delta/2.$$

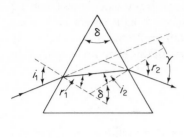
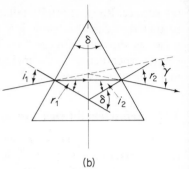

(a)   (b)

Figure 3.12.   Refraction through a prism. (*a*) General case. (*b*) Condition for minimum deviation

Hence

$$\gamma = 2(i_1 - r_1) = 2\left(\sin^{-1} n_{12} \sin\frac{\delta}{2} - \frac{\delta}{2}\right),$$

from which

$$\sin\tfrac{1}{2}(\delta + \gamma) = n_{12} \sin\frac{\delta}{2}$$

and therefore

$$n_{12} = \frac{\sin\tfrac{1}{2}(\delta + \gamma)}{\sin\tfrac{1}{2}\delta}.$$

This provides a convenient method for determining the refractive index of a material.

3.14   Dispersion

Since the index of refraction varies with the wavelength, the angle of deviation of a ray of light which passes through a prism will also vary with the wavelength. Hence if a beam of collimated white light is incident on a prism as shown in Figure 3.13, the constituent waves of different wavelengths will have slightly different deviations and the light will be resolved into its separate constituent colours. Since the wavelength of red light is the longest and that of violet light the shortest in the visible spectrum it follows from equation (3.16) that the red will suffer the least deviation and the violet the greatest. If, after passing through the prism, the rays of light fall on a screen placed as shown, in general a continuous spectrum will be observed with the colours appearing in the order indicated. The resolution of white light into its separate monochromatic components is called dispersion.

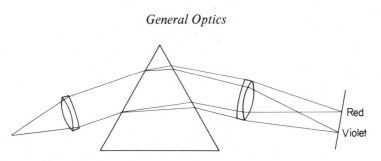

Figure 3.13. Dispersion of light by a prism

3.15 Huygens' principle

Huygens suggested that every point on a wave front acts as a source of secondary waves. Successive positions of the wave front are then given by the envelopes of the secondary waves. Let us consider a spherical wave front W_1, Figure 3.14, such as is obtained with a point source in an isotropic medium.

Figure 3.14. Propagation of a spherical wave front according to Huygens' principle

The wavelets from each point on W_1 are represented in the diagram by circular arcs the radii of which are proportional to the time which has elapsed since the wave front occupied the position W_1. The envelope of these arcs, i.e. W_2, represents the new wave front. This wave front is also spherical so that successive wave fronts in this case form a series of concentric spheres. This method of determining wave fronts is known as Huygens' construction.

3.16 Wave theory of reflection and refraction

Huygens' construction provides a simple interpretation of reflection and refraction in terms of the wave theory. Let us suppose a parallel beam of light falls on the surface of separation between two media 1 and 2. As previously mentioned, the wave theory assumes that the velocity of light depends on the medium through which it is transmitted. Let v_1 and v_2 be the velocities of light in the respective media. We begin by considering the formation of the reflected wave, referring to Figure 3.15. If we assume the wave front AB to reach position CD at time $t = 0$, the light travelling along the ray BE will

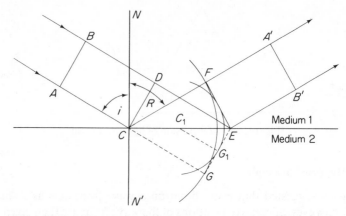

Figure 3.15.   Huygens' construction demonstrating reflection at a
plane surface

reach E after a time t such that

$$v_1 t = DE = CE \sin i = CG$$

where i is the angle of incidence, which is also equal to $D\hat{C}E$.

Thus, when the secondary wave is just starting at E, that from C will have formed a hemisphere of radius CG. At the same instant the secondary wave from any intermediate point such as C_1 will have formed a hemisphere of radius $C_1 G_1$ and it may easily be verified that this hemisphere touches the tangent plane EF. EF is therefore the reflected wave front. The reflected rays make an angle R with the normal to the surface. Observing that $R = C\hat{E}F$ we obtain

$$\sin R = CF/CE = v_1 t/CE = \sin i$$

as required.

Considering the refracted rays in a similar way, when the secondary wave is just starting at E, Figure 3.16, that from C will have formed a hemisphere of radius $v_2 t$. The tangent plane EF_1 is found to touch the hemispheres from intermediate points such as C_1 so that EF_1 is the refracted wave front. Since the angle of refraction $r = C\hat{E}F_1$ we have

$$\sin r = \frac{CF_1}{CE} = \frac{v_2 t}{CE}$$

and since $\sin i = v_1 t/CE$,

$$\sin r = v_2/v_1 \sin i,$$

which agrees with equation (3.12) since $v_2/v_1 = n_1/n_2 = 1/n_{12}$.

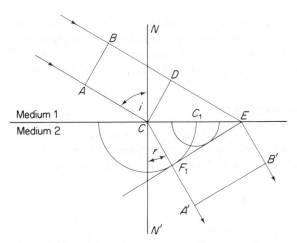

Figure 3.16. Huygens' construction demonstrating refraction at a plane surface

3.17 Polarization by reflection and refraction

If a beam of natural light is incident on the surface of a transparent medium at an angle of about 57°, it is found that the reflected beam is plane polarized, the plane of vibration being perpendicular to the plane of incidence as indicated in Figure 3.17. The plane of polarization is conventionally taken to

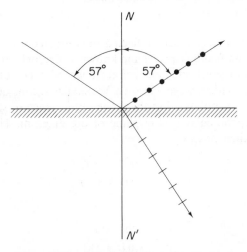

Figure 3.17. Polarization by reflection and refraction

Photoelastic Stress Analysis

be perpendicular to the plane of vibration. The refracted ray is also partially plane polarized but with the vibrations lying in the plane of incidence. The angle of incidence giving the greatest degree of polarization depends on the refractive index of the material and is called the polarizing angle.

3.18   Fresnel's equations

We now consider a plane polarized wave, represented by the ray *AO* which is incident at an angle *i* on the plane surface of a transparent medium, Figure 3.18. *OB* and *OC* are the resulting reflected and refracted rays. Let *a*, *b* and *c*

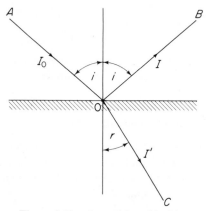

Figure 3.18.   Intensities of incident,
reflected and refracted waves

be the amplitudes of the incident, reflected and refracted waves. The following formulae, first obtained by Fresnel and known as Fresnel's equations, enable the amplitudes of the reflected and refracted waves to be obtained when the amplitude of the incident wave is known. We shall first consider the case when the incident wave is polarized in the plane of incidence so that the plane of vibration is perpendicular to the plane of the diagram. The ratios of the amplitudes are then given by

$$\frac{b}{a} = -\frac{\sin(i-r)}{\sin(i+r)}, \tag{3.17}$$

and

$$\frac{c}{a} = \frac{2\cos i \sin r}{\sin(i+r)}. \tag{3.18}$$

For light plane polarized in the perpendicular direction so that the plane of vibration coincides with the plane of the diagram,

$$\frac{b}{a} = \frac{\tan(i-r)}{\tan(i+r)},$$ (3.19)

and

$$\frac{c}{a} = \frac{2\cos i \sin r}{\sin(i+r)\cos(i-r)}.$$ (3.20)

The square of each of the above equations gives the ratio of the corresponding intensities. The quantity b^2/a^2, giving the relative intensity of the reflected light to the incident light, is known as the reflective power or coefficient of reflection of the surface.

For the present purposes, natural or unpolarized light can be assumed to consist of equal amounts of two components plane polarized in mutually perpendicular azimuths. Hence, the intensity J of natural light reflected at the surface is given by

$$\frac{J}{I_0} = \frac{\sin^2(i-r)}{2\sin^2(i+r)} + \frac{\tan^2(i-r)}{2\tan^2(i+r)},$$ (3.21)

where I_0 is the intensity of the incident light. The intensity I of the refracted wave is given by

$$\frac{I}{I_0} = \frac{\cos^2 i \sin^2 r}{\sin^2(i+r)} + \frac{\cos^2 i \sin^2 r}{\sin^2(i+r)\cos^2(i-r)}.$$ (3.22)

We shall now consider some particular applications of Fresnel's equations. When the angle of incidence i is small, equations (3.17) to (3.19) become

$$\frac{b}{a} = -\frac{i-r}{i+r} = -\frac{n-1}{n+1},$$ (3.23)

$$\frac{c}{a} = \frac{2r}{i+r} = \frac{2}{n+1},$$ (3.24)

$$\frac{b}{a} = \frac{i-r}{i+r} = \frac{n-1}{n+1},$$ (3.25)

$$\frac{c}{a} = \frac{2r}{i+r} = \frac{2}{n+1},$$ (3.26)

so that, for normal incidence with either direction of polarization or with natural light,

$$\frac{J}{I_0} = \left(\frac{n-1}{n+1}\right)^2,$$ (3.27)

and

$$\frac{I}{I_0} = \left(\frac{2}{n + 1}\right)^2. \tag{3.28}$$

If we assume an air-glass surface and a refractive index of 1·5 for the glass, we find from equation (3.27) that about 4% of the incident light is reflected. When reflection and refraction occur from glass to air, n in the above formulae must be replaced by $1/n$. Hence we find that when light is transmitted through a glass plate for which $n = 1·5$, about 4% of the light is lost at each surface and about 92% is transmitted.

At grazing incidence ($i = 90°$) both terms on the right hand side of equation (3.21) reduce to $\frac{1}{2}$ while those in equation (3.22) become zero, showing that total reflection occurs.

As the angle of incidence increases from normal to grazing incidence, the first term on the right of equation (3.21) increases continuously while the second diminishes to zero when $i + r = 90°$. The reflected light is therefore completely plane polarized in the plane of incidence so that the vibrations are parallel to the surface of the plate. If the incident light is plane polarized with the plane of vibration coinciding with the plane of incidence, no light is reflected.

When $i + r = 90°$, $\sin r = \cos i$ so that Snell's law gives

$$\tan i = n. \tag{3.29}$$

This relation is known as Brewster's law. The angle of incidence so defined is the polarizing angle and as we have already seen, has a value of about 57° for glass. In practice it is found that with incident natural light the reflected ray is not completely polarized at the polarizing angle due to surface contamination.

The second term on the right of equation (3.22) is obviously greater than the first so that some of the refracted light is plane polarized with the plane of vibration coinciding with the plane of incidence (except in the limiting case of normal incidence). In this case also the maximum degree of polarization occurs when the angle of incidence is equal to the polarizing angle but polarization is only partial. The degree of polarization can be greatly increased by employing a pile of parallel glass plates arranged so that the light is incident on them at the polarizing angle as shown in Figure 3.19. When light travels through such a pile some of the transmitted light is polarized by refraction in the first plate. This passes through the second plate in which a further proportion of the unpolarized light is polarized. In this way the degree of polarization increases with each plate and when seven or eight plates are used a high degree of polarization is obtained. Due to reflection and absorption there is of course a loss of intensity at each plate. Since, however, the intensity of the refracted beam from a single plate is about 80%

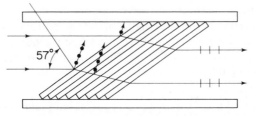

Figure 3.19.   Polarization by refraction through
a pile of glass plates

of that of the incident beam, the intensity at emergence from the pile can be approximately the same as the intensity obtained by reflection from a single plate.

In modern photoelastic practice, polarized light is obtained almost exclusively by means of Polaroid filters. These are described in Section 4.7.

3.19   Phase change due to reflection

Waves having amplitudes of opposite sign obviously differ in phase by π. When reflection occurs in the rarer medium so that $i > r$ the ratio b/a given by equation (3.17) is negative so that for light plane polarized in the plane of incidence a change of phase equal to π is introduced. On the other hand, if reflection occurs internally, i.e. in the medium of greater density, so that $i < r$, b/a is positive and there is no change in phase.

For waves plane polarized in planes perpendicular to the plane of incidence the ratio b/a given by equation (3.19) is positive with external reflection for small values of i but diminishes to zero and changes sign at the polarizing angle. Thus there is a phase change of π when the angle of incidence exceeds the polarizing angle but none for smaller angles of incidence. This is reversed when reflection occurs internally.

3.20   Path difference or relative retardation. Interference

We now consider two sinusoidal waves of the same frequency but different phase and amplitude travelling in the same direction along the same straight line. The distance measured in the direction of propagation between similar points on each wave is called the optical path difference or relative retardation. Since the wave undergoes a phase difference of 2π when the wave advances through a distance equal to one wavelength it follows that

$$\text{optical path difference} = \text{phase difference} \times \lambda/2\pi.$$

Since both path difference and phase difference can be expressed in wavelengths, the two terms are sometimes used synonymously.

The path difference between two waves originating at the same point source may be due to a difference in the lengths of the geometrical paths of the rays, or, as will be shown later, to differences in the refractive indices of the mediums through which the separate waves pass. The two waves can be expressed in the form of equation (3.3) by

$$y_1 = a_1 \sin\left[\omega\left(t - \frac{x}{v}\right) + \alpha_1\right]; \qquad y_2 = a_2 \sin\left[\omega\left(t - \frac{x}{v}\right) + \alpha_2\right].$$

It was shown in Section 3.3 that for a given value of x, these equations represent the vibrations at the point x. Further, when x is regarded as constant, the term $\omega x/v$ can be incorporated in the initial phase so that the above equations become

$$y_1 = a_1 \sin(\omega t + \alpha_1'); \quad y_2 = a_2 \sin(\omega t + \alpha_2'),$$

where

$$\alpha_1' = \alpha_1 - \omega x/v \quad \text{and} \quad \alpha_2' = \alpha_2 - \omega x/v.$$

From Section 3.5, the amplitude of the resultant vibration is

$$A = \sqrt{(a_1^2 + a_2^2 + 2a_1 a_2 \cos \alpha)}, \tag{3.7}$$

where $\alpha = \alpha_1' - \alpha_2' = \alpha_1 - \alpha_2$. Hence the resultant amplitude depends on the phase difference α. It will have a maximum value equal to $(a_1 + a_2)$ when $\cos \alpha = 1$, i.e. when $\alpha = 0$ or $2m\pi$. This corresponds to a path difference between the waves equal to zero or any number of complete wavelengths so that crests coincide with crests and troughs with troughs.

The minimum amplitude of the resultant vibration is $(a_1 - a_2)$ and occurs when $\alpha = (2m + 1)\pi$ where m is zero or any integer. In terms of path difference, this represents an odd number of half wavelengths so that crests coincide with troughs.

The energy transmitted or intensity of illumination produced by a light wave is proportional to the square of the amplitude so that the maximum intensity is proportional to $(a_1 + a_2)^2$ and the minimum intensity to $(a_1 - a_2)^2$.

When $a_1 = a_2$, the maximum amplitude is twice that of the separate waves and the intensity is four times as great. This is referred to as constructive interference. The minimum amplitude is zero, i.e. the intensity vanishes. This is called destructive interference.

It should be noted that in accordance with the principle of superposition each wave produces its own effect independently of the other. The combined effect, which is called interference, is a direct result of the superposition of the two waves.

The interference of two waves of equal amplitude and frequency for various phase differences is illustrated in Figure 3.20.

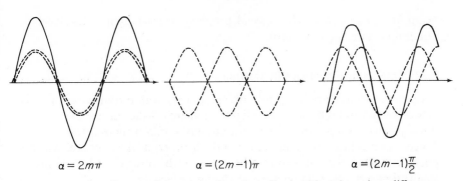

$$\alpha = 2m\pi \qquad\qquad \alpha = (2m-1)\pi \qquad\qquad \alpha = (2m-1)\frac{\pi}{2}$$

Figure 3.20.   Interference of two waves of equal amplitude with various phase differences

When light consisting of waves superimposed in the above manner inter-sects a surface such that the path difference varies continuously from point to point over the surface, it is possible under suitable conditions to observe a series of bright and dark bands. These are produced by interference and are known as interference fringes.

3.21   Coherent and non-coherent waves

With conventional light sources, the phenomenon of interference can only be observed when the individual waves are derived from a single primary source, in fact from the same atom. Waves originating from different sources can never interfere, even when identical monochromatic sources are used such that the conditions of equal wavelength and amplitude are satisfied. This is because the vibrating electrons which give rise to the radiation change phase discontinuously millions of times in one second. The waves are there-fore not produced as continuous trains but as successions of different trains with different phases, and random phase differences are continually being introduced between waves originating from different atoms. Each change in phase produces a change in the position of the interference fringes which shift so rapidly that they cannot be observed. Such waves are said to be non-coherent. It can be shown that the resultant intensity produced by non-coherent waves is always equal to the sum of the intensities produced by the individual waves when acting alone.

When waves emanate from the same atom source or from secondary sources derived from the same atom, the changes in phase are always identical and simultaneous. The phase difference then depends only on the difference in lengths of the optical paths. Such waves are known as coherent waves and the sources as coherent sources.

A different situation exists in the case of laser light, which possesses a high degree of spatial coherence. This means that the different waves in a laser beam are all in phase. This special property of laser light is the basis of holography (see Section 10.10).

3.22   Interference produced by thin films

Coloured fringes are frequently observed when a thin transparent film is viewed by natural reflected white light. Particular instances are the colours displayed by a soap bubble or by a thin film of oil on the roadway. These fringes are produced by interference of the form typified by the well known phenomenon of Newton's rings. The latter can be observed when a lens of small curvature is placed on a flat reflecting surface.

Consider a ray of light AP_1 which falls on the surface of a thin reflecting film, Figure 3.21. This gives rise to a reflected ray P_1B_1 and a refracted ray

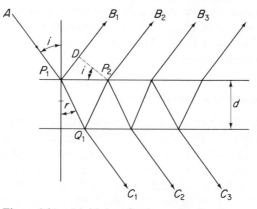

Figure 3.21.   Multiple reflection and refraction at
the surfaces of a thin film

P_1Q_1. The ray P_1Q_1 is in turn reflected and refracted at Q_1 to produce the internally reflected ray Q_1P_2 and the transmitted ray Q_1C_1. This process is continued, resulting in a series of reflected rays B_1, B_2, B_3 etc. and a series of transmitted rays C_1, C_2, C_3 etc.

We shall now consider the path difference between two successive rays, e.g. B_1 and B_2. For simplicity it will be assumed that the medium surrounding the film is air.

Let d be the thickness and n the refractive index of the material of the film and let the angles of incidence and refraction at P_1 be denoted by i and r respectively. The optical path difference between the two rays is then equal to $n(P_1Q_1 + Q_1P_2) - P_1D$.

Also

$$P_1Q_1 = Q_1P_2 = d/\cos r$$

and

$$P_1D = P_1P_2 \sin i = 2d \tan r \sin i = 2nd \sin^2 r/\cos r.$$

We therefore obtain

$$\text{optical path difference} = 2nd(1 - \sin^2 r)/\cos r$$

$$= 2nd \cos r.$$

The corresponding phase difference is $2\pi/\lambda \times 2nd \cos r$.

We have already seen (Section 3.19) that there is a change in phase equal to π when reflection occurs in the medium of lower refractive index but none when it occurs in the medium of higher refractive index. There is therefore an additional phase difference of π and a corresponding path difference of $\lambda/2$ introduced between the two rays. The total optical path difference is therefore $2nd \cos r \pm \lambda/2$, the sign chosen being immaterial since a path difference of λ will not affect the results.

Destructive interference will therefore occur when

$$2nd \cos r \pm \lambda/2 = (2m + 1)\lambda/2,$$

i.e. when

$$2nd \cos r = m\lambda, \tag{3.30}$$

where m is zero or any integer.

Similarly, constructive interference will occur when

$$2nd \cos r = (2m + 1)\lambda/2. \tag{3.31}$$

We will now consider the first two transmitted rays C_1 and C_2. Since the ray C_2 is reflected internally, there is no change in phase due to reflection. The total optical path difference is therefore $2nd \cos r$ so that destructive interference will occur when

$$2nd \cos r = (2m + 1)\lambda/2, \tag{3.32}$$

and constructive interference when

$$2nd \cos r = m\lambda. \tag{3.33}$$

In the above analysis, the first two reflected and transmitted rays only have been taken into account. To show the validity of this restriction, we assume that the angle of incidence is small. Equations (3.27) and (3.28) may then be applied to determine the relative intensities of successive reflected and transmitted rays, the refractive index n being replaced by $1/n$ when considering rays within the film.

Taking the intensity of the incident ray as unity, the intensity of the first reflected ray is

$$J_{B1} = \beta \tag{3.34}$$

where $\beta = ((n - 1)/(n + 1))^2$. It follows that the intensity of the refracted ray P_1Q_1 is $(1 - \beta)$. The intensity of the first transmitted ray Q_1C_1 is then

$$I_{C1} = (1 - \beta) \times \left(\frac{2}{(1/n) + 1}\right)^2 = (1 - \beta)^2, \qquad (3.35)$$

while that of the reflected ray Q_1P_2 is

$$(1 - \beta) \times \left|\frac{(1/n) - 1}{(1/n) + 1}\right|^2 = \beta(1 - \beta). \qquad (3.36)$$

The intensity of the second reflected ray P_2B_2 is therefore

$$J_{B2} = \beta(1 - \beta) \times \left(\frac{2}{(1/n) + 1}\right)^2 = \beta(1 - \beta)^2. \qquad (3.37)$$

Continuing in this way we obtain

$$I_{C2} = \beta^2(1 - \beta)^2 \qquad (3.38)$$

$$J_{B3} = \beta^3(1 - \beta)^2 \qquad (3.39)$$

and so on. Assuming a refractive index of 1·6, the intensities of the first three reflected rays taken in order and expressed as percentages of the incident intensity are found to be 5·33, 4·78 and 0·014. Thus we see that while the intensity of the ray B_2 is only slightly less than that of B_1, that of B_3 is greatly diminished. The effect of the third ray is therefore very small. The amplitudes of the first two rays, which are proportional to the square roots of the corresponding intensities, are as 1 : 0·95. Thus, the condition of equal amplitudes for complete interference is approximately satisfied and as a result the interference fringes formed when the rays are brought to a focus by a lens or the eye are well defined.

Considering now the transmitted rays, we find that the ratios between the intensities and the amplitudes of any two successive rays are $1/\beta^2$ and $1/\beta$ respectively. Since β is a small quantity (0·053 for $n = 1·6$) both the intensities and the amplitudes of the second and subsequent transmitted rays are small compared with those of the first. For this reason interference fringes viewed by transmitted light are weak and indistinct.

So far, we have considered films having a higher refractive index than the surrounding medium but equally well the film may have a lower refractive index as in the case of a film of air between glass plates. We will now consider the case shown in Figure 3.22 in which a plate of slightly varying thickness with a reflecting surface is viewed with normal incidence through an optically flat glass plate with a narrow film of air between the first plate and the glass. On substituting $n = 1$ and $r = 0$ in equation (3.30) we obtain the condition for destructive interference:

$$d = m\lambda/2. \qquad (3.40)$$

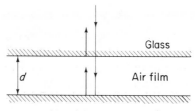

Figure 3.22. Interference by reflection
through an air film

Thus, interference fringes will be formed when the thickness of the air film is equal to an integral number of half wavelengths. These fringes are loci of points on the surface of equal distance from the glass plate and hence the contours of the surface can be plotted. This method has been successfully applied by Frocht[1] to determine the isopachic curves (i.e. curves of constant thickness or constant sum of principal stresses) for models with initially flat surfaces which were distorted under two-dimensional loading.

3.23 Interference with white and monochromatic light

Since equations (3.30) to (3.33) depend on the wavelength λ, it follows that the position of points of maximum and minimum intensity will differ according to the colour. Hence, if white light is used, the resultant effect is that of superposing interference patterns for all the constituent wavelengths.

For small path differences, only one or two wavelengths will completely interfere simultaneously and the resulting fringes assume characteristic colours as the different wavelengths successively interfere. When the path difference is great, however, many wavelengths interfere simultaneously; in fact all those of which the path difference is a simple multiple integral. Under these conditions, white light is practically restored and, except in special circumstances, the fringes cannot be observed.

Since the path difference is proportional to the thickness d of the film, it follows from the above that clearly distinguishable interference fringes will be obtained when using white light if the film is only a few wavelengths thick. With thicker films or plates such as are normally used in photoelastic work, the fringes can only be observed if monochromatic light having a high degree of purity is used. With thick plates the fringes lie very close together and it becomes necessary to observe them with the aid of a suitable telescope.

The interference between the rays reflected at the two surfaces of a photo-elastic model forms the basis of a method developed by Favre and Schumann (Section 10.8) in which measurements are made of both the absolute and relative retardations produced when the model is loaded.

3.24 Non-reflecting films

The reflection of light at surfaces between two media such as air and glass is very undesirable in certain photoelastic methods which rely on measurement of the intensity of transmitted light since some of the light reaches the plane of the image after multiple reflections. This reflection may be reduced from about 5 % to about 1 % of the incident light by coating such surfaces with a film of refractive index and thickness such that the waves reflected from, for example, the air/film and film/glass surfaces interfere destructively. The necessary conditions to be satisfied by the two sets of waves are that (1) the amplitudes shall be equal, (2) the phase difference shall be equal to π.

From equations (3.23) or (3.25) it is found that (1) is satisfied when the refractive index of the film is $\sqrt{n}$ where n is the refractive index of the glass.

Since both reflections take place in the medium of lower refractive index, both waves suffer a change of phase of π. For condition (2) to be satisfied it is therefore necessary that the path difference produced by the film shall be one half wavelength, i.e.

$$2d\sqrt{n} = \lambda/2,$$

or

$$d = \lambda/4\sqrt{n}.$$

The two reflected waves will then suffer complete destructive interference only for one particular wavelength and with normal incidence. The effectiveness of the film is not greatly reduced, however, for a considerable range of wavelengths and angles of incidence. For camera lenses, etc., the optical thickness of the film is usually made equal to one quarter the wavelength of green light. These lenses are therefore suitable when using the mercury green band. The film is deposited on the surface of the glass by evaporating crystals, usually of cryolite or magnesium fluoride, in a vacuum.

3.25 Multiple beam interference

The reflection coefficient of the surfaces of a film or plate can be increased by coating them with thin metallic layers which partially reflect and partially transmit the incident light. The intensities of the various reflections are thereby substantially increased while the reduction in intensity between successive reflections is reduced. The fringes observed when the various rays are collected and brought to a focus by a lens are produced by the mutual interference of several rays so that the simple analysis given in Section 3.22, in which the effects of the first two rays only were considered, is no longer valid. With metallic films, the phase differences on reflection are not exactly equal to 0 and π. If T_1 and T_2 denote the fractions of the incident intensity transmitted at the air–film and film–air interfaces and R the fraction reflected at each interface, the intensities of the various rays are as shown in Figure 3.23. Due to absorption at the metallic layers, $T + R < 1$.

By this method, known as multiple beam interference, well defined interference patterns with sharp fringes are obtained by transmitted light as well as by reflected light.

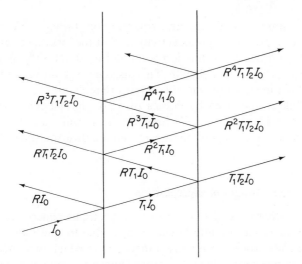

Figure 3.23. Intensity distribution for a plate having partially mirrored surfaces

3.26 Interference filters

An interference filter is a device based on the principle of multiple beam interference by means of which a narrow region of the spectrum can be isolated from a white light source much more effectively than by the use of coloured glass or gelatine filters. In its simplest form, it consists of a transparent dielectric spacer with both surfaces bounded by thin partially transparent metallic layers and protected externally by glass plates.

With normal incidence the transmitted rays reinforce when $m = 2d'$, where d' is the effective optical thickness of the spacer allowing for changes in phase at the dielectric/metal interface. Hence, for an optical thickness of $\lambda_0/2$, transmission peaks will occur at λ_0, $\lambda_0/2$, $\lambda_0/3$, etc., with other wavelengths more or less cutting out. The distance between peaks increases as the thickness is reduced, and when d' lies between about 2×10^{-5} and 6×10^{-5} cm, only one maximum within the visible spectrum is transmitted. For instance, if it is desired to isolate the mercury green line of 5460 Å and d' is made equal to 5.46×10^{-5} cm, peaks occur at 10,920, 5460, 3640 Å etc. of which only the required wavelength lies in the visible region. When wavelengths greater than about 6000 Å are required, unwanted visible peaks occur on the short wavelength side. These can be suppressed by incorporating a suitable cover glass in the filter.

The width of the transmitted band, which is measured by the distance between points of half light intensity on either side of peak transmission, is usually about 180 Å.

With the simple metal-dielectric construction, rather thick metal layers are required to produce high reflectivity so that the loss due to absorption is rather high and peak transmission is limited to about 30% of the incident light of that particular wavelength. This disadvantage is overcome in the all-dielectric filter in which the metallic layers are replaced by series of alternate layers of highly reflecting dielectrics (usually zinc sulphide and cryolite) having high and low refractive indices. By this means transmissions as high as 80% can be achieved and the transmitted bandwidth can be as narrow as 20 Å.

3.27   Brewster's fringes of superposition

We have already seen that, in ordinary circumstances, white light is unsuitable for observing interference fringes due to multiple reflections in thick plates. We now consider an arrangement in which two thick plates are employed by means of which low order fringes are obtained. These can be observed with white light, thus avoiding the difficulty of producing mono-chromatic light of extreme purity combined with sufficiently high intensity.

Assume that white light falls successively on two thick plates A and B with reflecting surfaces as shown in Figure 3.24. An incident ray may divide and pursue various paths before emerging from plate B, several of which are

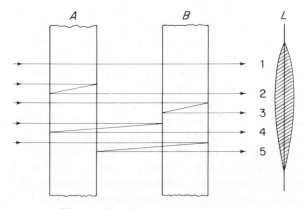

Figure 3.24.   Fringes of superposition

indicated. Ray 1 passes straight through both plates. Ray 2 is internally reflected twice in plate *A* and passes straight through plate *B*. Ray 3 passes without reflection through *A* but is twice reflected internally in *B*. Hence, if the optical thicknesses of *A* and *B* are equal, the rays 2 and 3 will be in phase when they arrive at the lens *L*. If the optical thicknesses are not quite equal, or if one of the plates is slightly inclined to the other, a small phase difference will be introduced between these rays. Under these conditions, interference fringes will be produced. Similarly, rays 4 and 5 pursue similar paths and can produce interference fringes.

For the sake of clarity, the paths of the rays 1, 2, 3, etc., are shown separately in Figure 3.24. All of these in fact originate from the same incident ray and are superimposed.

Fringes of the above type were first observed by Brewster and are known as Brewster's fringes or fringes of superposition. They are employed in several photometric procedures.

In practical applications the surfaces are provided with highly reflective coatings. The pairs of reflecting surfaces may be formed on the two surfaces of an optically flat glass plate or on adjacent surfaces of two glass plates with an air space between them. Pairs of such reflecting surfaces are known as etalons.

A variation of the arrangement shown in Figure 3.24 forms the basis of an interferometer developed by Post in which three reflecting surfaces are employed, the first reflecting coating on plate *A* being omitted. A photo-elastic model is inserted in the air space between *A* and *B*. If the optical thickness of plate *B* approximately the same as that of the model plus the remaining air gap, an interference pattern describing the variations in thickness of the model can be observed. This method is described in detail in Section 10.6.

4

TRANSMISSION OF
LIGHT THROUGH CRYSTALS

4.1 Double refraction

When a ray of light is incident on certain crystals it is split at entry into two components which in general are transmitted through the crystal in different directions. The refractive index and the velocity of transmission are therefore different for the two rays. This phenomenon is known as double refraction or birefringence and can be observed with all crystals other than those of the cubic class. If the emergent rays are observed through an analyser, it is found that they are plane polarized in mutually perpendicular planes.

The behaviour of light passing through a crystal can be conveniently studied with reference to a crystal of calcite or Iceland spar. The basic form of this crystal is a rhomb as illustrated in Figure 4.1. Two opposite·corners of this rhomb contain three obtuse angles of 101° 55'. The direction of a line which is equally inclined to the edges at these corners is called the principal axis of the crystal. A plane containing the axis and which is normal to one of the faces is called a principal section. There are thus three principal sections at every

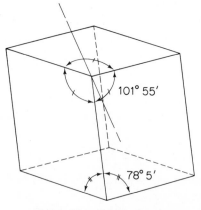

Figure 4.1.   Calcite rhomb

72

point in the crystal. These consist of parallelograms with angles of approximately 71° and 109°.

In Figure 4.2 *ABCD* represents a principal section of a calcite rhomb. When a ray of natural light falls with normal incidence on the face *AB* it is

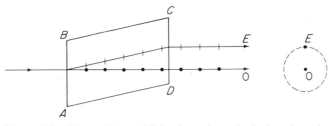

Figure 4.2.   Transmission of light through a principal section of a calcite rhomb

split into two components one of which, the ray *O*, travels without deviation through the crystal. This ray obeys the laws of refraction and is called the ordinary ray. The other ray *E* is deviated in the crystal and thus suffers a lateral displacement on emergence. This ray does not obey Snell's law and is called the extraordinary ray. If the rhomb is rotated about the ray *O* as axis, the ray *E* describes a circle around *O* showing that the direction of propagation of ray *E* is dependent and that of *O* independent of the orientation of the crystal. Further, if the emergent rays are examined through an analyser, it will be found that both rays are plane polarized, the extraordinary ray vibrating in the plane of the principal section and the ordinary ray in a plane perpendicular to this. If the incident ray is plane polarized, the ordinary ray is extinguished and the extraordinary ray has maximum brightness when the plane of vibration of the incident light is parallel to the principal section. This will be reversed when the plane of vibration is perpendicular to the principal section.

When the incident ray falls obliquely on a face of the rhomb, in general both the ordinary and extraordinary rays are refracted but by different amounts. The two components are transmitted with the same velocity and in the same direction for only one direction of transmission. This direction is known as the optic axis and in calcite this coincides with the principal crystallographic axis. Crystals of this type, having only one optic axis, are called uniaxial crystals. Many crystals have two optic axes and as such are known as biaxial crystals.

4.2   The index ellipsoid

The effects produced when a beam of light falls on a crystal for varying angles of incidence can be readily visualized by a geometrical representation known as the index ellipsoid or Fresnel's ellipsoid.

If from a point O (Figure 4.3) we imagine straight lines to be drawn in all directions such that the length of each line represents the reciprocal of the velocity, or the index or refraction, for a ray vibrating in the direction of that line, the end points of these lines lie on the surface of an ellipsoid. Figure 4.3a

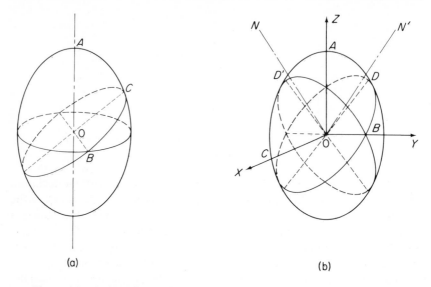

(a)   (b)

Figure 4.3.   Index ellipsoid for (a) a uniaxial crystal and (b) a biaxial crystal

shows the index ellipsoid for a positive uniaxial crystal. The principal axis OA corresponds to the optic axis of the crystal. The section through O normal to OA is circular so that when a ray passes through the crystal in the direction OA the component rays have equal velocities, i.e. the refractive index is the same for both. For any other direction of transmission, the normal section through O is an ellipse. The directions of polarization coincide with the directions of the principal semi-axes OB and OC of this ellipse. The length of one of these semi-axes, i.e. OB, is obviously always equal to the radius of the circular section showing that the velocity of the ordinary ray is independent of the direction of transmission. The velocity of the extraordinary ray can vary between $1/OA$ and $1/OB$ depending on the direction of transmission. The index of refraction for the extraordinary ray represented by OA, is known as the principal index of refraction for this ray and is denoted by n_E. That for any random direction of vibration will be denoted by n_E'. The index of refraction n_O for the ordinary ray is represented by the radius of the circular section.

Uniaxial crystals are classified as positive or negative according to whether the velocity v_E of the extraordinary ray is respectively smaller or greater than that of the ordinary ray, i.e. on whether n_E is greater or less than n_O.

For a positive uniaxial crystal the ellipsoid is a prolate spheroid of revolution as shown, while for a negative crystal it is oblate.

In the case of a biaxial crystal, the principal semi-axes of the ellipsoid all have different lengths. We denote the principal refractive indices by n_x, n_y and n_z and assume $n_x < n_y < n_z$ so that in Figure 4.3b, $OC < OB < OA$. There must then exist two central sections of the ellipsoid normal to the xz plane for which the semi-axes OD and OD' are equal to OB. These sections are circular and are equally inclined on either side of OA. Light passing in the directions normal to these sections behaves as if the crystal were isotropic. These directions, i.e. ON and ON', are the optic axes of the crystal. The lengths of both principal axes of the central section vary however with the direction of transmission so that no ordinary ray exists in a biaxial crystal.

4.3 Optical path difference or relative retardation in crystals

The refractive index of a transparent medium can be defined as in Section 3.10 by

$$n = c/v,$$

so that

$$c = nv,$$

where c is the velocity of light in a vacuum and v the velocity in the medium. It follows that when light travels a distance d in the medium it would travel a distance nd in a vacuum in the same time. The product nd is known as the optical path. The absolute retardation R_0 is equal to the difference between the lengths of the optical and geometrical paths, i.e.

$$R_0 = nd - d = (n - 1)d. \tag{4.1}$$

If we consider rays of light of the same wavelength transmitted normally through two plates of the same thickness d but having different refractive indices n_1 and n_2 $(n_1 > n_2)$ the optical paths are respectively $n_1 d$ and $n_2 d$. The optical path difference or relative retardation, as it is more commonly called in photoelasticity, is then given by

$$R = (n_1 - n_2)d. \tag{4.2}$$

In the transmission of light through a doubly refracting crystal plate, we take n_1 and n_2 to represent the indices of refraction for the two principal planes of vibration as indicated by the semi-axes of the corresponding central section of the index ellipsoid. Except when the direction of transmission is perpendicular to the principal axis in a uniaxial crystal or to two principal axes in a biaxial crystal, the rays along which the oppositely polarized waves are transmitted deviate within the crystal. This deviation, however, is extremely small in crystals of low birefringence and in all photoelastic model materials and can be disregarded for most practical purposes. Equation 4.2 then gives the retardation between oppositely polarized waves forming parallel wavefronts.

4.4 Ray velocity surfaces

As we have seen, the velocities of the two rays passing in any given direction through a crystal are inversely proportional to the lengths of the semi-axes of the central section of the index ellipsoid which is normal to that direction. If from a point O we imagine distances OP_1 and OP_2 representing the respective velocities of the two rays to be marked off along straight lines corresponding to every direction of transmission, all points such as P_1 and P_2 lie on a surface of two sheets. This surface is known as the ray velocity surface. Any plane wave passing through O will interest the appropriate surface in one point only, i.e. the wave front is tangential to the surface. The surface can therefore be regarded as the wave surface after unit time for a wave originating at O.

Biaxial crystals

It can be shown that the equation to the ray velocity surface is

$$\frac{a^2 x^2}{r^2 - a^2} + \frac{b^2 y^2}{r^2 - b^2} + \frac{c^2 z^2}{r^2 - c^2} = 0, \tag{4.3}$$

in which a, b and c are the principal ray velocities in the directions of the axes OX, OY and OZ respectively and r is the length of the radius vector from O to a point (x, y, z) on the surface. We assume, in conformity with Figure 4.3b, that $a > b > c$.

The equations to the curves in which the surface intersects the yz plane are obtained by substituting $x = 0$ in equation (4.3).

This gives

$$y^2 + z^2 = a^2,$$

and

$$\frac{y^2}{c^2} + \frac{z^2}{b^2} = 1.$$

These equations represent a circle of radius a and an ellipse with semi-axes c and b. Further, since $a > b > c$, the ellipse lies within the circle as shown in Figure 4.4a.

The intersections with the zx plane ($y = 0$) are given by

$$z^2 + x^2 = b^2,$$

and

$$\frac{y^2}{a^2} + \frac{x^2}{c^2} = 1,$$

i.e. a circle of radius b intersected by an ellipse of semi-axes a and c (Figure 4.4b).

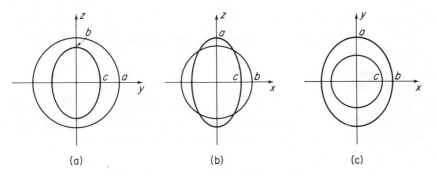

(a)   (b)   (c)

Figure 4.4.   Intersections of ray velocity surfaces with the principal planes

Similarly, the equations to the intersections with the plane $z = 0$ are

$$x^2 + y^2 = c^2,$$

and

$$\frac{x^2}{b^2} + \frac{y^2}{a^2} = 1,$$

which represent a circle of radius c within an ellipse of semi-axes b, a (Figure 4.4c).

One octant of the wave surface is represented in Figure 4.5. AB is the common tangent to the two surfaces in the xz plane. The radius OA is therefore an optic axis since it is normal to the corresponding wave front. The surfaces intersect in the point R in the xz plane. OR is a direction of equal ray velocities as distinct from the optic axis which is a direction of single wave velocity. These differ since the direction of one of the rays does not coincide with that of the wave normal.

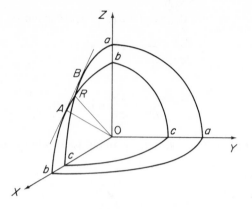

Figure 4.5. Wave surface for a biaxial crystal

Uniaxial crystals

If we assume $a = b$ to represent the velocity of the ordinary ray so that c represents that of the extraordinary ray in a uniaxial crystal, the optic axis coincides with the axis OZ. Equation (4.3) then reduces to

$$(r^2 - c^2)\left(\frac{x^2 + y^2}{a^2} + \frac{z^2}{a^2} - 1\right) = 0,$$

which represents a surface of two sheets, namely a sphere of radius a and a spheroid formed by the rotation of an ellipse of semi-axes a and c about the z axis. The intersection with any plane containing the z axis consists of a circle of radius a and an ellipse of semi-axes a and c which touch on the z

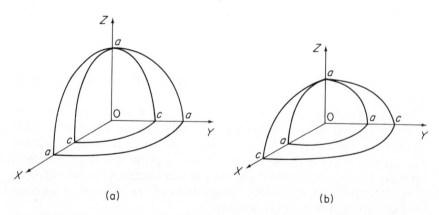

(a)   (b)

Figure 4.6. Wave surfaces for (a) positive and (b) negative uniaxial crystals

axis. If the crystal is positive, $a > c$ so that the circle includes the ellipse while if it is negative the ellipse includes the circle. Sections normal to the optic axis consist of two concentric circles. Figure 4.6 shows one octant of the wave surfaces for positive and negative uniaxial crystals.

4.5 Huygens' constructions for doubly refracting crystals

The directions of the rays and wavefronts in crystals can be traced using Huygens' principle. Figure 4.7a shows the construction for a plane wave front AB falling with oblique incidence on the face of a negative uniaxial crystal

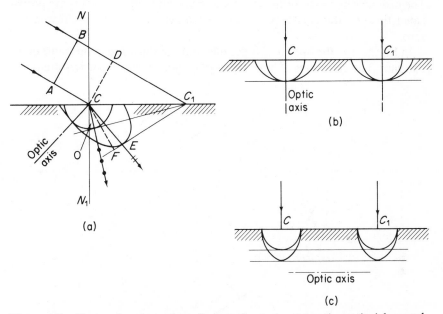

Figure 4.7. Huygens' constructions for wavefronts in a negative uniaxial crystal. (a) Oblique incidence. (b) Normal incidence parallel to the optic axis. (c) Normal incidence perpendicular to the optic axis

which is oriented such that the optic axis lies in the plane of incidence. At any instant the sections of the wavelets originating at C are respectively circular and elliptical in form and touch on the optic axis. We assume that the curves in Figure 4.7a represent the wavelets at the instant when the incident wave reaches C_1. The tangents C_1O and C_1E drawn from C_1 to the circle and ellipse respectively then represent the instantaneous traces of the ordinary and extraordinary wavefronts while CO and CE are the corresponding refracted rays. The extraordinary ray does not obey Snell's law since $\sin N\widehat{C}A/\sin N_1\widehat{C}E \neq DC_1/CE$. Considering the normal CF to the extraordinary wavefront, however, we obtain

$$\sin N\widehat{C}A/\sin N_1\widehat{C}F = DC_1/CF = ct/v_e t = c/v_e,$$

for light entering the crystal from a vacuum, where c is the velocity of light in a vacuum and v_e the velocity of the extraordinary wavefront in the crystal. This ratio will be denoted by n'_E and represents the extraordinary index of refraction for that particular direction of the ray.

Figure 4.7*b* illustrates the particular case when the incident light is normal to the surface of a uniaxial crystal cut perpendicular to the optic axis. The sections of the wavefronts, which can be derived from Figure 4.4, are as shown. The ordinary and extraordinary rays coincide and the wavefronts are propagated through the crystal with equal velocities so that the birefringence is zero.

In Figure 4.7*c* the optic axis is parallel to the surface and in the plane of the diagram. The directions of both rays again coincide but the waves are trans-

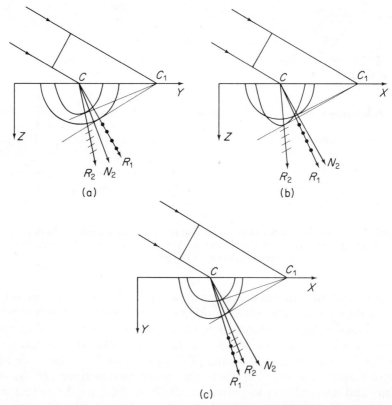

Figure 4.8. Huygens' constructions for wavefronts in a biaxial crystal with wavefronts normal to the axes (*a*) *OX*, (*b*) *OY*, (*c*) *OZ*

mitted with different velocities so that a phase difference is established between them.

In general, when a beam of light falls on the surface of a biaxial crystal, the incident ray and the two refracted rays are not all coplanar and so cannot be represented in a two dimensional drawing. Only when the plane of incidence contains two of the principal axes and hence is normal to the third are the three rays coplanar. Figure 4.8 shows Huygens' constructions for the particular cases when the plane of incidence is normal to the axes OX, OY and OZ respectively in Figure 4.3b. In each case the character of the secondary waves is as indicated in Figure 4.4. The directions of the refracted wavefronts are obtained by drawing tangents from C to the circular and elliptical sections of the secondary waves.

The ray CR_1, drawn from C through the point of tangency on the circular section, coincides with the wave normal. This ray is polarized in the plane of incidence, i.e. its plane of vibration is normal to the plane of incidence. The other ray CR_2, drawn through the point of tangency on the elliptical section, does not coincide with the corresponding wave normal CN_2. This ray is polarized in a plane normal to the plane of incidence.

With normal incidence, it is obvious that the rays and wave normals are all perpendicular to the surface of the crystal and that in all three cases the two wavefronts are propagated with different velocities through the crystal.

4.6 Polarizing prisms. The Nicol prism

The difference in directions of the ordinary and extraordinary rays in calcite is utilized in various prisms designed to produce a field of plane polarized light. The original and best known of these is the Nicol prism in which a calcite crystal is cut as shown in Figure 4.9, the two pieces being cemented together with a thin film of Canada balsam. The refractive index of Canada balsam is intermediate between the indices of calcite for the ordinary and extraordinary rays. As a result the extraordinary ray is trans-

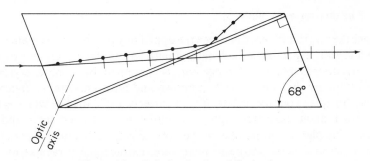

Figure 4.9.   Polarization by a Nicol prism

mitted through the film while the ordinary ray is totally reflected. Later improved types of polarizing prism such as the Glan–Thompson and Ahrens prisms are designed to permit the use of a wider cone of incidence thus increasing the amount of light passing through the prism and to eliminate the lateral displacement of the transmitted rays.

4.7 Dichroism. Polaroid

Certain crystals possess the property of absorbing one of the two plane polarized rays to a greater extent than the other. Thus, by using a crystal of suitable thickness, one of the rays can be almost completely absorbed while the other is transmitted to an appreciable extent. The emergent light is then practically plane polarized. This property, which is known as dichroism, is particularly strong in tourmaline. This material however is unsuitable for photoelastic work since the transmitted ray is heavily absorbed and also since the dichroism is greatly dependent on the wavelength of the light used.

In modern photoelastic practice, plane polarized light is most commonly obtained by the use of Polaroid sheet, manufacture by the Polaroid Corporation, U.S.A. In the earlier J-type sheet, microscopic needle-shaped crystals of herapathite are embedded in a plastic base and arranged to lie parallel to one another by an extrusion process. Herapathite is a synthetic crystal of a sulphate of iodoquinine and is highly dichroic. The residual light transmitted by crossed pieces of J-sheet is of a deep red colour.

The J-type sheet has been largely superseded by the H-type sheet. This is a molecular rather than a crystalline polarizer in which the dichroic agent is polymeric iodine formed by impregnating an initially transparent plastic sheet of polyvinyl alcohol with iodine and heating. The necessary orientation of the molecules is obtained by stretching. The extinction colour of the H-sheet is a deep blue. Crossed sheets transmit less than 0·01 % of the incident light over a wide range of the visible spectrum so that polarization is virtually complete.

4.8 The quarter-wave plate

Since the phase difference between the oppositely polarized waves emerging from a crystal depends on the length of path within the crystal, it is evident that any desired phase difference can be obtained by using a crystal of the appropriate thickness. The best known and most important application of this principle in photoelasticity is the quarter-wave ($\lambda/4$) plate which is used to convert plane polarized light into circularly polarized light and the reverse. The plate is cut parallel to the optic axis of such a thickness that the phase difference between the ordinary and extraordinary rays at exit for normal incidence is equal to one quarter of a wavelength for the particular monochromatic light used.

Quarter-wave plates are commonly made of mica which is available in fairly large sheets and can readily be split to the required thickness. They may also be manufactured from non-crystalline materials such as glass, cellophane or transparent plastics, utilizing the artificial anisotropy resulting from stresses introduced during manufacture or applied subsequently.

4.9   The half-wave plate

A half-wave plate, as the name implies, produces a phase difference of one half of the particular wavelength of light for which it is to be used. As shown in Section 5.9, a half-wave plate does not alter the state of polarization of a polarized light beam passing through it but merely rotates the plane of polarization through a certain angle. This property makes the half-wave plate a useful device for rotating the plane of polarization of a beam, for example, from a laser.

4.10   Isochromatic surfaces

The two waves associated with any given direction of a wave normal in a crystal travel in general with different velocities so that a path or phase difference is introduced between them. The path difference increases with the distance travelled by the waves through the crystal and depends on the direction of transmission and hence on the angle of incidence of the incident waves. From equation (4.2) the path difference R after travelling a distance r in the crystal is given by

$$R = r(n_1 - n_2),$$

where n_1 and n_2 are the refractive indices for the respective waves.

We consider first a plane parallel slice cut from a uniaxial crystal in the direction normal to the optic axis. Then $n_1 = n_0$, the index of refraction

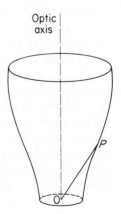

Figure 4.10.   Isochromatic surface for a uniaxial crystal

for the ordinary wave and $n_2 = n'_E$, the index for the extraordinary wave corresponding to the particular direction of the wave normal. If from a point O on the first surface of the slice, distances OP representing values of r corresponding to a given value of the path difference are marked off along each wave normal, all points such as P lie on a surface of the form shown in Figure 4.10. This is known as an isochromatic surface. By choosing different values for the path difference R a coaxial family of such surfaces is obtained.

We now consider a biaxial crystal and assume that a plane parallel slice is cut perpendicular to the direction bisecting the acute angle between the optic axes. The isochromatic surfaces in this case have the form shown in Figure 4.11. The intersection between an isochromatic surface and the second

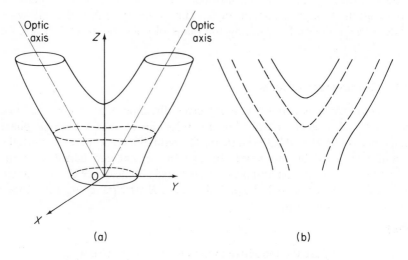

Figure 4.11.   Isochromatic surface for a biaxial crystal

surface of the crystal slice is a curve along which the waves incident at O have a constant phase difference on emergence. If the emergent rays are projected by means of a lens L as shown in Figure 4.12 on to a screen S placed in the focal plane of the lens after passing through an analyser A, an interference pattern can be observed. If monochromatic light is used this pattern contains a series of dark and bright lines while with white light they are coloured. These lines are called isochromatics. Isochromatic surfaces can obviously be described for every point on the first surface of the crystal and rays from all of these points which are parallel in the crystal are brought to a focus at a point on the screen. Each point on the screen thus corresponds to a particular direction in the crystal and not to a particular location on its surface.

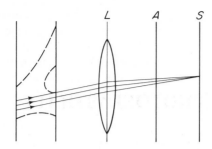

Figure 4.12.   Projection of isochromatic
lines

For a uniaxial crystal cut normal to the optic axis, the curves of inter-section and the isochromatic lines consist of concentric rings. In general, the isochromatics observed are different from the curves of intersection due to refraction of the rays on emergence from the crystal. They are, however, of similar form so that the isochromatic lines provide a qualitative indication of the curves of intersection for a slice cut in any particular direction relative to the optic axis. The various possible forms of the curve of intersection of the isochromatic surface for a plane parallel slice of a biaxial crystal cut per-pendicular to the bisector of the optic axes are indicated in Figure 4.11*a*. Those for a section parallel to the optic axes are shown in Figure 4.11*b*.

Interference patterns of the form described are of importance in certain photoelastic techniques in which convergent light is employed (see Section 14.4).

5

OPTICS OF PHOTOELASTICITY

5.1 Temporary or artificial double refraction

Many non-crystalline transparent materials which are ordinarily optically isotropic become anisotropic and display optical characteristics similar to crystals when they are stressed. This effect normally persists while the loads are maintained but vanishes, almost instantaneously or after some interval of time depending on the material and the conditions of loading, when they are removed. This phenomenon is known as temporary or artificial double refraction and was first observed by Sir David Brewster in 1816. It is this physical characteristic of these materials on which the science of photo-elasticity is based.

As we have seen in Chapter 1, when a body is subjected to any three dimensional stress system there exist at every point within the body three mutually perpendicular planes across which the resultant stress is normal, i.e. there is no shear stress. These planes are the principal planes of stress and correspond, in a body displaying temporary double refraction, to the principal sections of the crystal represented by an element of the material at that point. The normal stresses acting on these planes are the principal stresses and their directions correspond to the principal crystallographic axes. There is thus a direct relationship between the ellipsoid of stress and the index ellipsoid.

In general, the three principal stresses at a point are all different. An element of the material at such a point behaves like a crystal of the orthorhombic system in which the three principal axes are all mutually perpendicular and unequal. If two of the principal stresses are equal or zero as in the case of a thin rod subjected to a simple longitudinal tension or compression, the material at that point has the characteristics of a uniaxial crystal of the tetragonal system. If the three principal stresses are all equal as in the case of pure hydrostatic compression or zero as in the unloaded model, the material may be regarded as corresponding to a crystal in the cubic or isometric system and the birefringence produced is zero.

5.2 Stress optic law

The optical properties of the material at any point can be represented by an index ellipsoid, the principal axes of which coincide with the principal axes of stress at the point. Let n_1, n_2 and n_3 be the principal refractive indices for waves vibrating parallel to the principal stresses σ_1, σ_2, and σ_3 respectively at any point and let n_0 be the index of refraction for the unstressed material. The following equations, formulated by Maxwell in 1852, express the relationships between the principal refractive indices and the principal stresses:

$$(n_1 - n_0) = a\sigma_1 + b(\sigma_2 + \sigma_3), \tag{5.1a}$$

$$(n_2 - n_0) = a\sigma_2 + b(\sigma_3 + \sigma_1), \tag{5.1b}$$

$$(n_3 - n_0) = a\sigma_3 + b(\sigma_1 + \sigma_2), \tag{5.1c}$$

in which a and b are constants depending on the material.
Subtracting we obtain

$$n_1 - n_2 = a(\sigma_1 - \sigma_2) - b(\sigma_1 - \sigma_2) = (a - b)(\sigma_1 - \sigma_2),$$

i.e.

$$n_1 - n_2 = C(\sigma_1 - \sigma_2). \tag{5.2a}$$

Similarly,

$$n_1 - n_3 = C(\sigma_1 - \sigma_3), \tag{5.2b}$$

and

$$n_2 - n_3 = C(\sigma_2 - \sigma_3), \tag{5.2c}$$

where $C = a - b$ is known as the stress-optic coefficient. Equations (5.2) express the stress-optic law for a material under any general triaxial stress system. We next consider a material which is subjected to particular states of stress.

Uniaxial stress

A material which is in a state of simple uniaxial stress behaves as a uniaxial crystal with its optic axis coinciding with the direction of the applied stress (Figure 5.1).
On substituting $\sigma_2 = \sigma_3 = 0$ in equations (5.2) we obtain

$$n_2 = n_3,$$

corresponding to the index of refraction for the ordinary wave and

$$n_1 - n_2 = C\sigma_1.$$

A plane polarized light wave vibrating in a plane containing the axis of stress, i.e. in a principal section of the ellipsoid, emerges from the material

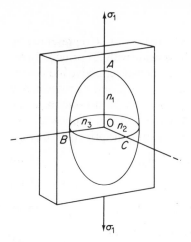

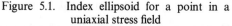

Figure 5.1. Index ellipsoid for a point in a
uniaxial stress field

unchanged except for a loss of intensity. In this case, no ordinary wave is produced. The same applies if the plane of vibration is perpendicular to the axis of stress in which case there is no extraordinary wave. For any other direction of vibration and with normal incidence, the wave is split into two components vibrating in the planes AOC and BOC respectively. These are transmitted with different velocities and on emergence the relative retardation is

$$R = (n_1 - n_2)d = C\sigma_1 d,$$

where d is the thickness of the plate.

Since the central section of the ellipsoid perpendicular to the axis of stress is circular, it follows that the birefringence is constant and equal to $n_1 - n_2$ for all directions of transmission perpendicular to the direction of the stress and is zero for transmission parallel to the stress.

Biaxial stress system

We now consider a plate of temporary doubly refracting material in a two-dimensional or biaxial state of stress. This state is approximately realized in a thin plate or one of only moderate thickness which is loaded in its own plane by forces applied uniformly over the thickness at its edges. Under these conditions we can assume that the principal stresses σ_1 and σ_2 in the plane of the plate are constant throughout the thickness and that the principal stress σ_3 normal to this plane is everywhere zero.

If $\sigma_1 \neq \sigma_2$, the principal stresses are all different and in this condition, as we have already seen, the material behaves like a biaxial crystal. The optical properties at any point can thus be represented by a triaxial ellipsoid with principal axes coinciding with the principal stresses as shown in Figure 5.2.

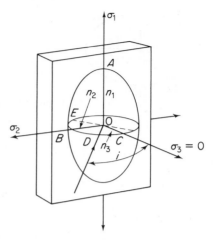

Figure 5.2.   Index ellipsoid for a point in
a biaxial stress field

On substituting $\sigma_3 = 0$, equations (5.2) reduce to

$$n_1 - n_2 = C(\sigma_1 - \sigma_2), \tag{5.3a}$$

$$n_1 - n_3 = C\sigma_1, \tag{5.3b}$$

$$n_2 - n_3 = C\sigma_2. \tag{5.3c}$$

A plane polarized wave vibrating in the plane of any of the three principal sections is transmitted unchanged through the plate except for loss of intensity. For normal incidence and any other direction of vibration, the incident wave is split into components vibrating in the planes AOC and BOC. If the stresses are constant throughout the thickness these components emerge with a relative retardation

$$R = (n_1 - n_2)d = C(\sigma_1 - \sigma_2)d. \tag{5.4}$$

For any other direction of transmission inclined at an angle i to the normal and in the plane perpendicular to σ_1, the wave is split into two components vibrating in the planes AOD and EOD. In this case the two components corresponding to a particular incident wave are in fact transmitted along rays which deviate within the crystal. Since the birefringence of all temporary doubly refracting materials is small, however, this deviation is small and can be neglected. The birefringence is $n_1 - n'$, where n' is the index of refraction for the wave vibrating parallel to OE and can be obtained from the equation to the elliptical section normal to OA (Figure 5.3). This equation is

$$\frac{n'^2 \cos^2 i}{n_2^2} + \frac{n'^2 \sin^2 i}{n_3^2} = 1,$$

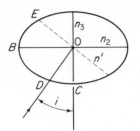

Figure 5.3.   Transmission of light in an arbitrary direction within the plane normal to σ_1

i.e.

$$\frac{1}{n'^2} = \frac{\cos^2 i}{n_2^2} + \frac{\sin^2 i}{n_3^2}.$$

Hence,

$$\frac{1}{n'^2} - \frac{1}{n_2^2} = \left(\frac{1}{n_3^2} - \frac{1}{n_2^2}\right)\sin^2 i,$$

i.e.

$$\frac{n_2^2 - n'^2}{n'^2} = \frac{(n_2^2 - n_3^2)}{n_3^2}\sin^2 i,$$

or

$$n_2 - n' = \frac{n'^2(n_2 + n_3)}{n_3^2(n_2 + n')} \times (n_2 - n_3)\sin^2 i. \tag{5.5}$$

In photoelastic materials, the differences in the refractive indices are small compared with their absolute values so that $n_1' \simeq n_2 \simeq n_3$ and $(n_2 + n_3) \simeq (n_2 + n')$. With these approximations, equation (5.5) reduces to

$$n_2 - n' = (n_2 - n_3)\sin^2 i. \tag{5.6}$$

From equations (5.3a), (5.3c), and (5.6) we obtain

$$n_1 - n' = C(\sigma_1 - \sigma_2) + C\sigma_2 \sin^2 i,$$

i.e.

$$n_1 - n' = C(\sigma_1 - \sigma_2 \cos^2 i). \tag{5.7}$$

The length of the geometrical path travelled in the plate is $d \sec i$. The relative retardation on emergence is therefore

$$R = C(\sigma_1 - \sigma_2 \cos^2 i)d \sec i. \tag{5.8}$$

It can be seen from equation (5.8) that if the relative retardation is measured for two different values of i, the principal stresses σ_1 and σ_2 can be determined. This is the basis of the oblique incidence method (Section 14.7) for the separation of the principal stresses.

5.3 The plane polariscope

The polariscope is an instrument used to measure the relative retardations or phase differences produced when polarized light passes through a stressed photoelastic model. It can have a variety of forms depending on the technique best suited to the type of problem being investigated and also to some extent on the personal preferences of the investigator.

In its simplest form the polariscope consists of a suitable light source and two polarizers. The first polarizer converts the natural light from the source into a field of plane polarized light in which the model is placed. The second polarizer, which is called the analyser, resolves the component waves emerging from the model into one plane so that the effects produced by the model can be measured from the resulting interference of the waves. Such an arrangement is known as a plane polariscope. The polarizer and analyser are frequently referred to collectively as the 'polaroids'.

In the plane polariscope the polaroids are usually set with their axes 'crossed' or perpendicular to one another. In the absence of a model, none of the light emerging from the polarizer is then transmitted by the analyser, Figure 5.4a. A model inserted into the field of a crossed plane polariscope therefore appears against a dark background.

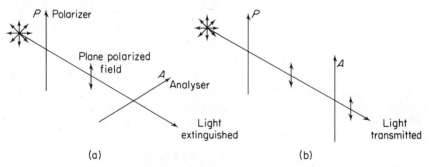

Figure 5.4.   The plane polariscope, (a) Crossed polaroids. (b) Parallel polaroids

If the axes of the polaroids are set parallel to one another, all of the light transmitted by the polarizer is transmitted also by the analyser except for losses due to absorption and reflection, Figure 5.4b. The background in this case is therefore bright. This is advantageous in obtaining accurate alignment of the model in the polariscope.

5.4 Effect of a stressed plate in the plane polariscope

We now consider the effects produced when a plate of temporary doubly refracting material in a two-dimensional or plane state of stress is placed in the field of a plane polariscope.

Let the axis of the polarizer be vertical and let the vector OP (Figure 5.5) represent a plane polarized wave emerging from it. This wave can be represented by

$$x = 0; \qquad y = a \sin \omega t$$

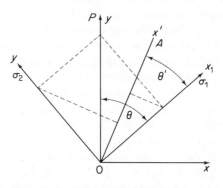

Figure 5.5.   Effect of a stressed plate in a plane polariscope

We assume that this wave falls with normal incidence on the plate at a point where the principal stresses are σ_1 and σ_2 and that its plane of vibration is inclined at an angle θ to the direction of σ_1. On entering the plate, the wave is split into components vibrating in the directions of the axes ox_1 and oy_1 parallel to σ_1 and σ_2 respectively. These components are given by

$$x_1 = y \cos \theta = a \cos \theta \sin \omega t,$$

$$y_1 = y \sin \theta = a \sin \theta \sin \omega t.$$

These are transmitted through the plate with different velocities so that on emergence a phase difference exists between them. Assuming ox_1 to be the 'fast' axis and denoting the phase difference by α, the emergent components are

$$x_1 = a \cos \theta \sin (\omega t + \alpha),$$

$$y_1 = a \sin \theta \sin \omega t.$$

As shown in Section 3.6, these equations represent an elliptical vibration. In general, therefore, the effect of the plate is to convert the incident plane polarized light into elliptically polarized light.

The analyser transmits only the component parallel to its axis of the resultant vibration emerging from the plate. If the axis OA of the analyser includes an angle θ' with the direction of σ_1, the transmitted wave is

$x' = x_1 \cos \theta' + y_1 \sin \theta'$

$= a \cos \theta \cos \theta' \sin (\omega t + \alpha) + \sin \theta \sin \theta' \sin \omega t$

$= (a \cos \theta \cos \theta' \cos \alpha + a \sin \theta \sin \theta') \sin \omega t + (a \cos \theta \cos \theta' \sin \alpha) \cos \omega t$

$= A \cos (\omega t + \beta)$,

of which the amplitude A is given by

$$A = \sqrt{[(a \cos \theta \cos \theta' \cos \alpha + a \sin \theta \sin \theta')^2 + (a \cos \theta \cos \theta' \sin \alpha)^2]},$$

which reduces to

$$A = \frac{a}{\sqrt{2}} \sqrt{(1 + \cos 2\theta \cos 2\theta' + \sin 2\theta \sin 2\theta' \cos \alpha)}.$$

The intensity of the light transmitted by the analyser is proportional to the square of the amplitude, i.e.

$$I_{\theta'} = ka^2(1 + \cos 2\theta \cos 2\theta' + \sin 2\theta \sin 2\theta' \cos \alpha), \tag{5.9}$$

where k is a constant.

If the analyser is now rotated through 90°, we obtain on replacing θ' by $\theta' \pm 90°$,

$$I_{\theta' \pm 90°} = ka^2(1 - \cos 2\theta \cos 2\theta' - \sin 2\theta \sin 2\theta' \cos \alpha).$$

Adding the last two equations, we obtain

$$I_{\theta'} + I_{\theta' \pm 90°} = 2ka^2 = I_0, \tag{5.10}$$

i.e. the sum of the intensities transmitted by the analyser when set in turn in any two mutually perpendicular directions is a constant. This constant, denoted by I_0, may be regarded as the total intensity and is the intensity of the light that would be transmitted by the analyser if it were optically isotropic. Equation (5.9) may now be written

$$I_{\theta'} = \tfrac{1}{2}I_0(1 + \cos 2\theta \cos 2\theta' + \sin 2\theta \sin 2\theta' \cos \alpha). \tag{5.11}$$

The direction of the analyser in which the intensity transmitted will be a maximum or a minimum is obtained from the condition

$$\frac{dI_{\theta'}}{d\theta'} = 0,$$

which produces

$$\tan 2\theta' = \tan 2\theta \cos \alpha. \tag{5.12}$$

Comparing this result with equation (3.10), we see that for maximum or minimum intensity, $\theta' = \psi$. Thus, the directions defined by equation (5.12) coincide with those of the principal axes of the light ellipse, i.e. the ellipse which characterizes the state of vibration of the light emerging from the plate. Substituting the corresponding values of $\sin 2\theta'$ and $\cos 2\theta'$ in equation (5.11), we obtain for the maximum and minimum intensities transmitted:

$$I_1, I_2 = \tfrac{1}{2}I_0[1 \pm \sqrt{(\cos^2 2\theta + \sin^2 2\theta \cos^2 \alpha)}]. \tag{5.13}$$

Intensity transmitted in the crossed plane polariscope

If the analyser is set in the crossed position, i.e. with its axis perpendicular to that of the polarizer, $\theta' = \theta - 90°$. Inserting this value in equation (5.11) we obtain for the intensity transmitted:

$$I = \frac{I_0}{2}(1 - \cos^2 2\theta - \sin^2 2\theta \cos \alpha)$$

$$= \frac{I_0}{2}[\sin^2 2\theta(1 - \cos \alpha)],$$

or

$$I = I_0 \sin^2 2\theta \sin^2 \frac{\alpha}{2}. \tag{5.14}$$

Equation (5.14) shows that there are two separate conditions under which extinction of the light will be obtained. One condition is that $\theta = 0$ or $90°$. This is satisfied by all points on the plate where the directions of the principal stresses are parallel to the axes of the polaroids and such points appear dark. In general, these points lie on continuous curves forming a system of dark bands known as isoclinics. For any given setting of the crossed polaroids a corresponding isoclinic pattern can be observed. As the crossed polaroids are rotated, this pattern changes as the condition that the directions of the principal stresses and the axes of the polaroids coincide is satisfied in turn by different points within the plate.

The isoclinics corresponding to any given inclination φ of the axes of the polaroids, usually measured anticlockwise from the vertical and horizontal, are referred to as the $\varphi°$ isoclinics or the isoclinics of parameter $\varphi°$.

The isoclinic pattern is independent of the wavelength and so is the same for any wavelength of monochromatic light and for white light. It is also independent of the phase difference and hence of the magnitude of the applied loads and the stress optic coefficient for the material. These facts can be used to advantage in obtaining the most satisfactory isoclinic patterns as will be described later.

The second condition from equation (5.14) under which extinction is obtained is that $\sin^2 \tfrac{1}{2}\alpha = 0$, i.e. $\alpha = 2n\pi$ radians $= n$ cycles where n is zero or any integer. This is equivalent to a relative retardation of n wavelengths. Thus, all points on the plate at which the difference $(\sigma_1 - \sigma_2)$ in the principal stresses is such that the relative retardation produced is equal to a whole number of wavelengths will appear dark. In general, the difference in the principal stresses varies continuously within the plate so that the loci of such points are smooth curves. These are known as isochromatics and are classified in terms of their fringe order. For instance, the isochromatic representing the locus of all points at which the relative retardation is equal to one wavelength ($n = 1$) is called the first order fringe and so on. The complete pattern of the isochromatics is frequently referred to as the stress pattern.

Since the phase difference α depends on the wavelength, it follows that the points of extinction for light of different colours lie on different curves. Thus, when white light is used, all wavelengths of which the relative retardation at a given point is an integral number will be extinguished. The colour observed is then due to the remaining wavelengths which are transmitted to a greater or lesser degree depending on their proximity to the wavelengths absorbed. For small values of the retardation, only one colour is extinguished along any given isochromatic line which thus has a characteristic colour. The fringes observed therefore appear as bands of varying colour consisting of the isochromatic lines for all wavelengths as they are progressively extinguished.

With greater values of the retardation, two or more wavelengths of the light may be extinguished simultaneously so that isochromatics of different order are not identical in appearance. This variation in colour can be used to determine the order of the isochromatics provided this is low. With still higher retardation the fringes become progressively paler until eventually the number of wavelengths extinguished is so great that the residual light appears practically white.

In the particular case when $\alpha = 0$, all wavelengths are extinguished so that the zero order fringe is always black.

Intensity transmitted in the parallel plane polariscope

If the axis of the analyser is set parallel to that of the polarizer, it follows from equations (5.10) and (5.14) that the intensity transmitted is

$$I = I_0 \left(1 - \sin^2 2\theta \sin^2 \frac{\alpha}{2} \right). \tag{5.15}$$

The conditions for maximum intensity of the transmitted light are now the same as those for extinction when the polaroids are crossed.

5.5 Distinction between isoclinic and isochromatic fringes

From the above analysis we see that when a stressed plate is placed in the field of a plane polariscope, two different systems of bands or fringes appear simultaneously, namely, the isoclinics and the isochromatics. If white light is used with the usual crossed arrangement, these can easily be distinguished from one another since the isoclinics are black while the isochromatics, with the exception of fringes of zero order, are coloured. With monochromatic light, both sets of fringes are black. The isochromatics are nevertheless usually distinguishable since they are much more sharply defined than the isoclinics.

The individual fringe patterns can also be identified by varying the conditions under which they are observed. Thus, if the polaroids are rotated in the crossed position while the load is kept constant the isoclinic pattern varies while the isochromatic pattern remains unchanged. Conversely, if the load on the model is varied for a fixed setting of the polaroids, the isochromatic pattern will change while the isoclinic pattern remains unchanged.

It frequently happens that all the information required can be obtained from the isochromatic pattern alone. The isoclinics are then undesirable since they tend to obscure the stress pattern. In such cases the isoclinics can be eliminated by using circularly polarized light. The directional characteristic of the light incident on the model, on which the formation of isoclinics depends, is thus removed.

5.6   Effect of quarter-wave plates

We consider a plane polarized wave $x = 0$; $y = \sqrt{(2)}\, a \sin \omega t$ represented by the vector OP in Figure 5.6 which falls on a quarter-wave plate set with its

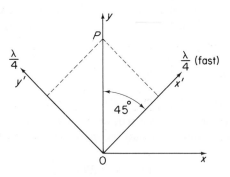

Figure 5.6.   Effect of a quarter-wave plate
on a plane polarized wave

principal axes at 45° to OP. The directions of these axes are represented by ox' and oy' and we assume in the first instance that ox' corresponds to the 'fast' axis.

On entering the quarter-wave plate, the initial wave is split into two components of equal amplitude vibrating in the directions ox' and oy'.

These are

$$x' = a \sin \omega t; \qquad y' = a \sin \omega t.$$

The phase difference introduced between these components by the quarter-wave plate is $\pi/2$ so that, on emergence,

$$x' = a \sin\left(\omega t + \frac{\pi}{2}\right) = a \cos \omega t,$$

$$y' = a \sin \omega t.$$

These equations represent a circular vibration. To obtain the sense in which it is described we have

$$\frac{\mathrm{d}y'}{\mathrm{d}t} = a\omega \cos \omega t.$$

When $x' = a$, $\mathrm{d}y'/\mathrm{d}t = a\omega$ which is positive, showing that the vibration is counterclockwise. In the same way, it can be shown that if ox' is the 'slow' axis, the quarter-wave plate converts the initial plane polarized wave into a clockwise circularly polarized wave.

We now consider the effect produced when a circularly polarized wave is incident on a quarter-wave plate set with its principal axes at 45° to the horizontal and vertical. Referring again to Figure 5.6 we assume that the incident wave is circularly polarized in the counterclockwise direction and given by

$$x' = a \cos \omega t; \qquad y' = a \sin \omega t.$$

After transmission through the quarter-wave plate these become

$$x' = a \cos (\omega t + \tfrac{1}{2}\pi) = -a \sin \omega t,$$

$$y' = a \sin \omega t.$$

The resultant is a plane polarized wave $x = \sqrt{(2)}\, a \cos (\omega t + \tfrac{1}{2}\pi)$, $y = 0$ of amplitude $\sqrt{(2)}a$ vibrating in the horizontal plane (Figure 5.7a). Combining this result with the preceding one, we see that the effect of two quarter-wave plates in series with their corresponding axes parallel is to rotate the plane of vibration through 90°. In other words, the two quarter-wave plates are equivalent to a single half-wave plate.

If we now suppose ox' to be the slow axis of the quarter-wave plate, then on emergence

$$x' = a \cos \omega t,$$

$$y' = a \sin (\omega t + \tfrac{1}{2}\pi) = a \cos \omega t.$$

The resultant is a plane polarized wave $x = 0$, $y = \sqrt{(2)}a \cos \omega t$ vibrating in the vertical plane (Figure 5.7b). In the same way, it can be shown that

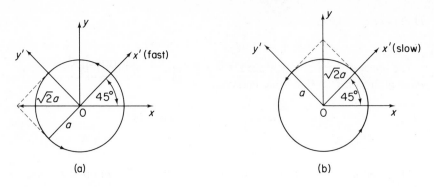

Figure 5.7.   Effect of a quarter-wave plate on a positive circularly polarized wave
when (a) *ox′* is the fast axis and, (b) *ox′* is the slow axis

if the entering wave is circularly polarized in the clockwise direction the
resultant is a plane polarized wave of amplitude $\sqrt{(2)}a$ vibrating in either the
vertical or horizontal plane depending on whether *ox′* is the fast or slow axis
respectively of the quarter-wave plate. These various effects are illustrated in
Figure 5.8.

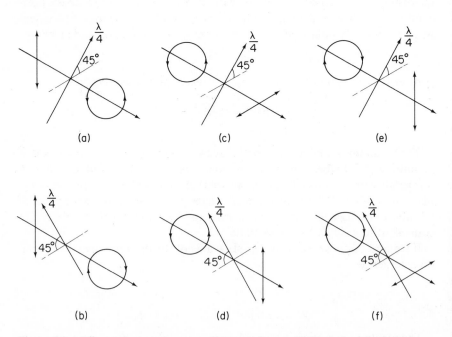

Figure 5.8.   Effects of a quarter-wave plate on plane and circularly polarized waves
for different directions of the fast axis.

5.7 The circular polariscope

The circular polariscope is obtained by inserting two quarter-wave plates between the polaroids in a plane polariscope with their principal axes at 45° to those of the polaroids. The axes of the polaroids may be either crossed or parallel and so also may those of the quarter-wave plates. The model is placed in the field of circularly polarized light between the quarter-wave plates.

The arrangement shown in Figure 5.9*a*, in which both the polaroids and the quarter-wave plates are crossed, is known as the crossed or standard circular

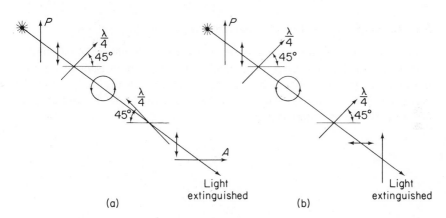

Figure 5.9. The circular polariscope. (*a*) Crossed or standard arrangement. (*b*) Parallel arrangement

polariscope. From the results obtained in the preceding section we see that a plane polarized wave emerging from the polarizer is converted by the first quarter-wave plate into a counterclockwise circularly polarized wave. The second quarter-wave plate reconverts this into a plane polarized wave vibrating in the vertical plane identical with that emerging from the polarizer except for some loss of intensity. With the axis of the analyser horizontal, the light is therefore extinguished. Thus, when a model is inserted in this polariscope it appears against a dark background. If the analyser is rotated through 90° from the position shown so that it is then parallel to the polarizer, the light emerging from the second quarter-wave plate is transmitted and the background will be bright.

The arrangement shown in Figure 5.9*b* is known as the parallel circular polariscope. Here the axes of the polaroids are parallel to one another and so also are those of the quarter-wave plates. The effect of the two quarter-wave plates is to rotate the plane of vibration of the initial plane polarized wave through 90°, i.e. from the vertical into the horizontal plane. The axis of the analyser being vertical, this wave is therefore extinguished and the background is dark. If the analyser is rotated through 90° from the position shown so that its axis lies in the horizontal plane, the final wave is transmitted and the background appears bright.

From the above, we see therefore that in the circular polariscope, if the polaroids and the quarter-wave plates are both crossed or both parallel, the background is dark. If the polaroids are crossed and the quarter-wave plates parallel or if the polaroids are parallel and the quarter-wave plates crossed, the background is bright.

5.8   Effect of a stressed plate in the circular polariscope

Let us consider a photoelastic model consisting of a plate in a two dimensional state of stress inserted in the field of a circular polariscope such that the plane of the plate is normal to the direction of transmission. Let the vertical axis Oy (Figure 5.10) represent the axis of the polarizer, and the axes Ox' and Oy' the principal axes of the quarter-wave plates.

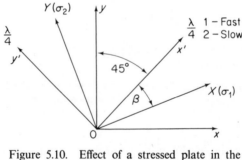

Figure 5.10.   Effect of a stressed plate in the
circular polariscope

If Ox' is the 'fast' axis of the first quarter-wave plate then, as shown in Section 5.6, the wave emerging from it will be circularly polarized in the counterclockwise sense. On entering the model, this wave is split into two components vibrating in the directions OX, OY of the principal stresses σ_1, σ_2 at the point of incidence. It is immaterial to which axes the incident circular vibration is referred; for convenience the principal axes of stress will be chosen. The components entering the model can then be expressed by

$$X = \frac{a}{\sqrt{2}} \cos \omega t; \qquad Y = \frac{a}{\sqrt{2}} \sin \omega t.$$

If OX is the fast axis and α is the phase difference introduced between the components due to the difference $(\sigma_1 - \sigma_2)$ of the principal stresses, then leaving the model we have

$$X = \frac{a}{\sqrt{2}} \cos(\omega t + \alpha); \qquad Y = \frac{a}{\sqrt{2}} \sin \omega t.$$

This elliptical vibration when referred to the axes Ox', Oy' of the second quarter-wave plate which include angles β with the principal axes of stress becomes

$$x' = X \cos \beta + Y \sin \beta,$$

$$y' = Y \cos \beta - X \sin \beta.$$

If Oy' is the fast axis of this quarter-wave plate, the emergent wave is

$$x' = \frac{a}{\sqrt{2}}[\cos \beta \cos(\omega t + \alpha) + \sin \beta \sin \omega t], \qquad (5.16)$$

$$y' = \frac{a}{\sqrt{2}}[\cos \beta \sin(\omega t + \tfrac{1}{2}\pi) - \sin \beta \cos(\omega t + \tfrac{1}{2}\pi + \alpha)]$$

$$= \frac{a}{\sqrt{2}}[\cos \beta \cos \omega t + \sin \beta \sin(\omega t + \alpha)]. \qquad (5.17)$$

If the axis of the analyser is set at 45° clockwise to Ox', corresponding to a dark field setup, the component transmitted is

$$x = \frac{x' - y'}{\sqrt{2}} = \frac{a}{2}[\cos(\beta + \omega t + \alpha) - \cos(\beta + \omega t)]$$

$$= \frac{a}{2}[\cos(\beta' + \alpha) - \cos \beta']$$

$$= \frac{a}{2}[(\cos \alpha - 1) \cos \beta' - \sin \alpha \sin \beta'],$$

where $\beta' = \beta + \omega t$. The amplitude of this wave is

$$\frac{a}{\sqrt{2}}\sqrt{[(\cos \alpha - 1)^2 + \sin^2 \alpha]} = \frac{a}{\sqrt{2}}\sqrt{[2(1 - \cos \alpha)]}$$

The intensity is proportional to the square of the amplitude:

$$I = \frac{ka^2}{2}(1 - \cos \alpha) = ka^2 \sin^2 \frac{\alpha}{2},$$

or

$$I = I_0 \sin^2 \frac{\alpha}{2}. \qquad (5.18)$$

If the axis of the analyser is at an angle of 45° counterclockwise from Ox', representing a bright field setup, the component transmitted is

$$y = \frac{x' + y'}{\sqrt{2}}$$

Proceeding as before, we find that the intensity of the light transmitted is now given by

$$I = I_0 \cos^2 \frac{\alpha}{2}. \tag{5.19}$$

Equation (5.18) shows that, when the circular polariscope is set to give a dark background, the intensity transmitted is zero when $\sin^2 \alpha/2 = 0$, i.e. when $\alpha = 2n\pi$, corresponding to a relative retardation equal to a whole number of wavelengths. This is the same as the second of the conditions for extinction obtained in the plane polariscope so that the isochromatic pattern is the same in each. This, however, is the only condition for extinction in the circular polariscope so that no isoclinics appear. The orders of the isochromatic fringes correspond to the number n of complete wavelengths of retardation which they each represent.

With the polariscope set for a bright background, we see from equation (5.19) that extinction occurs when $\cos^2 \alpha/2 = 0$, i.e. when $\alpha = (2n + 1)\pi/2$ where n is zero or any integer. This corresponds to a relative retardation equal to an odd number of half wavelengths. The dark fringes therefore now represent the loci of points at which the retardation $R = \lambda/2, 3\lambda/2, 5\lambda/2$, etc., and have the orders $\frac{1}{2}, \frac{3}{2}, \frac{5}{2}$, etc. Thus, in a circular polariscope, the number of points at which the relative retardation can be determined directly from the isochromatics is doubled by viewing the model with the analyser set in turn to give a dark and a bright background. With modern photoelastic materials of high optical sensitivity, this procedure is not usually necessary but it is particularly useful with models having regions where the isochromatics are widely spaced. The circular polariscope is usually preferred in those set ups which produce a dark background since the fringes then represent an integral number of wavelengths of retardation. The bright background is advantageous, however, in obtaining accurate alignment of the model and positioning of the loads.

5.9 Effect of a half-wave plate

Let the fast axis oy' (Figure 5.11) of a half-wave plate be inclined at an angle β to the direction of vibration of the incident plane polarized wave $x = 0, y = a \sin \omega t$. At entry, this wave is split into components:

$$x' = a \sin \beta \sin \omega t,$$

$$y' = a \cos \beta \sin \omega t,$$

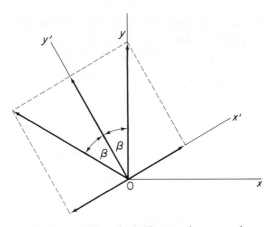

Figure 5.11. Effect of a half-wave plate on a plane
polarized wave

vibrating parallel to the principal axes ox', oy'. During passage through the half-wave plate, a relative retardation of one half-wavelength or phase difference π is introduced between the components so that on emergence they are represented by

$$x' = a \sin \beta \sin (\omega t + \pi) = -a \sin \beta \sin \omega t,$$

$$y' = a \cos \beta \sin \omega t.$$

The resultant is a plane polarized wave vibrating in a plane inclined at an angle 2β to that of the initial wave. The half-wave plate thus merely rotates the plane of vibration and may be used for this purpose, for example, with a laser beam.

5.10 Material fringe value

The retardation in terms of fringe order n and wavelength λ is given by

$$R = n\lambda.$$

The stress optic law for two dimensions, equation (5.4), can then be written

$$n\lambda = C(\sigma_1 - \sigma_2)d,$$

from which

$$(\sigma_1 - \sigma_2) = \frac{nf}{d} = \frac{\alpha f}{2\pi d}, \tag{5.20}$$

where $f = \lambda/C$ is a constant depending on the material and the wavelength of the light used. This constant is called the material fringe value or fringe-stress coefficient and represents the principal stress difference necessary to produce unit change in the fringe order in a model of unit thickness. Typical units of f are N/m fringe and lbf./in. fringe.

Since the maximum shear stress τ_{max} is equal to $(\sigma_1 - \sigma_2)/2$, equation (5.20) can be written in the form

$$\tau_{max} = \frac{nf_\tau}{d},\tag{5.21}$$

where $f_\tau = \frac{1}{2}f$ is the material fringe value for shear stress.

6

MEASUREMENT AND INTERPRETATION OF PHOTOELASTIC DATA

6.1 Determination of order of isochromatics

Equation (5.20) shows that the difference in the principal stresses or the maximum shear stress at any point in a photoelastic model is directly proportional to the isochromatic fringe order n. At any point on a load-free boundary, the principal stress normal to the boundary is zero and the fringe order gives the value of the other principal stress directly. At other points, the individual principal stresses can be determined using the methods described in Chapter 9.

In many cases the distribution of the principal stresses along a load-free boundary or of the maximum shear stress in the interior of the model can be determined with sufficient accuracy using the integral order isochromatic fringes only if these are produced in sufficient number and are reasonably well distributed over the model. In the circular polariscope the number of fringes can be doubled by setting for a dark and a bright background in turn. The number of fringes can also be increased using the method of fringe multiplication developed by Post and described in Section 6.9. The required distribution is obtained by measuring, usually from a photograph of the fringe pattern, the positions of the points in which the fringes of different order intersect chosen lines within the model.

The simplest method of determining the orders of the fringes is by counting in the direction of increasing order from a point of known zero order. With white light, points of zero order are easily recognized since they appear black. Such points may be isolated or form lines or zones. The direction of increasing fringe order can be recognized from the colours of the fringes which vary from reds to blues or greens in this direction.

With monochromatic light it is often possible to count from a point of known zero stress such as a projecting load-free corner. A more general method is to observe the growth of the stress pattern as the load is gradually applied. The fringe order at any observed point can then be determined by counting the number of fringes which successively appear at the point. Since the fringes originate at points of maximum order and are displaced towards points of lower order as the load is increased, the directions of increasing or diminishing fringe order can also be established.

When the number or distribution of the fringes is such that the stress pattern by itself does not form a satisfactory basis for the determination of the stress distribution, it becomes necessary to measure non-integral fringe orders at intermediate points. Several different methods by which this can be accomplished will now be described.

6.2 Colour matching

As we have seen previously, with white light the isochromatics in a field of non-uniform stress vary in colour as the different constituent wavelengths are successively extinguished. From equation (5.14), the intensity transmitted of any given wavelength λ in the crossed polariscope is proportional to $\sin^2 \alpha/2$ or $\sin^2 \pi R/\lambda$ where R is the relative retardation. Each wavelength or colour is extinguished in turn whenever $R = n\lambda$. The remaining colours are transmitted with different intensities depending on the wavelength and the colour observed is the complementary colour to that extinguished. Thus, any value of the retardation produces a characteristic colour of the transmitted light.

At points where the retardation is zero, all wavelengths are extinguished so that the zero order fringe is black. As the retardation gradually increases, it is at first less than the shortest wavelength in the visible spectrum and in this region the transmitted light changes from black through grey to white at about 200 mμ. Clearly defined colours appear at about 400 mμ, when the wavelength of violet light is reached and this colour is then extinguished. This is followed in turn by the other colours in the spectrum in the order indigo, blue, green, yellow, orange and red. The complementary colours vary in the sequence yellow, orange, red, indigo, blue and green. At about 800 mμ, the extinction of the first order of colours is complete.

When retardation is further increased it becomes equal in turn to twice the wavelengths in the visible spectrum and the colours are extinguished for the second time. The colours of the second and higher orders are not identical, however, with those observed in the first. Differences occur whenever the retardation becomes a simple multiple of more than one wavelength. Thus, the first extinction of deep red (770 mμ) lies close to the second extinction of violet (400 mμ). Similarly, the second extinction of colours between yellow and deep red (590–770 mμ) coincides with the third extinction of colours between violet and green (400–520 mμ). At higher retardations the number of wavelengths which are extinguished simultaneously increases. The colour observed is a mixture of the complementary colours of each wavelength extinguished. As a result, the colours become progressively paler and show less variation. With orders above the third (about 2000 mμ)

the colour variation reduces to an alternation between rose pink and pale green. At still higher orders the mixture of complementary colours is so great that white light is virtually restored. In such regions, the extinction of particular wavelengths can be recognized using a spectroscope.

To use the method, a simple calibration test is carried out, for instance on a uniform strip loaded in simple tension over the central region of which the colour will be more or less uniform. A chart or table is then prepared of the colours observed as the load is gradually increased. Table 6.1 gives corresponding values of the retardation in millimicrons and the colour transmitted

Table 6.1   Values of retardation and corresponding colours of transmitted light in crossed polariscope with white light source

retardation (mμ)	colour transmitted	retardation (mμ)	colour transmitted
0	black	843	yellow green
40	iron grey	866	green yellow
57	lavender grey	910	clear yellow
158	grey blue	948	orange
218	grey	998	brilliant orange red
234	green white	1101	dark violet red
259	off white	1128	bright blue violet
267	yellow white	1151	indigo
275	pale straw yellow	1258	green blue
281	straw yellow	1334	sea green
308	bright yellow	1376	brilliant green
332	brilliant yellow	1426	green yellow
430	brown yellow	1495	flesh colour
505	red orange	1534	crimson
536	red	1621	dull purple
551	deep red	1652	violet grey
565	purple	1682	grey blue
575	violet	1711	dull sea green
583	indigo	1744	blue green
664	sky blue	1811	bright green
728	green blue	1927	bright green grey
747	green	2007	white green
826	bright green	2048	flesh red

in a crossed polariscope with a white light source giving a continuous spectrum. Similar tables can be compiled for other sources such as the mercury vapour lamp, the light from which is composed of a few narrow wavebands only.

The method is not a particularly accurate one since it depends to a large extent on the sensitivity of the eye of the observer to small colour changes.

It is nevertheless of value in determining isochromatic orders in regions where the retardation changes slowly from point to point and is frequently favoured in methods employing photoelastic coatings (see Section 11.5) in which the restriction on the thickness of coating limits the order of the isochromatics produced.

6.3   The Babinet and Babinet–Soleil compensators

Small relative retardations less than one wavelength or larger retardations at any desired point in a photoelastic model can be measured to a high degree of accuracy by means of an instrument known as a compensator. This consists essentially of a doubly refracting crystal plate the effective thickness of which can be varied so as to introduce between the component vibrations leaving the plate a phase difference equal but opposite in sign to that produced by the model.

The best known of such instruments is the Babinet compensator. This consists of two quartz wedges of the same small angle (2°–3°). One of the wedges is fixed to the frame of the compensator while the other can be moved relative to it by rotating a micrometer screw. In this way, the overall thickness of the two wedges can be varied. The wedges are cut with their optic axes parallel to the surfaces but perpendicular to each other as indicated in Figure 6.1a.

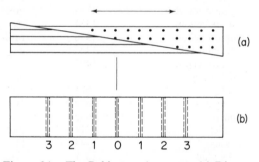

Figure 6.1.   The Babinet compensator. (*a*) Directions of optic axes. (*b*) Isochromatic pattern with the compensator set in the zero position

If plane polarized light falls with normal incidence on the surface of the compensator, the direction of vibration not coinciding with the optic axis of either wedge, it is split at entry into the first wedge into components vibrating parallel and perpendicular to the optic axis. These components are transmitted through the wedge with different velocities. Since the optic axes of the wedges are perpendicular to each other, the ordinary wave in the first wedge becomes the extraordinary wave in the second and vice versa so that their velocities are exchanged. The component which is advanced relative to the other in the first wedge is thus retarded in the second. The effective thickness of the plate is therefore equal to the difference in thickness of the wedges at any point. At the section where the thicknesses are equal, the relative retardation is zero for all wavelengths. This results in the appearance of a black band or fringe across the field of the compensator at this section when it is viewed between crossed polaroids. The central section of the fixed wedge is indicated by a cross wire or fine line engraved on its surface and the instrument is set in the zero position by moving the other wedge by means of the micrometer screw until the zero order fringe is bisected by this line.

On either side of the zero fringe, the effective thickness and hence the retardation increase in proportion to the distance from the centre line so that the field is crossed by other equally spaced fringes of increasing order representing retardations equal to an integral number of wavelengths as shown in Figure 6.1b. With white light, these fringes are coloured so that the zero order fringe can be easily identified. Since the retardation varies with the wavelength, however, the actual measurements should be made with monochromatic light.

The instrument is calibrated by measuring the distance s through which the moveable wedge must be shifted to make successive fringes coincide with the centre line. This movement represents a change in the retardation of one wavelength.

To measure the retardation at any point in a photoelastic model, the model and compensator are placed between crossed polaroids so that the light passes normally through each in turn. The compensator is rotated about its axis parallel to the direction of propagation until its principal axes are parallel to those in the model at the desired point as determined from the isoclinics. The crossed polaroids are set with their axes at about 45° to those of the compensator since in this position the fringes show maximum contrast. If the compensator is set initially in the zero position with the central cross line passing through the point under observation, any retardation produced by the model causes the zero fringe to be displaced to one side or other of this line. The micrometer screw is then rotated to return the zero order fringe to the centre line at the point of observation and the corresponding movement x of the wedge is noted. The retardation produced by the model is equal to x/s wavelengths.

In some applications it is a disadvantage that the retardation produced by the compensator varies according to the effective thickness at the point of entry. This difficulty is overcome in the Soleil modification of the Babinet compensator. In this, the wedges are cut with their optic axes parallel to each other. As in the Babinet compensator, one of the wedges can be moved laterally relative to the other. A separate quartz plate of uniform thickness is placed with its optic axis perpendicular to those of the wedges as shown in

Figure 6.2. The effective thickness is now equal to the difference between the thickness of the uniform plate and the overall thickness of the composite plate formed by the two wedges. This difference and hence the retardation

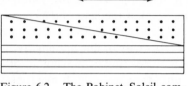

Figure 6.2.   The Babinet–Soleil compensator

produced are constant over the whole field of the compensator which therefore appears uniformly bright or dark when viewed between crossed polaroids.

The Babinet and Babinet–Soleil compensators enable fringe orders to be measured to within 0·01 fringe.

6.4   Simple compensators

Precision compensators of the Babinet type are expensive. Simple forms of compensator which, however, are sufficiently accurate for many purposes can be readily made in the laboratory.

The tensile strip compensator consists of a strip of photoelastic material of uniform section provided with the means of applying known variable axial loads and usually also for rotating the strip about an axis perpendicular to its plane. The compensator is placed either in front of or behind the model and parallel to it with the central region of the strip, over which the stress may be regarded as uniform, over the point at which compensation is to be carried out. The compensator is rotated about its axis until the principal axes in the strip are parallel to those in the model at the observed point. The load is then adjusted until the point appears black. The retardation produced by the model is then equal and opposite to that produced by the strip. Compensation is only possible when the compensator is in one of the two such possible positions, namely, when the axis of stress in the strip is parallel to the direction of the minor principal stress in the model. Thus, after a first trial, it may be found necessary to rotate the compensator through 90°. If the tension strip is replaced by a compression strip, which is sometimes preferred, compensation is only possible when the axis of stress in the strip is parallel to the major principal stress in the model. If the strip is made of the same material and thickness as the model, the required principal stress difference in the model is equal to the known applied stress in the strip when compensation is achieved. If the material or thickness is different, a calibration test can be carried out on the strip to determine the relationship between the applied stress and the retardation. The principal stress difference in the model can then be calculated from the retardation and the material fringe value and thickness of the model material.

In other simple forms of compensator, such as the wedge and beam type, advantage is taken of the frozen stress method. By this method, which is described in Chapter 14, the stress pattern is 'frozen' or locked in the material and persists after the load is removed. To make a wedge compensator, an axial tensile or compressive stress of suitable magnitude is frozen into a uniform strip. The strip is then cut to form a wedge of small angle with its edge perpendicular to the axis of the strip. Cutting must be performed with care to avoid disturbing the frozen stress pattern. The fringe order then varies linearly along the wedge from zero at the edge to a maximum at its thickest part. When the wedge is viewed in the polariscope, it is thus seen to be crossed by a series of equidistant fringes increasing in order from zero at the tip. The effect of refraction of the light passing through the wedge can be eliminated by cementing to it a wedge of equal angle made from the unstressed material in such a way that a compound strip of uniform thickness is formed.

In use, the wedge is set with its axis of stress parallel to either the minor or major principal stress in the model, depending on whether the frozen stress is tensile or compressive. The zero order fringe is then displaced from the edge to some other location along the wedge through a distance which is directly proportional to the retardation produced by the model.

In the beam type compensator a stress pattern produced by pure bending is frozen into the material. The zero order fringe then coincides with the neutral surface and equidistant fringes appear on each side with orders increasing towards the extreme fibres. The beam is placed with its axis aligned with the desired point in the model and parallel to one of the principal stresses at the point. The zero order fringe is then displaced from the axis towards the extreme fibres on one side or the other depending on whether the axis is parallel to the major or the minor principal stress. This displacement is proportional to the retardation produced by the model.

Since some alteration in the frozen stress pattern can occur with time, such compensators should be calibrated periodically. Compensators of the wedge and beam type are particularly useful for determining the integral orders of fringes when it is not possible to do this by counting.

6.5 Compensation by means of quarter-wave plates

In modern practice, fractional fringe orders are most frequently determined using either the Tardy[2] or Senarmont[3] quarter-wave plate methods. These combine a high degree of accuracy with the advantage that they utilize only the normal equipment of the polariscope.

6.5.1 *Tardy method*

The arrangement is that shown in Figure 5.9*a* for the crossed circular polariscope. The axes of the polaroids are first set parallel to the principal axes of stress in the model at the point at which the retardation is to be measured. This is done by rotating crossed polaroids with the quarter-wave plates removed until an isoclinic passes through the point under observation. The quarter-wave plates are then inserted with their axes crossed and at 45° to those of the polaroids and the principal axes in the model.

The state of polarization at exit from the second quarter-wave plate is defined by equations (5.16) and (5.17). With $\beta = 45°$ as in the present arrangement, these equations become

$$x' = \tfrac{1}{2}a[\cos(\omega t + \alpha) + \sin \omega t],$$

$$y' = \tfrac{1}{2}a[\cos \omega t + \sin(\omega t + \alpha)].$$

If the axis OX' of the analyser is inclined at an angle θ'' to the crossed position OX, Figure 6.3, the component transmitted is

$$X' = x' \cos(45 - \theta'') - y' \sin(45 - \theta'')$$

$$= \frac{a}{2\sqrt{(2)}}\{[\cos(\omega t + \alpha) + \sin \omega t][\cos \theta'' + \sin \theta'']$$

$$- [\cos \omega t + \sin(\omega t + \alpha)][\cos \theta'' - \sin \theta'']\},$$

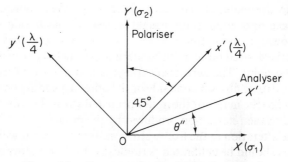

Figure 6.3. Compensation by the Tardy method

which reduces to

$$X' = \frac{a}{\sqrt{2}} \sin\left(\theta'' - \frac{\alpha}{2}\right)\left[\sin\left(\omega t + \frac{\alpha}{2}\right) + \cos\left(\omega t + \frac{\alpha}{2}\right)\right].$$

The amplitude of this wave is

$$\frac{a}{\sqrt{2}} \sin\left(\theta'' - \frac{\alpha}{2}\right), \tag{6.1}$$

and hence the intensity of the light transmitted by the analyser will be zero when $\sin(\theta'' - \alpha/2) = 0$.

Compensation is effected by measuring the angle through which the analyser must be rotated from the crossed position to produce extinction. This angle, which will be referred to as the angle of compensation, can have values between $+\pi$ and $-\pi$. Extinction can obviously be obtained at any point by rotating the analyser through a certain angle in one direction or through the complementary angle in the opposite direction. The phase difference α can be expressed in the form

$$\alpha = 2n'\pi + \Delta\alpha$$

where $2n'\pi$ and $\Delta\alpha$ represent the integral and fractional parts respectively and n' is zero or any integer. The condition for extinction is therefore

$$\sin(\theta'' - n'\pi - \tfrac{1}{2}\Delta\alpha) = 0,$$

i.e.

$$\theta'' = \tfrac{1}{2}\Delta\alpha \quad \text{or} \quad \theta'' = -(\pi - \tfrac{1}{2}\Delta\alpha).$$

Thus, extinction is obtained when the analyser is rotated clockwise through an angle $\tfrac{1}{2}\Delta\alpha$ or counterclockwise through $\pi - \tfrac{1}{2}\Delta\alpha$. The angle of compensation is therefore equal to one half of the amount by which the phase difference in circular measure is greater or less than an integral number of cycles.

In the above analysis, it has been assumed that the fast axis in the model is perpendicular to the axis of the polarizer. If these axes are parallel or, alternatively, if the axes of both quarter-wave plates are at 90° to the directions assumed, we obtain for the amplitude

$$\sqrt{(2)}\, a \sin(\theta'' + \tfrac{1}{2}\alpha) = \sqrt{(2)}\, a \sin(\theta'' + n'\pi + \tfrac{1}{2}\Delta\alpha), \tag{6.2}$$

and the directions of rotation corresponding to the two possible values of θ'' which give extinction are reversed.

The integral fringe order n' can usually be determined by counting from a point of zero order. The appropriate angle to be used can be ascertained by observing whether it is the fringe of lower or higher order which moves to produce extinction at the point as the analyser is rotated from the crossed position. If we denote by m the compensation angle measured in the direction which causes the fringe of lower order n' to move to the observed point, then the phase difference is

$$\alpha = 2n'\pi + 2m,$$

and the retardation in wavelengths or fringe order is

$$R = n = \frac{\alpha}{2\pi} = n' + \frac{m}{\pi} = n' + \frac{m°}{180°}.$$

If the complementary angle $(180° - m)$ is measured, then

$$n = (n' + 1) - \frac{(180° - m)}{180°}.$$

When the fringe order is everywhere non-integral so that no fringe pattern appears with the polaroids crossed, the value of n' can be determined with the aid of a simple improvised compensator of the wedge or beam type. The direction of rotation of the analyser which corresponds to an increase or a reduction in stress difference can be determined from a simple tensile test carried out on a specimen of the material.

6.5.2 *Senarmont method*

In the Senarmont method, only the analyser quarter-wave plate is used. This is set with its axes parallel to those of the crossed polaroids and at 45° to the principal axes of stress at the particular point in the model (Figure 6.4).

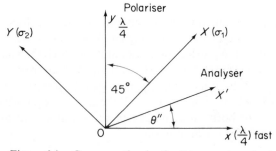

Figure 6.4. Compensation by the Senarmont method

The initial wave emerging from the polarizer may be represented by $x = 0$; $y = \sqrt{(2)}a \sin \omega t$. On entering the model, this wave is split into components $X = a \sin \omega t$, $Y = a \sin \omega t$ vibrating parallel to the principal stresses σ_1 and σ_2 respectively. If OX represents the fast axis, then leaving the model we have

$$X = a \sin (\omega t + \alpha); \qquad Y = a \sin \omega t.$$

On entering the quarter-wave plate, this vibration is resolved into the components

$$x = (X - Y)/\sqrt{2} = \frac{a}{\sqrt{2}}[\sin (\omega t + \alpha) - \sin \omega t]$$

$$= \frac{a}{\sqrt{2}}\left[2 \cos \left(\omega t + \frac{\alpha}{2}\right) \sin \frac{\alpha}{2}\right],$$

$$y = (X + Y)/\sqrt{2} = \frac{a}{\sqrt{2}}[\sin(\omega t + \alpha) + \sin \omega t]$$

$$= \frac{a}{\sqrt{2}}\left[2 \sin\left(\omega t + \frac{\alpha}{2}\right) \cos \frac{\alpha}{2}\right].$$

If ox represents the fast axis, then emerging from the quarter-wave plate we have

$$x = \sqrt{(2)}a \cos\left(\omega t + \frac{\alpha}{2} + \frac{\pi}{2}\right) \sin \frac{\alpha}{2} = -\sqrt{(2)}a \sin\left(\omega t + \frac{\alpha}{2}\right) \sin \frac{\alpha}{2},$$

$$y = \sqrt{(2)}a \sin\left(\omega t + \frac{\alpha}{2}\right) \cos \frac{\alpha}{2}.$$

If the analyser is set with its axis at an angle θ'' to the crossed position ox, the transmitted wave is given by

$$X' = x \cos \theta'' + y \sin \theta''$$

$$= \sqrt{(2)}a \sin\left(\omega t + \frac{\alpha}{2}\right)\left[\cos \frac{\alpha}{2} \sin \theta'' - \sin \frac{\alpha}{2} \cos \theta''\right]$$

$$= \sqrt{(2)}a \sin\left(\theta'' - \frac{\alpha}{2}\right) \sin\left(\omega t + \frac{\alpha}{2}\right).$$

The amplitude of this wave is $\sqrt{(2)}a \sin(\theta'' - \frac{1}{2}\alpha)$. If the principal axes in either the model or the quarter-wave plate are at 90° to the directions assumed, the amplitude is $\sqrt{(2)} a \sin(\theta'' + \frac{1}{2}\alpha)$. These values are the same as those given by equations (6.1) and (6.2). The conditions for extinction and the procedure for determining the required fringe order are therefore the same as described for the Tardy method.

As a practical example of compensation by means of quarter-wave plates, Figure 6.5 shows two views of a Y-shaped tube which was subjected to internal pressure. Due to the thinness of certain parts of the model, isochromatic patterns could not be observed in the usual way.

First, the isoclinics were drawn directly on the model. Next, the optical elements were set with their axes at 45° to the directions of the principal stresses on one of the isoclinics in the normal arrangement for compensation by the Senarmont method. The analyser was then rotated from its initial position in turn through 18°, 36°, 54°, etc., and the corresponding points of intersection of the zero order isochromatic were marked on the isoclinic. In this way, the points of isochromatic order 0·1, 0·2, etc., on the isoclinic were obtained. This procedure was repeated for each isoclinic in turn. Finally, the complete isochromatic pattern was obtained by drawing curves through points of equal order.

A similar procedure employing the Tardy method of compensation could equally well be applied. In this case, with circularly polarized light incident

Figure 6.5. Y-shaped tube illustrating method of plotting isochromatics of fractional orders by the Senarmont method

on the model, the second quarter wave plate is set at 45° to the directions of the principal stresses corresponding to a particular isoclinic, i.e. for an isoclinic of parameter φ the axis of the quarter wave plate is inclined at $\varphi + 45°$ to the reference direction. The analyser is initially set parallel to the reference direction and is rotated from this position in turn through 18°, 36°, etc. The points of intersection of the zero order isochromatic and the isoclinic are marked as before.

By means of a similar procedure, lines τ_{xy} = constant (see Section 8.5) may be drawn on the model. Denoting the parameters of the lines of constant shear stress and the isochromatics by n_τ and n_σ respectively, the relation

$$\tau_{xy} = \frac{\sigma_1 - \sigma_2}{2} \sin 2\varphi,$$

can be expressed in the form

$$n_\tau = \tfrac{1}{2} n_\sigma \sin 2\varphi,$$

or

$$n_\sigma = \frac{2n_\tau}{\sin 2\varphi}.$$

The value of the angle α through which the analyser should be rotated to determine the point of intersection of a line $\tau_{xy} = n_\tau$ with an isoclinic of parameter φ is therefore given by

$$\alpha = 180 n_\sigma = \frac{360 n_\tau}{\sin 2\varphi}.$$

6.6   Photometric method

A disadvantage of the Tardy and Senarmont methods of compensation is that they involve a sequence of operations and are therefore suitable only when the state of stress is static or varies to a negligible extent in the interval of time necessary to perform these operations. By means of the photometric or intensity method it is possible to determine both the phase difference and the directions of the principal stresses at a point from a single operation. If only the phase difference is required, no movement of the elements of the polariscope at all is necessary. The information required is obtained from measurements made with a photocell or photomultiplier of the intensity of light transmitted by the analyser. Several different arrangements of the polariscope are possible, in which the light incident on the model may be either plane or circularly polarized.

Plane polarized light

The intensities of the light transmitted in the plane polariscope when the polaroids are respectively crossed and parallel are given by

$$I = I_0 \sin^2 2\theta \sin^2 \frac{\alpha}{2}, \tag{5.14}$$

$$I = I_0\left(1 - \sin^2 2\theta \sin^2 \frac{\alpha}{2}\right). \tag{5.15}$$

Maximum intensity with crossed polaroids and minimum intensity with parallel polaroids are obtained when $\theta = 45°$. Denoting the corresponding intensities by I_1 and I_2 we obtain

$$\frac{I_1}{I_0} = \sin^2 \frac{\alpha}{2} = \tfrac{1}{2}(1 - \cos \alpha), \tag{6.3}$$

$$\frac{I_2}{I_0} = \cos^2 \frac{\alpha}{2} = \tfrac{1}{2}(1 + \cos \alpha), \tag{6.4}$$

from which

$$\cos \alpha = \frac{I_2 - I_1}{I_0} = 1 - \frac{2I_1}{I_0} = \frac{2I_2}{I_0} - 1, \tag{6.5}$$

where $I_0 = I_1 + I_2$.

To apply the method, a travelling microscope is first focused on the desired point on the model or its image. The eyepiece is then removed and the photomultiplier head substituted in the eyepiece tube. Crossed polaroids are now rotated until the intensity transmitted is a maximum and the corresponding value of I_1 is noted. The analyser is then rotated independently through 90° to obtain I_2. The phase difference is then obtained from equation (6.5) and the directions of the principal stresses are at 45° to those of the axes of the polaroids. As shown by equation (5.10), the total intensity I is equal to the sum of the intensities measured with the analyser in any two mutually perpendicular positions. With a stable light source, this can be determined from a preliminary observation. The only operation necessary is then the synchronous rotation of the polaroids in either the crossed or parallel positions until the intensity transmitted is respectively a maximum or a minimum.

The directions of the principal stresses cannot be obtained by the above method at points where the phase difference α is a whole number of wavelengths when using monochromatic light. In this case extinction occurs for all positions with crossed polaroids and maximum intensity for all positions with parallel polaroids. The photomultiplier is sufficiently sensitive, however, to distinguish between the isoclinics and the isochromatics if white light is used or, with a mercury vapour lamp, if the filters are removed.

Circularly polarized light

If a quarter-wave plate is inserted between the polarizer and the model with its axis at 45° to that of the polarizer, the light entering the model is circularly polarized. Irrespective of the directions of the principal stresses in the model, the total phase difference between the emergent components is then $\alpha \pm \frac{1}{2}\pi$ and equation (6.5) becomes

$$\pm \sin \alpha = \frac{I_2 - I_1}{I_0} = 1 - \frac{2I_1}{I_0} = \frac{2I_2}{I_0} - 1. \qquad (6.6)$$

As before, the maximum and minimum intensities are transmitted when the axis of the analyser is at 45° to the principal stresses in the model. If the total intensity I_0 is known, all the necessary information can be obtained by rotating the analyser only into the position giving maximum or minimum intensity. This overcomes the difficulty in applying the previous method when the polariscope is not equipped with the means for rotating both polaroids simultaneously.

The directions of the principal stresses cannot be determined by the procedure described at points where the light emerging from the model is circularly polarized since the intensity is then the same for all directions of the analyser. This occurs when the phase difference is equal to an integral number of half wavelengths. At such points the directions can be obtained by removing the quarter-wave plate and following the procedure for plane polarized light or by interpolation. At neighbouring points, the eccentricity of the light ellipse is small and the directions are best obtained by noting two settings of the analyser, taken on opposite sides of the maximum or minimum position, which give the same reading on the photometer. The bisectors of the angles between these two directions are parallel to the axes of the light ellipse and at 45° to the directions of the principal stresses.

In the procedure just described, only the polarizer quarter-wave plate is used. If the arrangement is that of the ordinary crossed or parallel circular polariscope employing both quarter-wave plates, the intensity transmitted is given by equation (5.18) or (5.19) depending on whether the polariscope is set to give a dark or a bright background. These equations are

$$I = I_0 \sin^2 \frac{\alpha}{2},$$

$$I = I_0 \cos^2 \frac{\alpha}{2},$$

and are the same as those obtained for the plane polariscope when $\theta = 45°$. The phase difference α is therefore again given by equation (6.5) but with I_1

and I_2 now representing the intensities transmitted in the circular polariscope when set to give a dark and a bright background respectively.

In the circular polariscope the intensity transmitted is independent of the directions of the principal stresses in the model so that these cannot be determined. The phase difference, however, can be obtained with no movement at all of the elements of the polariscope. This method is therefore particularly suitable when the state of stress varies rapidly with time if the directions of the principal stresses are not required. Rapid changes in the directions of the principal stresses can be determined using the method described in Section 19.12.

In order to obtain reliable results by the intensity method, care must be taken to reduce or if possible eliminate the various effects which can cause inaccuracies. The major sources of error are

(*a*) the measurement of the intensity over a finite area instead of at a point,

(*b*) the effects of stray light,

(*c*) the use of inaccurate quarter-wave plates.

The magnitude of the error due to (*a*) depends on the relative sizes of the model or its image and the field of view of the photometer and these must be suitably proportioned to give the requisite degree of accuracy. The field of view can be limited by inserting a stop in the focal plane of the microscope. The size of aperture used depends on the intensity of the light source, the magnification and the sensitivity of the photometer. Using a source of only moderate intensity and a nine-stage photomultiplier, satisfactory results can be obtained from measurements of the light passing through a circle of about 0·004 in. diameter on the model.

The presence of stray light is indicated by the absence of complete extinction when the phase difference is such that this should be achieved. Assuming that external stray light has been excluded, the principal source is the reflection which occurs at the surface of each element in the polariscope. Thus, some of the light which is reflected in the reverse direction from one surface, e.g. that of the model, is again reflected in the forward direction when it meets another surface, such as that of a quarter-wave plate. This effect obviously increases with the number of surfaces and thus is greater in a lens type polariscope using a point source than in a diffuse light polariscope with no lenses. It is also greater in a circular polariscope, due to the quarter-wave plates, than in a plane polariscope. The effect can be practically eliminated by using a narrow beam of light isolated from a field of diffuse light by means of two stops of small aperture one being placed a short distance in front of the model and the other in front of the polarizer. Alternatively, the intensity of the stray light can be measured under 'extinction' conditions and deducted from the test readings.

When using the single quarter-wave plate set up, the error in the measured phase difference due to an error ε in the quarter-wave plate varies from zero at points where the principal stresses are at 45° to the directions of the axes in the quarter-wave plate to ε at points where these axes are parallel. The effect on the indicated directions of the principal stresses is much more serious, however, and any departure from circularity of the incident light causes these to oscillate through $\pm 45°$ about the true direction. It is therefore essential that the incident light has a very high degree of circularity. This can be obtained by using two or more birefringent plates suitably oriented in series as described in Section 13.6.

6.7   Sign of boundary stresses

In many cases, consideration of the loading will indicate regions of the edges of a model where the sign of the stress is known with certainty. The stresses must vary continuously along the edge and can only change sign at a point where the isochromatic order is zero. Such zero points are not always well defined in the stress pattern, however, although their existence can often be deduced from the variation of isochromatic order along the edge. It sometimes happens however that the sign cannot be determined with certainty from inspection of the fringe pattern alone. In such cases, the so called 'nail test' provides a simple method of determining the sign. Local compression is applied normal to the edge at the point where the sign is to be determined using the finger nail or other sharp edged instrument. The required sign is determined by observing the resulting displacement of the isochromatics in the region of the superimposed load. If σ_1 denotes the stress tangential to the edge and σ_2 the compressive stress applied normal to it, then clearly if σ_1 is tensile,

$$\sigma_1 - \sigma_2 > \sigma_1.$$

The fringe order will thus increase and higher order isochromatics will appear at the edge. If the fringe order is increasing towards the edge, the isochromatics of lower order will be displaced towards the interior. On the other hand, if σ_1 is compressive then

$$\sigma_1 - \sigma_2 < \sigma_1.$$

The fringe order at the edge will now be reduced and the isochromatics of lower order will be displaced towards the edge. The technique is illustrated by its application to a beam in pure bending, Figure 6.6, in which isochromatic displacements in the directions mentioned are clearly visible. If the fringe order diminishes towards the edge, the isochromatics are displaced in the reverse direction.

In cases where local variations of fringe order are too small for the above effects to be observed, it is possible to recognize an increase or a reduction in the fringe order by observing the colour sequence using white light.

Figure 6.6.   Determination of the sign of boundary
stresses by the nail test

The sign of the edge stresses can also be determined in other ways, for example by means of a simple compensator aligned either parallel or perpendicular to the edge. Of course, if a precision instrument such as the Babinet–Soleil compensator is used, both the magnitude and the sign of the edge stress can be determined.

It is shown in Section 8.3 that if the direction of the greater principal stress is known at one point of a model, the signs of the stresses at all points on the free boundary can be deduced from the system of stress trajectories.

6.8   Construction of polariscopes

The polariscope consists basically of a light source and two rotatable polarizers between which the model is placed; for observation in circularly polarized light, two quarter-wave plates are introduced between the polarizers, one on each side of the model. The quarter-wave plates may be mounted on the same stands as the corresponding polarizers but should be capable of independent rotation. In some polariscopes, the quarter-wave plates are demountable to allow the observation of isoclinics in plane polarized light; in others their effect can be eliminated by rotating them until their principal axes are parallel to those of the polarizers or the polarizer-quarter-wave plate assemblies may be reversed so that they lie outside the polarized light field. Circular polarizers consisting of plane polarizers and quarter-wave plates cemented together are also available. Alternative light sources are usually provided, white light being employed for the observation of isoclinics or for the identification of isochromatic fringe orders from the colour, while monochromatic light is used for the production of isochromatic patterns in circularly polarized light. The retardation of the quarter-wave plates must of course correspond with the particular wavelength of the monochromatic light used. For the observation of isoclinics, it is an advantage if the two polarizers can be rotated synchronously.

Polariscopes for orthodox observations by transmitted light may be classified into two groups—the lens type or the diffuse light type—according to the system of illumination of the model. Figure 6.7 shows the arrangement

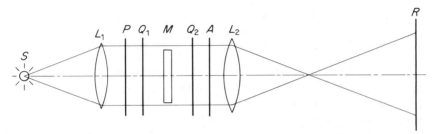

Figure 6.7.   Basic optical system of the lens type polariscope

of the lens type polariscope in its simplest form. An approximate point source S is placed at the focus of a collimating lens L_1 so that a beam of more or less parallel rays passes through the polarizers P, A, quarter-wave plates Q_1, Q_2 and the model M. The light emerging from the second polarizer (the analyser) is projected by the second field lens L_2 to form an image of the model on a screen R. Alternatively, the fringe pattern may be photographed by placing a camera so that the light from L_2 enters the camera lens. In some polariscopes, a second smaller lens is introduced between the lens L_2 and the screen in order to reduce the distance between the model and screen. For the production of monochromatic light from a mercury vapour lamp, suitable colour filters must be inserted in the system. With the object of reducing the various abberations associated with lenses, other elements are frequently included such as an additional condensing lens and iris diaphragm stop between the source and the collimating lens L_1, and a second stop between the lens L_2 and the screen.

The principal advantage of the lens polariscope is that images with large magnification may be produced so that the size of the field and of the model can be comparatively small. Disadvantages are the high cost of lenses of suitable size and quality, some discomfort to the eye due to glare in direct observation, and increased internal reflection resulting in ghost images and flare.

The diffuse light polariscope employs no lenses other than the camera lens. With this system, light from the source falls on a diffuser D such as a ground glass screen, Figure 6.8. With large field polariscopes, the source may consist of several lamps distributed within the housing in such a manner as to produce approximately uniform intensity of illumination over the diffuser.

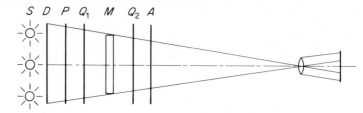

Figure 6.8.   Optical system of the diffuse light polariscope

Compared with the lens polariscope, the diffuse light polariscope has the advantages, (i) large field instruments can be produced at relatively low cost, (ii) since accurate alignment along an optical axis is not required, the elements which lie on opposite sides of the model can be formed as completely separate units so that the model loading frame is free from restrictions which might otherwise be imposed by the construction of a bench type polariscope, (iii) it is more satisfactory for direct viewing of the model. Disadvantages arise from the fact that the model is not illuminated by a collimated beam of light at normal incidence. Thus, rays from every point of the diffuser pass through each point of the model. These rays have different directions and those which enter the eye or the camera lens will lie within the cone subtended by the lens aperture at the point on the model as shown in Figure 6.9. The bire-fringence observed represents some mean value corresponding to all direc-tions of transmission within this cone. The effect obviously depends on the

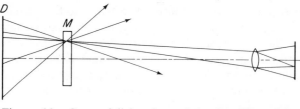

Figure 6.9.   Cone of light observed in the diffuse light
polariscope

size of aperture and on the distance of the lens from the model which in turn depends on its focal length. By a suitable choice of these dimensions the effect can be made negligible. Another effect resulting from the obliquity of the rays is that parallel rays passing through the front and rear edges of the model do not coincide. As a result, the corresponding images do not coincide and this produces an edge shadow effect which may conceal the isochromatic pattern in this important region. Here again the effect can be greatly reduced by employing a camera lens of large focal length and small aperture. Alternatively, a lens may be inserted between the camera lens and the analyser as close to the model as possible as shown in Figure 6.10 so that the image is

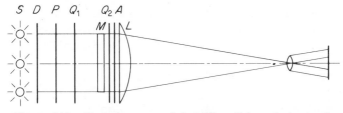

Figure 6.10.   Optical system of the diffuse light polariscope for reduced cone effect

formed essentially by rays which are parallel in the model. Edge shadow effect may also be eliminated locally from a region of interest by locating the model so that this region is in line with the axis of the camera.

Figure 6.11 illustrates a small bench type lens polariscope from a range manufactured by Norwood Instruments Ltd.* The lamp housing contains

Figure 6.11.   A lens type polariscope. Courtesy Norwood Instruments Ltd

* Norwood Instruments Ltd., New Mill Road, Honley, Nr. Huddersfield, HD7 2QD, England.

both white and monochromatic sources and can be swung about its supporting pillar to bring the source selected into position. The quarter-wave plates
are attached by bayonet type couplings to the supporting rings of their
respective polarizers and may therefore be removed for observations in
plane polarized light.

Two diffuse light polariscopes manufactured by the German firm of
Dr H. Schneider, Optotechnische Fabrik* are illustrated in Figures 6.12 and
6.13. That shown in Figure 6.12 is of the bench type and employs a condenser

Figure 6.12.   A bench type diffuse light polariscope. Courtesy Dr H. Schneider

lens to convert the divergent light from a point source into a collimated beam
which illuminates the ground glass diffuser screen. Alternative white and
monochromatic sources may be rotated into position at the principal focus
of the condenser lens. The various components of the polariscope shown in
Figure 6.13 are combined into two separate units which can be set up
independently of one another at any convenient point and distance. Either
white or monochromatic (sodium vapour) light can be selected and switched
on from the control panel at the base of the analyser. In both of these
polariscopes, the polarizer and analyser are provided with electric synchronous drive so that one follows the rotation of the other automatically. The
models illustrated have fields of 300 mm diameter but others up to 450 mm
diameter are manufactured by the same company.

Figure 6.14 shows a polariscope of 15 in. diameter field designed by
Sharples Photomechanics Ltd† for maximum versatility of the illuminating

*Dr H. Schneider, Optotechnische Fabrik, 655 Bad Kreuznach, W. Germany.
†Sharples Photomechanics Ltd, Europa Works, Wesley Street, Bamber Bridge, Preston,
   PR5 4PB, England.

Figure 6.13.   A large field diffuse light polariscope constructed as separate units.
Courtesy Dr H. Schneider

Figure 6.14.   A large field universal polariscope and straining frame. Courtesy Sharples
Photomechanics Limited

system. As illustrated, the polariscope is set up as a lens type employing a point source of light and a system of lenses. It may be converted into a diffuse light polariscope by removing the lenses and replacing the point source by a diffuse source which, when not in use, is stored in one of the cabinets forming the base of the instrument. The polariscope is also designed to accommodate a microflash lamp for the investigation of problems of dynamic stresses. Both units of the polariscope are fitted with wheels which may be mounted on rails as shown to allow the distance between the units to be varied. The straining frame is also provided with wheels running on a transverse track which allow it to be moved clear of the polariscope, for example for convenience when setting up the model and also allow horizontal adjustment of the model position. The vertical position of the straining frame is controlled by means of the large handwheel shown.

In the United States, a large range of polariscopes is manufactured by Photolastic Inc.,* of which one specially designed for the application of the fringe sharpening and fringe multiplication techniques is illustrated in Figure 6.20.

In addition to the types described above, other polariscopes have been developed for specific applications, such as the reflection polariscope and the scattered light polariscope described in Sections 11.2 and 16.7 respectively.

6.9   Isochromatic fringe multiplication and fringe sharpening

In regions of low fringe order or low fringe gradient, the ordinary isochromatic pattern may be inadequate to allow an accurate evaluation of the stresses. This is not only because of the sparsity of the fringes but also because the fringes in such regions consist of broad bands within which the lines of maximum darkness or brightness are difficult to locate. The first of these difficulties can be overcome by additional measurements of non-integral fringe orders by compensation or by the photometric method as described in the preceding sections; the second can be overcome by employing a photometric device to locate the lines of maximum and minimum intensity. Both of these are point by point procedures, however, so that the 'whole field' advantage of the photoelastic method is lost. Two experimental procedures possessing whole field characteristics which improve the isochromatic pattern in such cases and thus allow a more accurate analysis of the stresses have been developed by D. Post.[4,5,6] One procedure, known as fringe multiplication, increases the fringe order at every point and produces an isochromatic pattern containing several times as many fringes as the ordinary pattern. The other procedure, known as fringe sharpening, converts ordinarily diffuse isochromatic fringes into narrow black and white bands

---

* Photolastic Inc., 67 Lincoln Highway, Malvern, Pa., U.S.A.

which allow the whole and half order isochromatic lines to be located much more precisely.

In both procedures, the effects are produced by causing the light to pass through the model several times. This is achieved by multiple reflection of the light between two partial mirrors which are inserted into the collimated field of the ordinary lens type polariscope, one on each side of the model.

For fringe multiplication, the partial mirrors are inclined to each other at a small angle (in practice, about 1/200 radian). The resulting behaviour of the light is shown in Figure 6.15, where, for clarity, the inclination is greatly

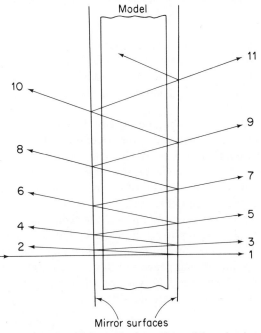

Figure 6.15.   Light path through model and mirror
assembly for fringe multiplication. After Post

exaggerated. Each time the light arrives at a mirror surface, part of it is transmitted and part is reflected. Rays which have travelled through the model a different number of times emerge in different directions. All rays which have passed through the model the same number of times emerge in the same direction to form parallel beams of light. As shown in Figure 6.16, the forward transmitted beams are converged by the second field lens to form a row of discrete images of the source in its focal plane; a similar row of images is formed by the backward directed beams in the focal plane of the first field lens which of course coincides with the plane of the source aperture.

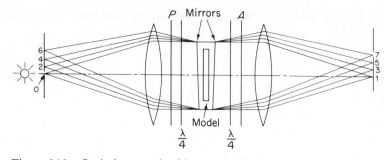

Figure 6.16. Optical system for fringe multiplication by transmitted light.
After Post.

The isochromatic fringe order at every point is multiplied by the number of times the light observed has passed through the model. As can be read from Figure 6.15, the multiplication factor is an odd or an even number depending on whether the beam observed emerges in the forward or the backward direction respectively.

It is clear from Figure 6.15 that, because of the lateral travel or spread of the rays over the model between the point of entry and the point of exit, the birefringence accumulated by a ray does not correspond to that at a unique point of the model. The lateral travel obviously increases with the distance between the mirrors and their angle of inclination so that these quantities should be kept as small as possible. As shown by Post, the lateral travel can be considerably reduced by arranging that the light shall retrace its path. This can be achieved using slightly oblique instead of normal incidence. The optimum angle of incidence for minimum lateral travel is such that the point of entry and the point of emergence are directly opposite one another or coincide according to whether a forward or a backward beam is observed. The optimum paths for multiplications of 7 and 8 are depicted in Figure 6.17.

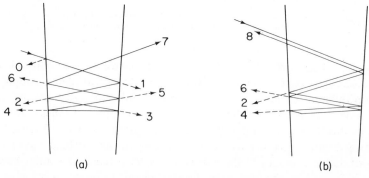

Figure 6.17. Optimum light paths for multiplication by (a) 7 and (b) 8.
After Post

Post suggests that the adjustment of the system for the observation of a beam giving the desired multiplication at the optimum angle of incidence be carried out with the aid of the images of the source. The source aperture is located by means of the images of the source formed by the backward directed beams. If a forward directed beam (odd multiplication factor) is to be observed, the aperture is moved (or the complete mirror assembly is rotated) until it is midway between the images corresponding to the next lower and the next higher even multiplication factors as indicated in Figure 6.18*a*.

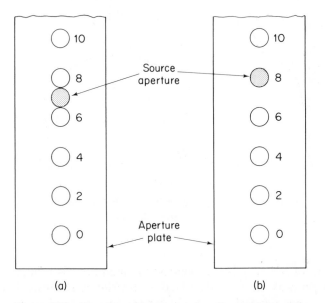

Figure 6.18.   Location of source aperture for optimum angle of incidence for multiplication by (*a*) 7 and (*b*) 8. After Post

For observation of a backward directed beam (even multiplication factor) the aperture is moved into coincidence with the image corresponding to the desired multiplication, Figure 6.18*b*. For correct positioning of the camera, the images are brought to a focus on an aperture plate placed before the camera lens. The plate and camera are then moved laterally until the aperture allows the desired beam to enter the camera while all other light is intercepted.

Fringe multiplication has the disadvantage that it greatly reduces the intensity of the light—to about 1 % of that of ordinary isochromatic fringe patterns.

For fringe sharpening, normal incidence is used and both partial mirrors are set parallel to the model. The behaviour of the light is basically the same as with fringe multiplication but all reflected and transmitted rays are now collinear with the incident ray. All of the forward directed beams enter the camera and their intensities are superimposed. The resultant intensity distribution is represented in Figure 6.19. As compared with the $\sin^2$ intensity

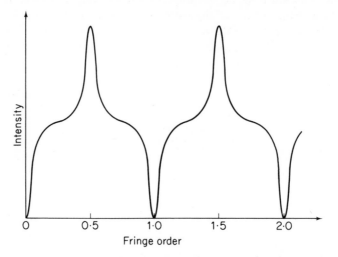

Figure 6.19.   Dark field intensity distribution for sharpened isochromatics. After Post

distribution of ordinary isochromatics, the modified distribution displays much sharper peaks at integral and half-order fringes. The result is a pattern consisting of narrow dark and bright fringes superimposed on a grey background from which the isochromatic lines can be more precisely determined.

With both procedures and particularly that of fringe multiplication the intensity is greatly reduced as compared with that of ordinary isochromatics. This disadvantage can usually be overcome by using a brighter light source, faster photographic materials and, with static problems, a longer exposure time.

While reasonably good results are possible using mirrors formed from ordinary polished glass plate, Post recommends optical flats for best results. Multilayer dielectric coatings are practically non-absorbing and for this reason are preferred to evaporated metal coatings; reflectance should be about 85%.

Figure 6.20 illustrates a commercially available fringe multiplication and fringe sharpening apparatus mounted on a polariscope specially designed to facilitate these applications. Two versions of the apparatus are produced, one

Figure 6.20.   Optical bench with mirror assembly for fringe multiplication and fringe
sharpening. Courtesy Photolastic Inc.

intended for two-dimensional models and the other for slices cut from three-dimensional frozen stress models (see Chapter 14). The mirror housings are connected by three micrometer screws reading to 0·001 in. which are used to vary the separation and the angle between the mirrors. The polariscope illustrated is provided with spherical pivot bearings allowing both tracks to rotate about vertical and horizontal axes to facilitate the alignment for observations at the optimum angle of incidence.

Photographs illustrating fringe multiplication and fringe sharpening kindly provided by D. Post are reproduced in Figures 6.21 and 6.22. Although these show applications to slices cut from three-dimensional models used in investigations carried out by the frozen stress method, the effects produced are exactly the same as with two-dimensional models. Figure 6.21 shows the ordinary isochromatic pattern (1 ×) and patterns with multiplication factors of 9 and 17 of a slice from a bolted flange section of a nuclear reactor closure assembly from an investigation by M. M. Leven and R. L. Johnson. From the enlargement of the region of stress concentration around a hole, it can be seen that resolution is excellent even with a multiplication factor as high as 17. Also visible in Figure 6.21 is a thin wire fitted into a slot cut in the slice to support it in the field of view ; local stresses which this produced did not affect the stresses in remote areas. Fringe sharpening is demonstrated in Figure 6.22 which shows ordinary and fringe sharpened isochromatic patterns for a slice from a model of a nuclear reactor tube sheet from an investigation by T. Slot.

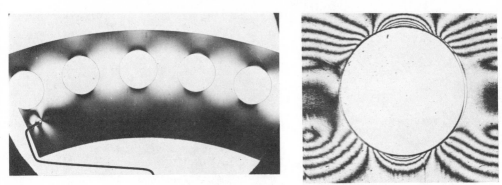

0·5×

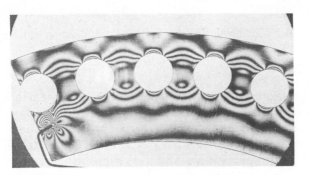

4·5×

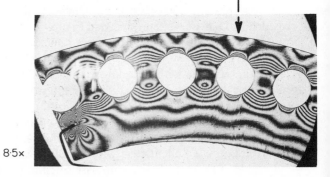

8·5×

Figure 6.21.   Ordinary and fringe multiplied isochromatic patterns of a slice from a bolted flange section of a nuclear reactor closure assembly. Courtesy D. Post

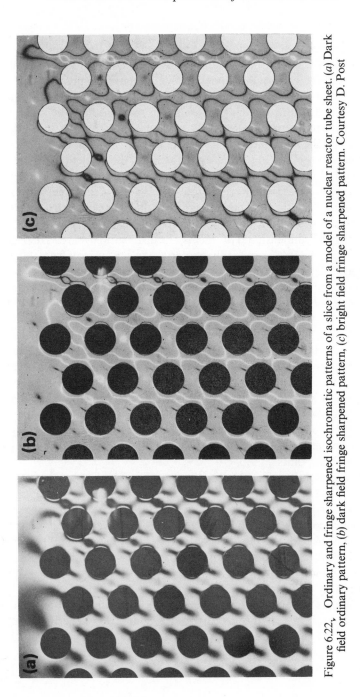

Figure 6.22. Ordinary and fringe sharpened isochromatic patterns of a slice from a model of a nuclear reactor tube sheet. (*a*) Dark field ordinary pattern, (*b*) dark field fringe sharpened pattern, (*c*) bright field fringe sharpened pattern. Courtesy D. Post

6.10 Application of model results to the prototype

It follows from the equations of equilibrium and compatibility that the distribution of stress in the elastic state is independent of the magnitude of the loads and the scale of the model. In two-dimensional problems, it is also independent of the elastic constants if the body forces are zero or uniform. For the investigation of such problems, the material and the scale of the model and the magnitudes of the loads may therefore be chosen as convenient. Provided that the model and prototype are geometrically similar and that the distribution of the loading is the same for both, the model results can then be transferred to the prototype with certain exceptions following the laws of similarity. In a plate which is in a state of plane stress, the stress distribution is constant through the thickness; the scale of the thickness need not therefore be the same as that of the dimensions in the plane of the plate.

In certain two-dimensional problems, the distribution of stress depends on the value of Poisson's ratio for the material. As can be read from equations (2.28) and (2.29), this occurs when the body forces vary over the field of the plate. An example of this type of problem is the distribution of centrifugal stresses in a rotating disc. The stress distribution also depends on Poisson's ratio in multiply connected bodies (i.e. bodies with holes) if a resultant force acts at the boundary of a hole. For such cases, the distribution of stress in the model will differ to some extent from that in the prototype unless Poisson's ratio is the same for both materials. Theoretical and experimental evidence indicates, however, that the difference will usually be small. The laws of similarity may also not apply near the loading points where the stresses may not be truly two-dimensional and where the area of contact may alter considerably under load. It follows from St Venant's principle, however, that the effect on the stress distribution will be negligible except in the immediate vicinity of such points.

In three-dimensional problems, the distribution of stress in general is dependent on the value of Poisson's ratio. The difference in the values of Poisson's ratio for the usual hard model plastics and the prototype material in room temperature investigations is small, however, and its effect can usually be neglected. Investigations by Fessler and Lewin,[7] and others have indicated that even when applying the frozen stress method when Poisson's ratio for the model material is approximately 0·5, errors arising from this effect are likely to be comparatively small.

In problems of plane stress and in three-dimensional problems with $v_m = v_p$, the stresses in the prototype are related to those in the model through

$$\sigma_p = \sigma_m \frac{F_p}{F_m} \left(\frac{L_m}{L_p}\right)^2,$$ (6.7)

where F denotes force and L some linear dimension while the subscripts m and p refer to the model and prototype respectively. For loading by pressure p, the corresponding relation is

$$\sigma_p = \sigma_m \frac{p_p}{p_m}. \tag{6.8}$$

For plane stress problems with $d_m/d_p \neq L_m/L_p$ where d is the thickness, equation (6.7) is replaced by

$$\sigma_p = \sigma_m \frac{F_p}{F_m} \frac{L_m}{L_p} \frac{d_m}{d_p}.$$

For stresses due to rotation with $v_p = v_m$,

$$\sigma_p = \sigma_m \left(\frac{L_p}{L_m}\right)^2 \frac{\rho_p}{\rho_m} \left(\frac{\omega_p}{\omega_m}\right)^2,$$

where ρ is the density and ω the angular velocity.

For gravitational stresses in dams, etc.,

$$\sigma_p = \sigma_m \frac{\rho_p}{\rho_m} \frac{L_p}{L_m}.$$

Similar relations for other forms of loading can be written down from elementary considerations.

For exact similarity, the model and prototype should be geometrically similar when deformed by their respective loads. If they are geometrically similar before loading, this means that corresponding strains in the model and prototype should be equal, i.e. $\epsilon_m = \epsilon_p$. Thus, if $v_m = v_p$,

$$\frac{\epsilon_m}{\epsilon_p} = 1 = \frac{\sigma_m}{\sigma_p} \frac{E_p}{E_m} = \frac{F_m}{F_p} \left(\frac{L_p}{L_m}\right)^2 \frac{E_p}{E_m},$$

from which

$$F_m = F_p \left(\frac{L_m}{L_p}\right)^2 \frac{E_m}{E_p}. \tag{6.9}$$

The force scale thus depends on the scale of the model and the ratio of the elastic moduli of the materials. Similarly, for loading by pressure p we obtain

$$p_m = p_p \frac{E_m}{E_p}. \tag{6.10}$$

For true similarity, the loads applied to the model should conform with equation (6.9) or equation (6.10). Provided that the absolute deformations are small, however, the influence of different relative deformations in general is negligible. Appreciable differences may occur, however, when small deformations considerably alter the loading conditions such as when a loaded pin bears on a hole with small clearance. For the investigation of stresses due to a combination of shrinkage and some other loading, it is necessary that ϵ_m/ϵ_p due to shrinkage shall equal ϵ_m/ϵ_p due to the other loading.

6.11   Stress concentrations and fatigue

A problem frequently encountered is that of determining the stresses in the vicinity of a notch or similar discontinuity in the shape of a body. Such a discontinuity or stress raiser causes a local increase of stress which may be considerably greater than the nominal stress calculated from the loads and the net cross sectional area of the part. This effect is usually expressed in terms of a theoretical stress concentration factor K_t, which is defined as the ratio of the maximum stress σ_{max} to the nominal stress σ_n, i.e.

$$\sigma_{max} = K_t\sigma_n.$$

It is well known that plastic flow may reduce the stress concentration at the root of a notch, particularly with ductile materials. In such cases, the stress determined photoelastically may be compared with the elastic limit of the prototype material.

The effects of stress concentrations are particularly important in parts subjected to repeated loading or fatigue conditions. Since the fatigue limit of a material depends on several factors such as metallographic structure, surface roughness and the scale, the fatigue strength of a prototype cannot be determined directly from tests on a photoelastic model; such tests merely enable the maximum stress or range of stress and the stress gradient in the prototype to be determined.

Numerous experiments have shown that the reduction in fatigue limit caused by a notch is less than would be expected from the value of K_t. The influence of a notch on the fatigue strength or notch sensitivity is usually expressed in terms of a fatigue stress concentration factor K_f. This factor is the ratio of the endurance limit of a plain specimen to that of a specimen containing the notch. Various methods have been proposed for determining the value of K_f from that of K_t. Neuber[8] has suggested the formula

$$K_f = 1 + \frac{K_t - 1}{1 + \dfrac{\pi}{\pi - \omega}\sqrt{\dfrac{a}{r}}},$$

where ω is the flank angle and r the root radius of the notch, while a is a constant depending on the grain size of the material. The following relation proposed by Heywood[9] shows good agreement with many experimental results:

$$K_f = \frac{K_t}{1 + 2[(K_t - 1)/K_t]\sqrt{a'/r}},$$

where a' is an empirical constant for the material. Peterson[10] has suggested the relation

$$q = \frac{K_f - 1}{K_t - 1},$$

and has produced curves giving values of the notch sensitivity factor q for various steels and notch radii.

Siebel[11] has found that the factor $n_\chi = K_t/K_f$ is a function of the material properties and the relative stress gradient χ normal to the surface defined by

$$\chi = \frac{1}{\sigma_1} \frac{\partial \sigma_1}{\partial s_2}, \tag{6.11}$$

where σ_1 is the normal stress tangential to the surface and s_2 is the stress trajectory normal to the surface at the bottom of the notch. The value of χ can be determined photoelastically. Since $\sigma_2 = 0$, equation (6.11) may be written

$$\chi = \frac{\partial(\sigma_1 - \sigma_2)/\partial s_2}{\sigma_1 - \sigma_2} + \frac{\partial \sigma_2/\partial s_2}{\sigma_1 - \sigma_2}.$$

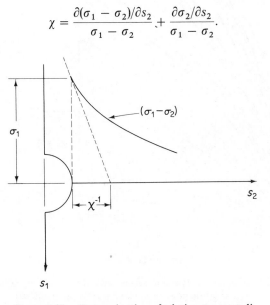

Figure 6.23.   Determination of relative stress gradient at a notch

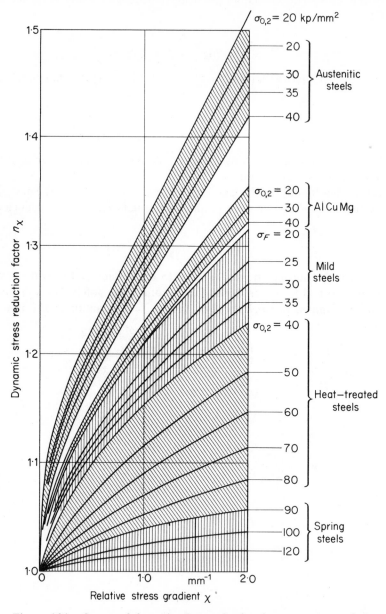

Figure 6.24. Curves of dynamic stress reduction factor n_χ versus relative stress gradient χ for various metals (after Siebel)

We also have from equations (1.16):

$$\frac{\partial \sigma_2}{\partial s_2} + \frac{\sigma_1 - \sigma_2}{\rho_1} = 0.$$

Hence,

$$\chi = \frac{\partial(\sigma_1 - \sigma_2)/\partial s_2}{\sigma_1 - \sigma_2} - \frac{1}{\rho_1},$$

where ρ_1 is the radius at the bottom of the notch. This equation allows the value of χ to be read from the isochromatics as indicated in Figure 6.23. Having determined the value of χ, that of n_χ for the material can be read from curves of the form shown in Figure 6.24 derived experimentally by Siebel.

When the loading consists of an alternating stress superimposed on a mean stress, two values—either the maximum and minimum stresses or one of these and the mean stress—are required to specify the fluctuation. These

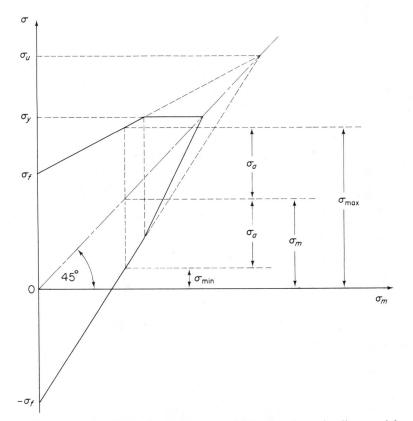

Figure 6.25. Modified Goodman or Smith diagram for a ductile material

values may be determined photoelastically and the results compared with the modified Goodman or Smith diagram for the propotype material, the construction of which is represented in Figure 6.25. In this diagram, from which the endurance strength corresponding to the measured fluctuation can be read, σ_m is the mean stress, σ_a the alternating stress, σ_y the yield stress and σ_u the ultimate tensile strength while σ_f is the endurance limit for complete reversals of stress.

Due to the effects of surface roughness, size, nature of loading, etc., the endurance limit of the prototype may differ from that measured in the laboratory using standard specimens of the material subjected, for example, to rotating bending. Consideration of these effects which have been the subject of numerous investigations and may be taken into account by introducing additional factors is outside the scope of this book.

7

PHOTOELASTIC MODEL MATERIALS

7.1 Mechanical behaviour of plastics

In common with many structural materials, plastics such as the epoxy and polyester resins which are used as model materials show a progressive increase in strain with time when subjected to a constant load for a prolonged period. This phenomenon is known as creep. Unlike the hard metals, however, which, with a few exceptions display measurable creep at elevated temperatures only, all plastic model materials are to some extent susceptible to creep at room temperature. The nature of the creep is also different in the two cases; in metals it involves plastic deformation while in plastics it is a damped elastic (viscoelastic) deformation. Thus, plastics exhibit recovery when the load is removed. In addition, the mechanical creep in photoelastic models is accompanied by a variation with time of the optical effects, known as optical creep.

The rate of strain creep in plastics depends on the temperature and normally increases as the temperature is raised. The strain does not increase indefinitely, however, but tends asymptotically towards a certain limiting or final value. This limiting strain is practically independent of the temperature.

Although most model plastics are hard at normal temperatures, they behave like rubber at high temperatures, becoming soft but remaining purely elastic within the elastic limit, which in most cases is practically identical with the ultimate stress. In this condition the limiting deformation is reached in a few seconds. The lower limit of temperature at which this occurs is known as the softening point or critical temperature. For rubber or soft plastics, room temperature corresponds to the 'high' temperature and these materials become hard only when the temperature is greatly reduced. Apart from this, there is no essential difference in the behaviour of soft and hard plastics.

The rate of recovery after unloading of model materials which are susceptible to creep is also dependent on the temperature. At room temperatures the recovery is very slow but at temperatures above the softening point it is practically instantaneous.

If a model is subjected to a load at a temperature above the softening point and this load is maintained while the model is subsequently cooled down to room temperature, the strain becomes 'frozen in'. When the load is removed, there is a comparatively small instantaneous recovery corresponding to the instantaneous strain produced by the same load when applied at room temperature.

While some of the properties of model materials described above are undesirable in many applications, they form the basis of certain others which will be described later.

7.2 Multiphase theory

The above phenomena, and some others to be observed with these materials can be explained by the multiphase (or diphase) theory. In this theory, the model material is assumed to be composed of an elastic phase and a plastic or viscous one—like a sponge impregnated with grease or wax.

The elastic phase is governed by Hooke's law and its modulus of elasticity is practically independent of the temperature. The plastic phase is assumed to consist of a multitude of components of different viscosity and temperature dependence of viscosity. For this reason, the designation 'multiphase' is considered to be more appropriate than 'diphase'.

While it is known from chemistry that the special properties of plastics are due to certain molecular linkages, van der Waals forces, etc., their mechanical and optical behaviour can be explained more easily by means of the analogy of the wax-impregnated sponge. This can be represented schematically by a series of springs and parallel dashpots (Kelvin or Voigt elements) as shown in Figure 7.1. The deformations of the springs $e_1 \ldots e_i$ (which may

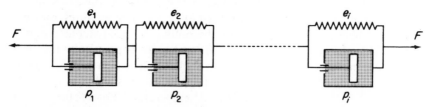

Figure 7.1.   Model representing the mechanical behaviour of a plastic according to the multiphase theory

or may not be identical) are proportional to the forces acting on them. The damping produced by the dashpot p_1 is assumed to be very low at room temperature while the effects of the others increase with increasing index. It is further assumed that the viscosities of the liquids in all the dashpots decrease with increasing temperature and that, at the softening point, even the viscosity of the liquid in p_i is fairly low.

By choosing appropriate viscous properties of the fluids in the dashpots, the above representation can be used to simulate the behaviour of any of the usual plastic model materials and is the most general possible of its kind. The differences as to viscosity with some materials may be greater or less than with others, not only with respect to the softening temperature but also to the temperature dependence of the viscosity. With some materials, the viscosity of some or all of the plastic components decreases gradually with increasing temperature while with others it changes suddenly at a certain temperature in a manner resembling melting. These differences are only quantitative, however, and the principles governing the behaviour of such materials are unchanged.

Since the viscosity of the liquid in the dashpot p_1 is negligible, spring e_1 carries the full load and reaches its full deformation immediately after loading. This is followed by yielding of the other dashpots, each at a different rate. As yielding proceeds, the force acting on any individual dashpot diminishes while that on the corresponding parallel spring increases by an equal amount since their sum is constant and equal to the applied load, Figure 7.2.

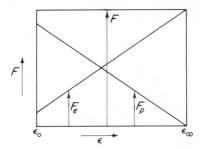

Figure 7.2. Variation of load distribution with strain in a spring/dashpot element

Ultimately, the load is transferred from all the dashpots to the springs and no further deformation will occur. The total deformation of the system then corresponds to the limiting deformation of the plastic. The total deformation at any instant can be regarded as being composed of the instantaneous deformation of e_1 and that resulting from creep up to that instant.

After unloading, there is an instantaneous recovery due to e_1, while the recoveries of the springs $e_2 \ldots e_i$ are damped by the parallel dashpots. The curves for recovery are therefore similar to those for the deformation, but inverted as shown in Figure 7.3.

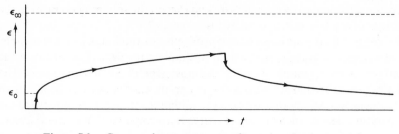

Figure 7.3. Creep and recovery curves for a viscoelastic material

It can easily be shown that the rate of deformation of a single spring-dashpot element at any instant is proportional to the difference between the limiting deformation and that at the instant considered, i.e. to $(\epsilon_\infty - \epsilon)$. This behaviour, if duplicated in plastics, would lead to the assumption of a homogeneous plastic phase. Experiments show, however, that with these materials the rate of deformation decreases faster than $(\epsilon_\infty - \epsilon)$, which means that some of the plastic components deform more rapidly than others.

It follows from the above considerations that the stress-strain diagrams will be straight lines up to fracture only when measurements are made at time t_0 immediately after loading, i.e. before the commencement of creep and after a time t_∞ when the limiting deformation has been reached, Figure 7.4.

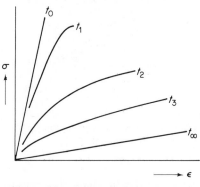

Figure 7.4. Stress–strain curves for a plastic relating to different instants after loading

From a practical viewpoint, the limiting deformation will not be realized at room temperature, but at high temperatures will be reached after a few seconds. For any other interval of time after loading, the stress-strain diagram will be curved as shown.

The model system of Figure 7.1 can also be used to explain the so-called 'memory' of plastics. When a model has been loaded for some time, part of the strain will vanish immediately the load is removed while that part which was caused by creep will persist for a period, the duration of which increases as the loading time, the stress and the temperature during loading are increased and as the temperature after unloading is decreased. If a model is loaded, then unloaded and subsequently loaded for a second time in the same or any other way, the deformation caused by the second load is superimposed on the residual strain resulting from the first. When the second load is removed, two different residual strains are superimposed.

This effect can be observed in practice. For instance, if a beam is bent in one direction, unloaded and then subjected to an equal bending moment in the opposite direction it is found that the strain and fringe orders at a given time after the second loading are less than those at the corresponding time after the first. Further, the residual values of strain and birefringence after the second unloading will diminish to zero after some time and then increase with opposite sign.

The phenomena of birefringence and the changes of phase differences which occur during creep, etc., are explained in a simple way by means of the multiphase analogy. The birefringence produced by a stressed model is partly due to the elastic phase and partly to the plastic phase. The birefringence produced by each phase is proportional to the stress optical coefficient for the phase and to the stress carried by the phase but not to the strain. The birefringence at any instant after loading is given by the sum of the partial birefringences:

$$(n_1 - n_2) = (n_1 - n_2)_e + (n_1 - n_2)_p$$
$$= C_e(\sigma_1 - \sigma_2)_e + C_p(\sigma_1 - \sigma_2)_p, \tag{7.1}$$

where C_e and C_p are the stress optical coefficients for the elastic and plastic phases respectively, $(\sigma_1 - \sigma_2)_e$ and $(\sigma_1 - \sigma_2)_p$ are the differences in principal stresses in the two phases.

The partial principal stress differences $(\sigma_1 - \sigma_2)_e$ and $(\sigma_1 - \sigma_2)_p$ can be found from the conditions that their sum is equal to the total stress difference $(\sigma - \sigma_2)$, i.e.

$$(\sigma_1 - \sigma_2)_e + (\sigma_1 - \sigma_2)_p = (\sigma_1 - \sigma_2), \tag{7.2}$$

and the stress in the elastic phase is proportional to the deformation or strain ϵ, i.e.

$$(\sigma_1 - \sigma_2)_e = \epsilon E_e, \tag{7.3}$$

where E_e is Young's modulus for the elastic phase. E_e can readily be obtained by carrying out a deformation test at high temperature.

The above relations, based on the assumptions of equal values of C_p for all components of the plastic phase and of C_e for all of the springs, appear to describe to a close approximation the behaviour of most model materials. Certain exceptions can be explained by expressing equation (7.1) in more general terms assuming different stress optical coefficients for each spring and dashpot in the analogous system. The birefringence is then given by

$$(n_1 - n_2) = C_{e1}(\sigma_1 - \sigma_2)_{e1} + C_{e2}(\sigma_1 - \sigma_2)_{e2} + \cdots + C_{ei}(\sigma_1 - \sigma_2)_{ei}$$

$$+ C_{p1}(\sigma_1 - \sigma_2)_{p1} + C_{p2}(\sigma_1 - \sigma_2)_{p2} + \cdots + C_{pi}(\sigma_1 - \sigma_2)_{pi},$$

where the subscripts are numbered to indicate the particular spring or dashpot to which these values pertain.

One of the most remarkable facts to be explained by means of the multi-phase analogy is that with some model materials the birefringence increases during creep while with others it diminishes, passes through zero and then increases with reversed sign. This can be explained with the assumption that the stress optical coefficients C_e and C_p are of the same sign with the first kind of material and of opposite sign with the second kind.

For all model materials, the absolute magnitude of C_e exceeds that of C_p. When a model is loaded at room temperature, however, the value of Young's modulus for the elastic phase is small so that initially the load is supported mainly by the plastic phase and the birefringence is mainly due to this phase. The stress, deformation and birefringence of the elastic phase then increase with time. Hence, if C_e and C_p are of the same sign, the total birefringence will increase, Figure 7.5a. If C_e and C_p are of opposite sign, however, the birefringence will diminish and reach zero when the deformation has reached the stage at which $|C_e(\sigma_1 - \sigma_2)_e| = |C_p(\sigma_1 - \sigma_2)_p|$. Thereafter, $|C_e(\sigma_1 - \sigma_2)_e|$ will exceed $|C_p(\sigma_1 - \sigma_2)_p|$ and the birefringence will increase with reversed sign until the limiting values of deformation and birefringence are reached, Figure 7.5(b); the reversed sign will be retained with frozen-in strain.

The stress optical coefficients C_e and C_p and the elastic modulus E_e for the elastic phase can be determined from simple tests. At temperatures above the softening point, the limiting ratio of stress to strain (which can be measured a few seconds after loading) is equal to E_e while the birefringence produced is due to the elastic phase only. For example, a simple tensile test carried out under these conditions will yield the values of both E_e and C_e.

At room temperature, the ratio of stress to instantaneous strain is conventionally called Young's modulus E for the material. The ratio E_e/E is equal to the fraction of the total stress carried by the elastic phase immediately after loading:

$$E_e/E = (\sigma_1 - \sigma_2)_e/(\sigma_1 - \sigma_2) = 1 - (\sigma_1 - \sigma_2)_p/(\sigma_1 - \sigma_2).$$

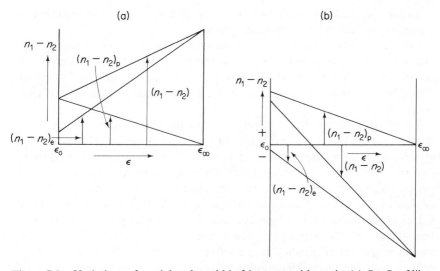

Figure 7.5. Variations of partial and total birefringences with strain, (a) C_e, C_p of like sign. (b) C_e, C_p of unlike sign

This equation, together with equation (7.2), allows the values of $(\sigma_1 - \sigma_2)_e$ and $(\sigma_1 - \sigma_2)_p$ to be determined.

Equation (7.1) can be rearranged in the form

$$C_p = [(n_1 - n_2) - C_e(\sigma_1 - \sigma_2)_e]/(\sigma_1 - \sigma_2)_p,$$

from which C_p can be calculated after inserting the values of the birefringence $(n_1 - n_2)$ observed immediately after loading and the partial principal stress differences.

Using equations (7.1) to (7.3), the birefringence observed at any time after loading can be expressed in terms of the total stress difference and the instantaneous deformation ϵ:

$$(n_1 - n_2) = C_e \epsilon E_e + C_p[(\sigma_1 - \sigma_2) - \epsilon E_e] = C_p(\sigma_1 - \sigma_2) + (C_e - C_p)\epsilon E_e.$$

Immediately after loading, ϵ is equal to $(\sigma_1 - \sigma_2)/E$. The birefringence resulting from the instantaneous deformation is therefore

$$(n_1 - n_2) = C_p(\sigma_1 - \sigma_2) + (C_e - C_p)(\sigma_1 - \sigma_2)E_e/E$$

$$= (\sigma_1 - \sigma_2)[C_p + (C_e - C_p)E_e/E],$$

from which it is seen that the conventional photoelastic constant C is composed of C_e and C_p in the following way:

$$C = C_p + (C_e - C_p)E_e/E.$$

The above analysis becomes more general if σ_1 and σ_2 are regarded as secondary instead of true principal stresses.

For the usual model materials, Poisson's ratio has values between 0·35 and 0·4 at room temperature. At the softening point it is approximately 0·5, which means that the material undergoes constant volume deformation. This effect is obviously due to the plastic phase which is liquid at this temperature.

Experiments show that the value of Poisson's ratio gradually increases during creep. This effect and some others indicate the possibility that the directions of the principal stresses in the elastic and plastic phases may differ during creep. With the additional assumption that a beam of light in a model passes alternately through elastic and plastic particles, this provides an explanation of the dispersion of birefringence which is observed with large values of creep strain and which can be utilized in the photoelastic investigation of elastoplastic stresses and strains (Section 20.4.2). In Section 13.8 it is shown by the *j*-circle method that a series of birefringent layers in which the principal stresses have different directions in general cause an effect identical with dispersion.

7.3   Choice of model material

The properties desirable in a photoelastic material differ according to the nature of the data required and the procedure to be employed. For the production of an isochromatic pattern, which is the most common photoelastic operation, the optical sensitivity (i.e. the stress optical coefficient) should obviously be as high as possible. Further, in order that the shape of the model when deformed under load shall not differ appreciably from that of the prototype, the elastic modulus should be as high as possible; this is particularly important with the frozen stress method when the residual deformations are generally much greater than those of the prototype.

The relative merits of photoelastic materials are sometimes compared using a quantity known as the figure of merit, which is the ratio of the elastic modulus to the material fringe value, and is usually denoted by Q. Thus,

$$Q = E/f.$$

The material fringe value f being inversely proportional to the stress optic coefficient, it follows from the above that the figure of merit should be as high as possible. However, this neglects certain other factors which influence the choice of material. For example, for the observation of isoclinics it is an advantage if the stress optic constant is small; for the insensitive material used with the sandwich technique it should if possible be zero while for one of the plates used with the two-sheet method it should be negative.

With respect to the mechanical properties, high strength, linear proportionality between stress and strain, and freedom from creep are the usual requirements. For photoplastic investigations, however, the material should have creep properties which simulate the plastic behaviour of the material of the prototype.

For all procedures, freedom from initial (or residual) stress and time-edge effect is desired.

7.4 Initial stresses. Annealing

Certain materials exhibit initial stresses resulting from the manufacturing process. These can be eliminated in some cases by annealing but not in others. Sheets may be annealed by placing on a flat surface and heating to a temperature above the softening point; this temperature is maintained for several hours and is then reduced to room temperature at a low rate to avoid freezing of thermal stresses. Models of other shapes may be supported on a bed of sand during annealing. Some authors prefer to use an oil bath as this allows greater uniformity of the rates of heating and cooling over the entire model and may also retard the formation of time-edge stresses.

For the production of castings in the laboratory, the materials and casting procedure should be chosen to avoid as far as possible the development of initial stresses during gelation and cure (see Section 7.6.1).

7.5 Time-edge effect

Most plastics suffer to a greater or less extent from the so-called time-edge effect. This is most commonly due to the evaporation or absorption of water which causes changes of volume and results in tensile or compressive stresses in a narrow zone bordering the edges of the model together with small equilibrating stresses distributed through the interior. The optical effects produced are superimposed on these due to the applied loading and distort the true fringe pattern. In severe cases, isochromatics of fairly high order may be produced near the edges which can completely obscure the behaviour of the true isochromatics in this important region.

The water content of the material is regulated by the atmospheric humidity. With a change of humidity, water can readily evaporate or be absorbed at the surfaces but diffuses very slowly through the interior so that stationary conditions are reached only after a considerable period of time. The rate of diffusion increases appreciably with temperature while the water solubility diminishes. The time required for stationary conditions to be achieved with consequent elimination of time-edge effect can therefore be reduced by moderate heating in an atmosphere of constant humidity. After removal from the oven, however, edge effect will reappear with any change of humidity.

Likewise, on account of the reduction of water content at the higher temperatures, models removed from the oven at the end of a stress freezing cycle will be susceptible to edge effect through absorption.

Some success in retarding the formation of edge effect has been achieved by other methods, e.g. smearing the surfaces with oil or silicone grease and covering with thin metal foil, and storing in a desiccator containing a suitable drying agent such as calcium chloride.

With certain plastics, edge effect may be due to the evaporation of solvents other than, or in addition to, water and the problem of its elimination is more complicated. When using a material susceptible to edge effect, it is obviously desirable that the model or frozen stress slices should be examined as soon as possible after manufacture.

7.6 Properties of model materials

Some photoelastic model materials are discussed individually in the following sections. Typical values of their properties at room temperature and at their critical temperatures are given in tables 7.1 and 7.2 respectively. Since considerable variation in these values is possible between different samples of the same material, the values to be applied in an investigation should be determined from calibration tests of the material actually used.

7.6.1 *Epoxy resins*

Epoxy resins are today the most widely used photoelastic materials, being suitable for the manufacture of models for both two- and three-dimensional investigations and for use as birefringent coatings. They possess good optical and mechanical properties, are only slightly susceptible to edge effect and are available at comparatively low cost. At room temperature the material has a slight but unimportant tendency to creep. Epoxies have good casting properties and parts may readily be cemented together at room temperature, allowing the fabrication of complicated models.

Although available as a fully polymerized material, usually in the form of flat sheets suitable for two-dimensional photoelastic work, epoxies are more commonly supplied as a basic resin and hardener or curing agent. The finished resin is formed by the chemical reaction which takes place between the resin and hardener when the two are mixed together at a suitable temperature. The basic resin is supplied in both solid and liquid forms but the type used has little effect on the properties of the finished product. The properties depend greatly, however, on the proportion and type of hardener used. For plates and other thin-walled castings, optimum properties are obtained using phthalic anhydride as hardener.

In the United Kingdom, the most commonly used epoxy for photoelastic work is Araldite CT200 manufactured by Ciba (A.R.L.) Limited. This is a

•

Table 7.1 Properties of photoelastic materials at room temperature

Type of Material	Trade Name	$C \times 10^7$ cm²/kp (m²/N × 10⁵)	in²/lbf	f^* kp/cm fr (N/m fr × 10⁻³)	lbf/in fr	E kp/mm² (N/m² × 10⁻⁷)	lbf/in²	Q^* fr/cm (fr/m × 10⁻²)	fr/in
Epoxy resin	Araldite CT200 30pph phthalic	56·1	3·94	10·5	58·8	320	455,000	3050	7740
	Araldite 6020† 50pph phthalic	56·6	3·98	10·4	58·2	317	450,000	3050	7730
	Bakelite ERL 2774† 50pph phthalic	57·2	4·0	10·3	57·8	335	475,000	3250	8200
	Standard HEX-phthalic†	52·1	3·66	11·3	63·0	335	475,000	2960	7540
Allyl diglycol	C.R.-39	34·2	2·4	17·2	96·0	200	284,000	1160	2960
Polymethacrylate	Perspex Plexiglas	−4·7– −4·5	−0·33– −0·31	−125– −130	−700– −730	280– 300	398,000– 427,000	220– 230	570– 580
Polycarbonate	Makrolon	78	5·5	7·5	42	260	370,000	3500	8800
Cellulose nitrate	Celluloid	2–20	0·14–1·4	30– 300	170– 1700	200– 250	284,000– 356,000	670– 80	1700– 200
Glass		−2– +1·5	−0·14– +0·10	−300– +400	−1700– +2200	7000	10,000,000	2300– 1700	6000– 4500
Polyurethane rubber	Photoflex	3500	240	0·17	0·97	0·07– 0·35	100– 500	40– 200	100– 500
	Hysol 4485‡	3500	240	0·17	0·97	0·30– 0·44	430– 620	180– 260	440– 640
Gelatine		20,000– 6,000	1400– 420	0·03– 0·09	0·17– 0·50	0·003– 0·03	4– 40	10– 33	24– 80

* Values for mercury green light (5461A).
† After Leven.
‡ After Durelli and Riley.

Table 7.2 Properties of epoxy resins at their critical temperatures

Material	$C \times 10^7$		f		E		Q		T
	cm²/kp (m²/N × 10⁵)	in²/lbf	kp/cm fr (N/m fr × 10⁻³)	lbf/in fr	kp/mm² (N/m² × 10⁻⁷)	lbf/in²	fr/cm (fr/m × 10⁻²)	fr/in	°C
Araldite CT200 30pph. phthalic	2560	180	0·23	1·28	1·3	1850	566	1400	140
Araldite 6020* 50pph. phthalic	1420	100	0·415	2·32	3·59	5100	866	2200	162
Bakelite ERL 2774* 50pph. phthalic	1330	93	0·444	2·48	3·67	5210	828	2100	160
Standard HEX-phthalic* 42pph. phthalic, 20pph. HEX.	1230	86	0·479	2·68	4·54	6450	945	2400	170

* After Leven.

hot-setting resin, the recommended hardener being phthalic anhydride HT901. The resin is supplied in solid lumps and the hardener as a powder or flake. A popular formulation is 30 parts by weight of HT901 to 100 parts of CT200. While casting procedures vary considerably between different laboratories, the following has been found to produce satisfactory results. The resin is first melted at a temperature of 120 to 130 °C. The hardener is then added slowly and the mixture is stirred thoroughly until all of the hardener has dissolved. Alternatively, the resin and hardener may be melted separately before mixing. Thorough stirring is essential in order to obtain a homogeneous product. It is also necessary to ensure that all air bubbles introduced during melting and stirring are allowed to escape before gelling begins. The mixture is poured into the mould, preheated to the same temperature. The duration of the curing cycle depends on the temperature, increasing as the temperature is reduced. For minimum shrinkage and residual stresses, however, it is desirable that the temperature should be fairly low, e.g. between 100 and 110 °C. The resin is allowed to gel for about 16 hr at this temperature. The material is then removed from the mould and heated slowly to 130 to 140 °C at which temperature it is held for a further 16 hr to complete the curing. The casting is finally cooled slowly to room temperature at a rate varying between $\frac{1}{2}$ and 3 °C/hr according to the thickness to avoid the freezing of thermal stresses. Some investigators prefer to carry out the entire curing cycle at the lower temperature but a considerably longer period of time must be allowed to ensure that curing is complete.

Metal moulds are generally to be preferred but other materials such as silicone rubber and plaster of Paris have been used successfully. With porous materials, however, it is necessary that the surfaces be sealed, for example by means of a cold curing epoxy resin. In all cases a release agent such as Releasil 7 or Dow Corning 7 must be applied to the surfaces to allow easy separation.

The relative merits of various epoxies and hardeners available in the United States have been studied in detail by Leven,[12] whose investigations indicate that optimum properties for frozen stress analyses are obtained with the following compositions:

(i)   100 parts by weight ERL2774 (supplied by the Bakelite Company) to 55 parts phthalic anhydride.

(ii)   100 parts by weight Araldite 6020 (supplied by Ciba Company, Inc.) to 50 parts phthalic anhydride.

In the production of thick-walled castings, considerable care is required to avoid residual stresses. These stresses arise from the exothermic nature of the reaction during gelation and cure. The rate of curing and consequently the rate of heat generation increase with the temperature; differential rates resulting from temperature gradients in the material introduce stresses which cannot be removed by later heat treatment. It is therefore desirable that the materials used and the casting procedure adopted shall involve slow gelation and cure. It has been found that the exothermic reaction can be reduced by adding hexahydrophthalic (HEX) anhydride to the phthalic anhydride. This reduces the susceptibility of the casting to the development of residual stresses but at the expense of some reduction in the figure of merit. Leven recommends the following standard composition:

> 100 parts by weight—ERL2774
> 42 parts by weight—phthalic anhydride
> 20 parts by weight—HEX anhydride.

The resin is heated to about 120 °C while the HEX anhydride is heated separately above its melting point to about 100 °C. The HEX is added to the resin after filtering. The phthalic anhydride is then added and the mixture is heated and stirred until the phthalic has dissolved. The mixture is poured into the mould preheated to 92 °C at which temperature gelation occurs in one to three days. The temperature is then gradually increased to 150 °C and maintained at this level for 8 to 10 days before finally cooling slowly to room temperature. A variation of this procedure recommended by Leven is to heat slowly to 100 °C after initial gelation, allow to cool to room temperature and then remove the casting from the mould. The casting is then replaced in the oven and heated fairly rapidly to 100 °C after which the curing cycle is completed as before.

For the production of large castings, McConnel[13] has found it advantageous to replace the phthalic anhydride entirely by HEX anhydride, and recommends the following formulation:

> 100 parts by weight—Araldite CT200
> 53 parts by weight—HEX anhydride (Shell HPA or Ciba HT907)

The resin and hardener, heated separately, are mixed at 130 °C. Gelation and cure require about 12 days at 90 °C with very large castings but the temperature may be increased up to 120 °C with a corresponding reduction in curing time with smaller castings.

7.6.2   *Columbia resin C.R.39*

This allyl type resin is a highly transparent, colourless material having good optical sensitivity. It is supplied by The Homalite Corporation and other firms in the form of flat plates of uniform thickness with highly polished surfaces and consequently is very popular for general two-dimensional work. C.R.39 is only slightly susceptible to time–edge effect but suffers to some extent from creep. Experiments by Durelli[14] and others have indicated, however, that over a considerable range the isochromatic order measured at a given time after loading is linearly related to the stress.

Although C.R.39 is virtually free from residual stresses in the plane of the plate, considerable residual stresses exist in the normal direction. These stresses influence the optical effect observed under oblique incidence so that the material is unsuitable for this application. C.R.39 is also unsuitable for the frozen stress method. '

The material is rather brittle and special care is required when shaping the model to avoid chipping at the edges.

7.6.3 *Polymethacrylate*

This material is marketed under the trade names of Plexiglas in Germany and the United States, and Perspex in the United Kingdom. It can be purchased at moderate cost in the form of flat sheet, round bar, etc. with highly polished surfaces and in a wide variety of sizes. The material is highly transparent, colourless and free from initial stress. It exhibits practically no creep or edge effect.

The optical sensitivity of Plexiglas is very low, being of the same order as that of glass. This precludes its use for the determination of isochromatics. On the other hand, the virtual absence of isochromatics makes the material particularly suitable for the determination of isoclinics. Its low optical sensitivity can also be used to advantage in investigations of dynamic stresses by the photometric method. At room temperature the stress–optic coefficient of Plexiglas is negative; when used in conjunction with another material having a positive coefficient for investigations by the two-sheet method, an increased optical effect is therefore produced.

With the frozen stress method, Plexiglas is used only for special investigations, e.g. in the determination of stress due to the self-weight of structures. Here again the double refraction is so small that no isochromatic pattern can be observed and compensation is necessary.

7.6.4 *Polycarbonate*

This plastic is manufactured in the form of flat sheet by Bayer Chemicals Ltd (U.K. distributors M. & B. Plastics Ltd) and marketed under the trade name Makrolon. It has a very high optical sensitivity combined with a reasonably high modulus of elasticity which gives it an exceptionally high figure of merit. Polycarbonate is practically free from time–edge effect and shows little creep at room temperature. Although some initial stresses resulting from the manufacturing process are usually present, these can be eliminated by careful heat treatment.

7.6.5 *Celluloid* (*cellulose nitrate*)

Celluloid was one of the earliest photoelastic materials but has been superseded for conventional work by modern materials having greater uniformity and higher optical sensitivity. Celluloid is susceptible to creep and is remarkable for the fact that its stress–strain diagram resembles that of different metals such as aluminium, mild steel and carbon steels according to its grade, the time of loading and, to some extent, the temperature and humidity of the environment. This property has led to a revival of interest in celluloid as a photoelastic material for the study of elastoplastic states of stress and strain (see Section 20.4.2).

7.6.6   *Polystyrene*

Polystyrene has a comparatively low optical sensitivity and is highly susceptible to creep. It is of interest mainly on account of its unusual optical behaviour in compression, resulting from the fact that the stress optical coefficients of the elastic and plastic phases are of opposite sign. The stress–strain and fringe–strain diagrams are linear up to a certain stress, the magnitude of which depends on the time which has elapsed since the load was applied. Beyond this stress, there is a sudden increase of strain while the fringe order diminishes, passes through zero and then increases with opposite sign. These properties make polystyrene suitable for certain photoplastic investigations (see Section 20.4.2) in which compressive stresses predominate.

7.6.7   *Glass*

Although glass was the first material to be used for photoelastic investigations, it is used nowadays only for special applications for which its properties render it particularly suitable. Its general use as a photoelastic material is precluded by its low optical sensitivity and the difficulty and cost of producing glass models.

Glass has several advantages as a photoelastic material. It has a high elastic modulus and figure of merit, linear stress-strain relation over a wide range, and is free from defects of modern plastics such as creep and time–edge effect. Residual stresses can be eliminated by careful annealing. It is chemically stable, unaffected by changes of humidity and insensitive to normal variations of atmospheric temperature. Glass can therefore be used to advantage for the measurement of suitable stress levels over a long period of time.

The stress-optic coefficient of glass depends greatly on its composition and may be positive or negative, or exceptionally zero.

Special applications of glass to photoelastic investigations are described in Sections 20.2 and 20.3.

7.6.8   *Polyurethane rubber*

Polyurethane (or urethane) rubber is a rubber-like plastic of very low elastic modulus and very high optical sensitivity. It is a highly transparent, amber coloured material and is practically free from time–edge effect. Optical and mechanical creep are negligible at room temperature after the load has been applied for a few seconds. The relation between stress and strain is linear over a wide range.

Urethane rubber can be purchased in the United Kingdom in the form of cured sheet under the trade name of Photoflex. Different grades having elastic moduli in the range 100 to 500 lbf/in^2 are available but the standard sheets have values within the range 400 to 500 lbf/in^2.

In the United States, urethane rubber is available in sheet form under the designation Hysol 4485 or it may be cast using 100 parts by weight of Hysol 2085 resin to 24 parts of Hysol 3562 hardener. Curing is carried out for 2 hr at 280 °C followed by 4 hr at 210 °C. A detailed investigation of the static and dynamic properties of this material has been reported by Durelli and Riley in their book 'Introduction to Photomechanics'.[15] In this investigation, the elastic modulus was found to vary between 430 and 620 lbf/in^2 at room temperature depending on the batch and the casting procedure. The stress–strain and stress-fringe relations were found to be linear up to a stress of 20 lbf/in^2, beyond which both the elastic modulus and Poisson's ratio appeared to increase slightly. Under dynamic loading, the material fringe stress coefficient was found to be practically independent of the strain rate within the test range 8 to 65 in/in sec, while the fringe–strain coefficient diminished and the initial elastic modulus increased with increasing strain rate.

Urethane rubber is a popular material for the investigation of problems of dynamic stresses. The velocity of propagation of stress waves (which is proportional to the square root of the elastic modulus) in urethane rubber is only about one-thirtieth of that in the conventional hard plastics. The problem of recording transient fringe patterns is thus greatly simplified as has been shown by Durelli and others. Urethane rubber is also suitable for the investigation of certain problems of stresses due to self weight. Although it is not sufficiently sensitive for isochromatics to form under the action of direct stresses due to self weight only, satisfactory patterns are produced in cases where the self weight results in high bending stresses. Since the material is available in grades covering a considerable range of elastic moduli and over which Poisson's ratio is approximately constant it may be used for the manufacture of composite models free from the complicating effects of differences of Poisson's ratio.

7.6.9 *Gelatine*

The optical sensitivity of gelatine greatly exceeds that of any other photo-elastic material and is so high, in fact, that isochromatics are produced by the stresses arising from the weight of the substance itself. For this reason, gelatine is suitable for the investigation of stress problems in which the influence of self weight is important, e.g. in determining the stress distribution in dams and around tunnels.

The material is obtained in the form in which it is used by dissolving dry gelatine in water at a temperature of 50 to 60 °C, with or without the addition of glycerine. The mechanical and optical properties depend greatly on the composition, the elastic modulus increasing and the stress-optic constant diminishing as the gelatine content is increased. In order that the photoelastic effect produced shall be as great as possible, the elastic modulus should be as low as possible; on the other hand, it must not be reduced to such an extent that the geometrical shape of the model alters appreciably under load.

Due to rapid evaporation of water from the surfaces, gelatine models are extremely susceptible to time–edge effect. This can be reduced by adding glycerine—a non evaporating solvent—to the mixture. A typical composition is 15 % gelatine, 25 % glycerine and 60 % water, which gives an elastic modulus of about 10^5 N/m^2 and a material fringe value of about 40 N/m^2 fringe. It is also an advantage to store and use the model in a constant cool and humid atmosphere.

Models may be produced by pouring the mixture to the required depth in a mould consisting of a frame having the contour of the model placed on a carefully levelled glass plate and allowing to set. To obtain satisfactory isochromatic patterns, model thicknesses of the order of 5 cm are usually required.

Because of its lateral instability when erected, the model requires to be supported, for example between two parallel glass plates. The model will of course then be in a state of plane strain in contrast to the plane stress condition existing in ordinary two-dimensional tests. Although the normal stresses thus introduced between the plates and the model do not influence the photoelastic effects observed under normal incidence, shear stresses between them must be eliminated as far as possible. The contacting surfaces must therefore be well lubricated, e.g. with light machine oil.

8

TWO-DIMENSIONAL PHOTOELASTICITY

8.1 Isoclinics

As shown in Section 5.4, the isoclinic lines are the loci of points at which the directions of the principal axes of stress are parallel to the axes of the crossed polaroids. If the polaroids are rotated in the crossed position, the isoclinic lines will move as parallelism is obtained in turn at other points within the field. One isoclinic must obviously pass through every point of the field when the polaroids are rotated through 90°; as will be shown in the following section, however, particular points may exist through which all or a wide range of isoclinics pass.

By combining in a single diagram isoclinic lines having parameters varying in increments of, say 5° or 10°, corresponding to successively different orientations of the polaroids, a system of isoclinics covering the entire field may be drawn. This can be accomplished by tracing the lines of different parameters from photographs or directly on tracing paper.

The isoclinic pattern is independent of the magnitude of the load employed and the fringe value of the material. Patterns which are not confused by the presence of isochromatics can therefore be obtained by applying small loads which produce only a few or no isochromatic fringes or by using model materials of low sensitivity such as Plexiglas or Perspex. If the first of these methods is employed, it is necessary to ensure that the applied loads are sufficiently great that the isoclinics will not be appreciably influenced by any initial or residual stresses present in the material. Confusion between the two sets of lines can further be avoided by using white light since the isoclinics will appear black in contrast to the isochromatics which, with the exception of the zero order fringe, will be coloured.

In regions where the directions of the principal stresses do not vary greatly from point to point, the isoclinics appear as wide diffuse bands. The true isoclinic lines are the lines of maximum darkness passing through these bands and can best be plotted with the aid of a photomultiplier.

It is frequently found that the isoclinics are vague or vanish altogether in certain regions of a plate; for example, near the edges. This may be due to variations of the state of stress through the thickness of the plate or the effects of initial stresses in regions where the applied stresses are low. Errors in plotting the isoclinics may be avoided if the following rules are observed:

1. Isoclinic lines do not intersect one another except at an isotropic point (Section 8.2).

2. An isoclinic intersects a load free boundary or one subjected to normal forces only at a point where the inclination of the tangent to the boundary is identical with the parameter of the isoclinic, the reference direction being the same for each. It follows that a straight shear-free edge is an isoclinic.

3. Isoclinics with parameters differing by 90°, 180°, etc., are identical. It follows that a rectangular shear-free corner is part of one isoclinic.

4. One isoclinic coincides with a section of symmetry. In plates containing such sections, i.e. plates which are symmetrical in form and loading, the isoclinic pattern is also symmetrical. An isoclinic of parameter φ on one side of a section of symmetry corresponds, however, with the isoclinic of parameter $90° - \varphi$ on the other side.

5. At a point on a shear free boundary where the stress parallel to the boundary has a maximum or a minimum value, the isoclinic intersects the boundary at right angles. This is proved in Section 8.4.

Some other properties of isoclinics are included in the following sections.

8.2   Isotropic and singular points

It is sometimes observed that all the isoclinics pass through a particular point. At such a point the principal stresses are equal in magnitude and sign so that $(\sigma_1 - \sigma_2) = 0$; the material is in a pure hydrostatic state of stress and behaves as if it were optically isotropic in the plane of the plate. Such points are known as isotropic points. If the stress in every direction is zero, the isotropic point is called a singular point.

An isotropic point situated at a free boundary must be a singular point; since the stress is zero in the direction normal to the boundary, it must be zero in all directions. Such points usually indicate a change of sign of the boundary stress.

Two different forms of isotropic point can be distinguished. As the crossed polaroids are rotated, the isoclinic lines appear to rotate about an isotropic point. In some cases it will be observed that the isoclinics rotate in the same sense as the polaroids while in others they rotate in the opposite sense; such points are referred to as positive and negative isotropic points respectively. The order of succession of the isoclinics about an isotropic point of each type is shown in Figure 8.1.

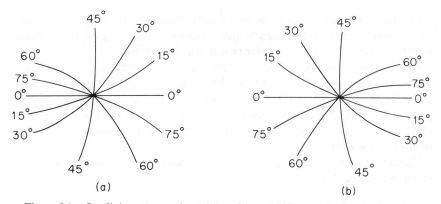

Figure 8.1.   Isoclinic patterns about (*a*) positive and (*b*) negative isotropic points

It can be shown that a singular point at a load free edge of a plate can only be of the negative type. A projecting load free corner of a plate is always a negative singular point.

When the isoclinics pass in turn through several isotropic points, these must be alternately positive and negative since the order of succession of the isoclinics at one point will be reversed at the next, Figure 8.2. If an

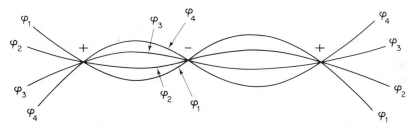

Figure 8.2.   Illustrating alternation of signs of successive isotropic points

isoclinic passing through an isotropic point on a free boundary also passes through another, the latter must be of the positive type. Further, if an isoclinic passes through two singular points on the free edges of a plate, it must also pass through a positive isotropic point in the interior.

The order of an isotropic point is defined by the number of isoclinic lines passing through it with any given orientation of the polaroids; one isoclinic of every parameter between 0° and 90°, passes through an internal isotropic point of the first order. In the case of a singular point occurring at the tip of an acute angled projection, some of the isoclinics between 0° and 90° will be missing.

Occasionally, two or more isoclinics of the same parameter pass through an isotropic point. Such higher order points can be regarded as being formed by two or more first order points which coincide.

Occasionally, isotropic lines or zones can be observed. For example, the neutral axis of a beam under pure bending is an isotropic line.

Isoclinics of all parameters radiate from a point on the boundary of a plate at which a concentrated load acts. Such a point is not isotropic in the sense of our previous definition, however, since the difference of principal stresses is not zero.

Further special rules which apply in the region of an isotropic point are given in Sections 8.3 and 8.6.

8.3 Stress trajectories

A stress trajectory, line of principal stress, or isostatic, is a line such that its direction at any point coincides with that of one of the principal stresses at the point. Since the two principal stresses at any point of a two-dimensional stress system are mutually perpendicular, it follows that a system of stress trajectories will consist of two orthogonal families of curves. One of these families indicates the directions of the σ_1 (algebraically greater) principal stresses and the other those of the σ_2 stresses.

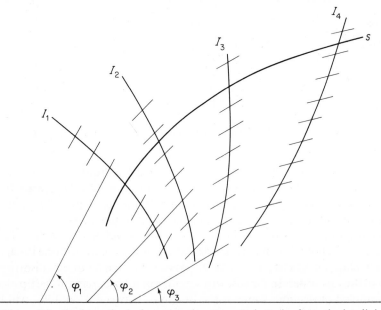

Figure 8.3. Basic method of constructing stress trajectories from the isoclinics

A system of stress trajectories covering the field of a plate may be drawn from the isoclinic pattern. The basic method of construction is indicated in Figure 8.3. At points on the isoclinics I_1, I_2, etc., small lines are drawn in a direction inclined to the reference direction at an angle equal to the parameters φ_1, φ_2, etc., of the respective isoclinics. The stress trajectories of one family are obtained by drawing smooth curves tangential to these lines. By drawing small lines in the perpendicular direction, the orthogonal family of stress trajectories can be obtained.

The stress trajectories obtained by the above method are sufficiently accurate in most cases. When the isoclinics are more widely spaced, the construction indicated in Figure 8.4 will produce more accurate results. From

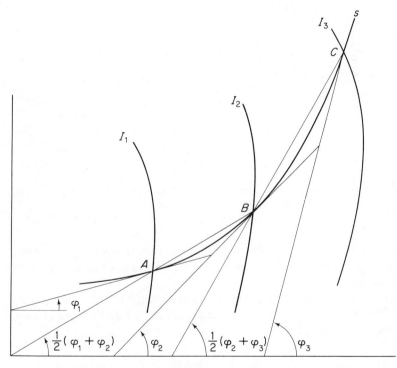

Figure 8.4. Method of constructing stress trajectories from widely spaced isoclinics

any point A on the isoclinic I_1, a straight line is drawn inclined at an angle $\frac{1}{2}(\varphi_1 + \varphi_2)$ to the reference direction to cut the isoclinic I_2 in the point B. From B, a line is drawn at an inclination of $\frac{1}{2}(\varphi_2 + \varphi_3)$ to cut I_3 in C and so on. Lines are then drawn through the points A, B, C, etc., at inclinations

$\varphi_1, \varphi_2, \varphi_3$, respectively. The curve passing through the points A, B, C tangential to the polygon formed by the second set of lines is a close approximation to the line of principal stress.

The accuracy of construction of the stress trajectories can frequently be improved by determining their radii of curvature at different points. The radius of curvature of a stress trajectory denoted by s_1, Figure 8.5, is

$$\rho_1 = \frac{ds_1}{d\varphi} \simeq \frac{\Delta s_1}{\Delta\varphi}, \tag{8.1}$$

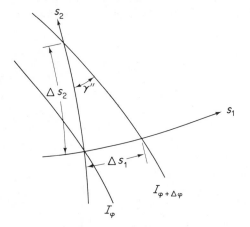

Figure 8.5. Determination of the radius of curvature of a stress trajectory

where Δs_1 represents the distance between two neighbouring isoclinics measured in the direction of s_1 and $\Delta\varphi$ is the difference of parameters of the two isoclinics.

If the angle between the isoclinic and the direction of the principal stress is very small, it is difficult to determine Δs_1 accurately. In this case, equation (8.1) may be replaced by

$$\rho_1 = \frac{\Delta s_2}{\Delta\varphi} \tan \gamma'',$$

where $\gamma'' = \tan^{-1}(\Delta s_1/\Delta s_2)$ is the angle between the isoclinic and the stress trajectory s_2.

In the neighbourhood of an isotropic point, the stress trajectories form a characteristic pattern depending on the sign of the isotropic point. Near a positive isotropic point, the system consists of two families of curves of parabolic appearance which partly enclose the isotropic point, Figure 8.6*a*,

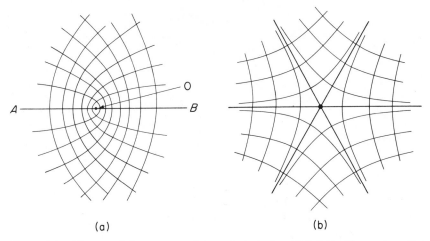

(a)   (b)

Figure 8.6.   Characteristic patterns of stress trajectories. (*a*) Interlocking system at a
positive isotropic point. (*b*) Non-interlocking system at a negative isotropic point

forming what is known as an interlocking system. In the vicinity of a negative
isotropic point, the system consists of families of hyperbolic type curves
separated by asymptotes, Figure 8.6*b*, and is known as a non-interlocking
system. Each asymptote separates two families of the same class.

While such systems of stress trajectories may vary in detail according to the
state of stress near the isotropic point, their general pattern of behaviour
always corresponds with one or the other of the above two types.

Several other important properties of stress trajectories are given in the
following section.

8.4   Properties of stress trajectories

1. Stress trajectories of one family never intersect each other or merge with
those of the other family.

2. At points on a load free boundary or one subjected to normal forces only,
the directions of the principal stresses are normal and tangential to the boundary. A stress trajectory of one family will therefore coincide with such a
boundary while those of the other family will intersect it orthogonally.

3. It can be shown that the distance between a load-free boundary and
a neighbouring stress trajectory of the same family varies inversely as the
tangential stress at the boundary, i.e. $\sigma_1 \propto 1/\Delta s_2$, Figure 8.7.

4. Since a section of symmetry is shear-free, it coincides with a stress trajectory. As stated previously, such a section also coincides with an isoclinic.

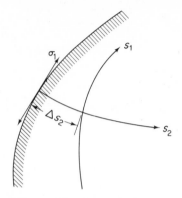

Figure 8.7. Stress trajectory
adjacent to a load-free boundary

5. In doubly or multiply connected plates such as rings, or in plates where loads or thermal stresses act within the field of the plate as well as at the edges, the stress trajectories may form closed loops but not spirals.

6. The sign of $(\sigma_1 - \sigma_2)$ is constant along each stress trajectory. Since it is possible to move from any one point of a plate to any other by following one or more stress trajectories which do not pass through an isotropic point, the sign of $(\sigma_1 - \sigma_2)$ is therefore constant throughout the plate. An apparent or pseudo change in the sign of $(\sigma_1 - \sigma_2)$ occurs when a stress trajectory passes through an isotropic point; the two parts, however, of such stress trajectories lying on opposite sides of an isotropic point, e.g. *AO* and *BO*, Figure 8.6*a*, are in fact parts of two different trajectories belonging to opposite families.

7. The principal stress pertaining to one stress trajectory has a maximum or a minimum value at a point where the orthogonal stress trajectory is straight or has a point of inflexion. This can be shown from the Lamé–Maxwell equations of equilibrium. Considering, for example, the first of equations (1.16), the necessary condition for a maximum or a minimum value of σ_1, i.e. $\partial\sigma_1/\partial s_1 = 0$, is obviously satisfied when $\rho_2 = 0$, i.e. when s_2 is straight or has a point of inflexion. Further, we then have

$$\frac{1}{\rho_2} = \frac{d\varphi}{ds_2} = 0,$$

which means that the direction of s_2 coincides with that of the isoclinic passing through the point. The isoclinic is therefore perpendicular to the line s_1.

Since a shear free edge is a stress trajectory, it follows from the above that the stress tangential to the edge is a maximum or a minimum at a point where an isoclinic intersects the edge at right angles.

A similar conclusion can be drawn with respect to a section of symmetry. Since a section of symmetry coincides with both an isoclinic and a stress trajectory, the stress normal to the section has everywhere a maximum or a minimum value.

8.5 Lines of constant shear stress

Lines can be drawn representing the loci of points at which the shear stress component τ_{xy} in a plate is constant. The system of lines obtained depends, of course, on the direction of the x, y axes which can be chosen arbitrarily.

Lines of shear stress may be used with the shear difference method for the separation of principal stresses. Their construction is described in detail in Section 9.1.2.

8.6 Stresses at a point in a two-dimensional system

8.6.1 *A general setup*

The stresses at a point in a two-dimensional system can be investigated using Airy's stress function defined in Section 2.9 or a system of stresses corresponding to the laws expressed by the stress function. This system is defined by

$$\tau_{xy} = ay + bx + \tau_0, \tag{8.2a}$$

$$\sigma_y = -by + cx, \tag{8.2b}$$

$$\sigma_x = -ax + dy + \sigma_0, \tag{8.2c}$$

in which $a, b, c, d, \sigma_0, \tau_0$ are constants which can assume different values. This setup satisfies the equations of equilibrium and compatibility as can readily be verified by substitution. It therefore expresses a possible state of stress. Since the equations are linear, the setup can be applied only to a point and its surroundings within a distance where the influence of higher order terms on the stress distribution can be neglected. Terms of higher order must be considered only if the linear terms vanish as in the case of isotropic points of higher order which are dealt with later.

8.6.2. *Linear isotropic points*

For convenience, the origin of co-ordinates is chosen to coincide with the isotropic point. We then have $\sigma_0 = \tau_0 = 0$. The directional angle of the stress trajectories can therefore be expressed, from equation (1.3), by

$$\tan 2\varphi = \frac{2\tau_{xy}}{\sigma_x - \sigma_y} = \frac{2(ay/x + b)}{(d + b)y/x - (a + c)} \tag{8.3}$$

At the origin this gives $\tan 2\varphi = 0/0$. Equation (8.3) shows, however, that $\tan 2\varphi$ is constant along each straight line passing through the isotropic point. The isoclinics in the region of the isotropic point are therefore straight. The laws which shall now be derived from the setup of stresses defined by equations (8.2) can thus be taken to apply with sufficient accuracy within a region surrounding the isotropic point in which the isoclinics approximate to straight lines.

The isoclinics satisfy the condition $\gamma = $ constant, where $\gamma = \tan^{-1} y/x$ is the directional angle relative to the x axis.

The asymptotes, by which we mean those stress trajectories which coincide with isoclinics, are expressed by

$$\tan 2\varphi = \tan 2\gamma.$$

For these, equation (8.3) gives

$$\tan 2\varphi = \tan 2\gamma = -\frac{2(b + a \tan \gamma)}{(a + c) - (d + b)\tan \gamma} = \frac{2 \tan \gamma}{1 - \tan^2 \gamma}. \tag{8.4}$$

With the substitution $\tan \gamma = C$, this produces the third order equation

$$C^3 + C^2[(2b + d)/a] - C[(2a + c)/a] - (b/a) = 0. \tag{8.5}$$

If the x axis is chosen to coincide with one of the asymptotes, then $b = 0$ and one of the three solutions of equation (8.5) vanishes. The remaining solutions are the roots of the quadratic

$$C^2 + \frac{d}{a}C - \frac{2a + c}{a} = 0.$$

Denoting these roots by C_1, C_2 we have from elementary theory,

$$C_1 + C_2 = -\frac{d}{a}; \qquad C_1 C_2 = -\frac{2a + c}{a}.$$

The ratio between the constants can therefore be determined from the directions of the asymptotes. Thus,

$$\frac{d}{a} = -(C_1 + C_2); \qquad \frac{c}{a} = -(C_1 C_2 + 2). \tag{8.6}$$

The direction of the isopachic passing through the isotropic point can also be expressed in terms of the directions of the asymptotes. The directional angle of the isopachic is given by

$$\tan \vartheta = -\frac{\partial(\sigma_x + \sigma_y)/\partial x}{\partial(\sigma_x + \sigma_y)/\partial y}.$$

Differentiating equations (8.2b) and (8.2c) partially with respect to x and y and substituting, this becomes

$$\tan \vartheta = \frac{a - c}{d - b}.$$ (8.7)

This equation shows that the isopachic passing through the isotropic point is a straight line within the surroundings of the isotropic point as defined previously. Also, within this region, the isopachics are parallel.

If we combine equations (8.6) and (8.7) we obtain

$$\tan \vartheta = \frac{(C_1 C_2 + 3)}{(C_1 + C_2)},$$ (8.8)

which becomes equal to 0/0 when $C_1, C_2 = \pm\sqrt{3}$, i.e. when the asymptotes include angles of 60°. This means that the isotropic point coincides with a saddle point of the isopachics.

The isochromatics can be expressed by $(\sigma_1 - \sigma_2) = $ constant. From equation (1.4) we obtain

$$(\sigma_1 - \sigma_2)^2 = 4\tau_{xy}^2 + (\sigma_x - \sigma_y)^2.$$

Substituting for the stress components defined by equations (8.2), this becomes

$$(\sigma_1 - \sigma_2)^2 = y^2[4a^2 + (d + b)^2] + x^2[4b^2 + (a + c)^2]$$
$$+ xy[8ab - 2(d + b)(a + c)].$$ (8.9)

This equation is of the form $Ax^2 + 2Hxy + By^2 + C = 0$ and, since obviously $H^2 < AB$, shows that the isochromatic surrounding the isotropic point is an ellipse. Rewriting equation (8.9) in polar co-ordinates using the substitutions $x = r \cos \theta$, $y = r \sin \theta$, we obtain

$$(\sigma_1 - \sigma_2) = r\sqrt{\{[4a^2 + (d + b)^2 \sin^2 \theta + [4ab - (d + b)(a + c)]}$$
$$\times \sin 2\theta + [4b^2 + (a + c)^2] \cos^2 \theta\}, \quad (8.10)$$

where r is the distance of a point from the origin, i.e. from the isotropic point and $\theta = \tan^{-1} y/x$ is the directional angle. Equation (8.10) shows that $(\sigma_1 - \sigma_2)$ is proportional to the distance r from the isotropic point in any given direction.

For simplicity, the x axis can be chosen to coincide with one of the axes of the ellipse. This requires that the coefficient of the term in xy in equation (8.9) is zero, which produces the following relation between the constants:

$$b = \frac{d(a + c)}{(3a - c)}.$$ (8.11)

From equations (8.3) and (8.11), the angle ϕ_0 between the stress trajectory and the axis of the ellipse ($y = 0$) is

$$\tan 2\varphi_0 = -\frac{2b}{a + c} = -\frac{2d}{3a - c}. \tag{8.12}$$

The relation between the inclination γ_0 of an isoclinic to the axis of the ellipse and its parameter can be expressed in the following way : The difference between the parameter of an isoclinic and that of the isoclinic which coincides with the axis is equal to ($\varphi - \varphi_0$). We have

$$\tan 2(\varphi - \varphi_0) = \frac{\tan 2\varphi - \tan 2\varphi_0}{1 + \tan 2\varphi \tan 2\varphi_0}.$$

If we substitute in this equation for $\tan 2\varphi$ from equation (8.3) replacing y/x by $\tan \gamma_0$ and for $\tan 2\varphi_0$ from equation (8.12) we obtain after simplifying

$$\tan 2(\varphi - \varphi_0) = -\frac{2a}{(a + c)} \tan \gamma_0. \tag{8.13}$$

From equation (8.9), the ratio between the lengths of the axes of the elliptical isochromatic surrounding the isotropic point is

$$\sqrt{\left(\frac{4a^2 + (d + b)^2}{4b^2 + (a + c)^2}\right)}.$$

Substituting for b from equation (8.11), this reduces to $2a/(a + c)$ which is the same as the coefficient of $\tan \gamma$ in equation (8.13). This result establishes the geometrical relation between the isoclinics and the isochromatics in the neighbourhood of the isotropic point shown in Figure 8.8. In this diagram, a circle of radius equal to the major semi-axis of the elliptical isochromatic is described about the isotropic point. The points of intersection between the isoclinics, represented by the full radial lines, and the ellipse are projected parallel to the minor axis on to the circle. The radii drawn through the points thus obtained on the circle include with the x axis angles equal to twice the corresponding parameter differences, i.e. $2(\varphi - \varphi_0)$. If the parameters of the isoclinics differ by equal amounts such as 30° as shown in Figure 8.8, the circle is divided into equal parts by the projected points of intersection. The isoclinics and the elliptical isochromatic then form an oblique projection of a regularly spoked wheel.

The directional angle of the isopachic passing through the isotropic point can now be determined. Inserting the value of b given by equation (8.11) into equation (8.7) we obtain for this angle

$$\tan \vartheta_0 = \frac{3a - c}{2d}. \tag{8.14}$$

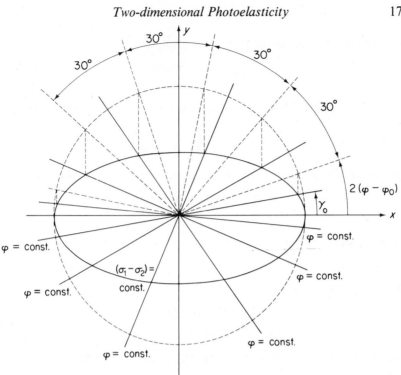

Figure 8.8.   Relation between the isoclinics and the isochromatic about an isotropic point

The right hand sides of equations (8.12) and (8.14) indicate two mutually perpendicular directions. Hence if the angle between the axis of the ellipse and one of the stress trajectories is doubled, the isopachic is normal to that direction.

8.6.3   *General points*

We now consider the state of stress at any random point and, as before, we choose the origin of co-ordinates to coincide with the point considered. In general, the stresses τ_0 and σ_0 at the origin now have finite values.

The direction of the principal stresses is given, from equations (1.3) and (8.2), by

$$\tan 2\varphi = \frac{2\tau_{xy}}{\sigma_x - \sigma_y} = \frac{2(ay + bx + \tau_0)}{(d + b)y - (a + c)x + \sigma_0}. \tag{8.15}$$

At the origin this produces for the direction of the stress trajectories:

$$\tan 2\varphi = \frac{2\tau_0}{\sigma_0}. \tag{8.16}$$

The directional angle of the isoclinic is given by

$$\tan \gamma = -\frac{\partial\varphi/\partial x}{\partial\varphi/\partial y}.$$

Applying this to equation (8.15) and substituting $x = 0$, $y = 0$ we obtain

$$\tan \gamma = \frac{\tau_0(a + c) + \sigma_0 b}{\tau_0(d + b) - \sigma_0 a}. \tag{8.17}$$

The directional angle β of the isochromatic through the point is given by

$$\tan \beta = \left[-\frac{\partial(\sigma_1 - \sigma_2)}{\partial x} \bigg/ \frac{\partial(\sigma_1 - \sigma_2)}{\partial y} \right]_0.$$

From equations (1.4) and (8.2) we obtain

$$(\sigma_1 - \sigma_2) = \sqrt{\{4(ay + bx + \tau_0)^2 + [(d + b)y - (a + c)x + \sigma_0]^2\}}.$$

Hence

$$\tan \beta = \frac{\sigma_0(a + c) - 4\tau_0 b}{\sigma_0(d + b) + 4\tau_0 a}. \tag{8.18}$$

For the directional angle α of the line $\tau_{xy} = \text{const}$, we have

$$\tan \alpha = -\frac{\partial\tau_{xy}}{\partial x} \bigg/ \frac{\partial\tau_{xy}}{\partial y},$$

where τ_{xy} is given by equation (8.2b). At the origin this produces

$$\tan \alpha = -\frac{b}{a}. \tag{8.19}$$

The directional angle ϑ of the isopachic is again given by equation (8.7). If the constants are eliminated from this equation by expressing them in terms of the angles φ, γ, β and α using equations (8.16)–(8.19) we obtain

$$\tan \vartheta = \frac{\tan^2 2\varphi \tan \gamma(\tan \beta - \tan \alpha) + \tan 2\varphi(\tan \beta - \tan \gamma) + \tan \beta(\tan \gamma - \tan \alpha)}{-\tan^2 2\varphi(\tan \beta - \tan \alpha) + \tan 2\varphi(\tan \beta - \tan \gamma) \tan \alpha - (\tan \gamma - \tan \alpha)} \tag{8.20}$$

At a saddle point of the isopachics, $\tan \vartheta = 0/0$ so that, from equation (8.7), $a = c$ and $d = b$. Substituting in equations (8.17) and (8.18) we find that

$$\tan \beta = -\frac{1}{\tan \gamma},$$

showing that the isochromatic and the isoclinic are mutually perpendicular.

If, moreover, $\tan \beta = 0$, equation (8.18) gives $2\sigma_0 a - 4\tau_0 b = 0$ or

$$\frac{2\tau_0}{\sigma_0} = \frac{a}{b}.$$

Comparing this result with equations (8.16) and (8.19) we see that the direction of the line $\tau_{xy} = $ constant is perpendicular to the direction obtained by doubling the angle between the stress trajectories and the x axis.

It is shown in Section 9.3.2 that at a saddle point of the isopachics the line $\tau_{xy} = $ constant is perpendicular to the isochromatic. One of the stress trajectories thus coincides in direction with the isochromatic and the other with the isoclinic at such a point. All of these conditions are necessary but not sufficient to indicate a saddle point.

8.6.4 *Isotropic points of second order*

At an isotropic point of the first order it was necessary to take linear terms only into account since the effect of higher order terms could be neglected in the immediate vicinity of the isotropic point. If, however, all of the linear terms of all three stress components or those of the shear stresses alone vanish, the second order terms must be considered. In such a case, the stress system defined by equations (8.2) can be replaced by the following setup:

$$\tau_{xy} = ax^2 + 2bxy + cy^2 \tag{8.21a}$$

$$\sigma_y = -2axy - by^2 + dx^2 + fx, \tag{8.21b}$$

$$\sigma_x = -2cxy - bx^2 + (2b - d)y^2 + ey. \tag{8.21c}$$

As can readily be verified, these equations satisfy the equations of equilibrium and compatibility.

The directional angle φ of the stress trajectories is given by:

$$\tan 2\varphi = \frac{2\tau_{xy}}{\sigma_x - \sigma_y} = \frac{2(a + 2by/x + cy^2/x^2)}{(4b - d)y^2/x^2 - (b + d) - 2(c + a)y/x + ey/x^2 - f/x} \tag{8.22}$$

This equation shows that, if $f = e = 0$, i.e. if no linear terms are contained in the setup, φ is constant with constant y/x; the isoclinics passing through the isotropic point are therefore again straight lines in its immediate vicinity. Otherwise, the isoclinics will be curved.

Since $y/x = \tan \gamma$ is of the second order in equation (8.22) where, as before, γ is the directional angle of the isoclinic, two values of γ are produced with a given value of φ. This means that for each setting of the crossed polaroids, two isoclinics pass through the isotropic point. The system of stress trajectories now has four asymptotes. The isochromatics surrounding the isotropic point are no longer ellipses but are curves of higher order. Isotropic points of order higher than the second can be treated using a setup similar to that of equations (8.21) but containing terms of correspondingly higher degree.

8.7 Photoelastic effect of superposition of two or more states of stress

The following method of determining the resultant optical effect of two or more superimposed states of stress finds application in several photoelastic procedures. In two-dimensional photoelasticity it can be used to eliminate the effects of initial strain or initial double refraction from the results obtained. In three-dimensional problems it allows the determination of the resultant birefringence necessary for plotting the path curves in the application of the *j*-circle method.

Since double refraction is proportional to the difference of the principal or secondary principal stresses, it is proportional in cases of superposition to the difference of principal stresses corresponding to the resultant state of stress. This also holds true when one of the component states of stress is frozen in or consists of 'initial' stress or 'initial' strain.

In applying the method, the initial or frozen-in double refraction is replaced by an equivalent or virtual state of stress. This state of stress is

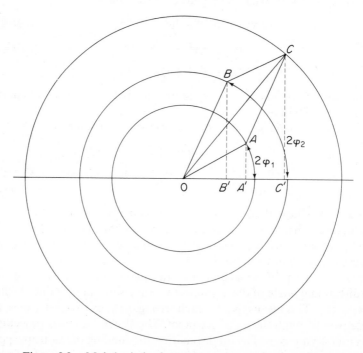

Figure 8.9. Mohr's circles for two superimposed states of stress

evaluated from the measured initial double refraction using the stress optic coefficient of the material at room temperature.

The difference of the resultant principal stresses and their directions are related to the stress components by

$$(\sigma_1 - \sigma_2) = \sqrt{[(\sigma_x - \sigma_y)^2 + 4\tau_{xy}^2]}, \tag{1.4}$$

$$\tan 2\varphi = \frac{2\tau_{xy}}{\sigma_x - \sigma_y}. \tag{1.3}$$

If two states of stress represented by the components $\sigma_{x1}, \sigma_{y1}, \tau_{xy1}$ and $\sigma_{x2}, \sigma_{y2}, \tau_{xy2}$ are superimposed, the resultant values of $(\sigma_x - \sigma_y)$ and τ_{xy} are

$$(\sigma_x - \sigma_y) = (\sigma_{x1} - \sigma_{y1}) + (\sigma_{x2} - \sigma_{y2}) = (\sigma_x - \sigma_y)_1 + (\sigma_x - \sigma_y)_2,$$

$$\tau_{xy} = \tau_{xy1} + \tau_{xy2},$$

since corresponding stresses act on the same faces of the element considered.

The resultant state of stress can be determined graphically using a method derived from Mohr's circle. The hydrostatic part $(\sigma_1 + \sigma_2)/2$ of a stress system does not influence the double refraction and can therefore be neglected here. The centres of the Mohr's circles for the two states of stress will then coincide as shown in Figure 8.9 where $A\hat{O}A' = 2\varphi_1, B\hat{O}B' = 2\varphi_2, OA' = (\sigma_x - \sigma_y)_1/2, OB' = (\sigma_x - \sigma_y)_2/2, AA' = \tau_{xy1}, BB' = \tau_{xy2}$.

Completing the parallelogram $OACB$, we see from the diagram that

$$OC' = OA' + A'C' = OA' + OB'$$

$$= \tfrac{1}{2}[(\sigma_x - \sigma_y)_1 + (\sigma_x - \sigma_y)_2].$$

Also, $CC' = AA' + BB' = \tau_{xy1} + \tau_{xy2}$.

It follows that OC is the radius of Mohr's circle for the resultant state of stress. Further, since

$$\tan C\hat{O}C' = \frac{2\tau_{xy}}{\sigma_x - \sigma_y} = \frac{2(\tau_{xy1} + \tau_{xy2})}{(\sigma_x - \sigma_y)_1 + (\sigma_x - \sigma_y)_2},$$

OC also indicates the direction of the resultant state of stress.

From the above, the resultant state of stress or the corresponding double refraction can be determined in the following way. The difference of the principal stresses (or the maximum shear stress) for each state of stress is represented by a vector. The direction of this vector is parallel to the algebraically greater principal stress and its length is proportional to the difference of principal stresses. The angle between one of these vectors and the other is then doubled after which they are added in the ordinary way. The length of the resulting vector is proportional to the difference of principal stresses of the resultant state of stress. The direction of the algebraically greater resultant principal stress is found by bisecting the angle between the resulting vector and that component vector which is drawn in its true direction.

The procedure given produces only the resultant shear stress or double refraction. If the resultant state of stress is to be completely determined, the hydrostatic parts of the component states of stress must be added to the results as scalars.

As a first example of the method we consider the effect of initial double refraction in a plate. If the initial double refraction corresponds to a virtual stress σ_1 (Figure 8.10) and the resultant double refraction observed when the plate is loaded corresponds to σ_R the applied stress is represented by σ_2.

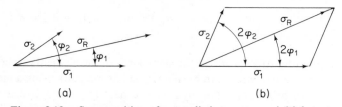

(a)   (b)

Figure 8.10.   Superposition of an applied stress on an initial stress

In Figure 8.10*a* the vectors are shown in their true directions while in Figure 8.10*b* the angles between them are doubled.

Our second example illustrates the superposition of membrane and bending stresses in a shell or plate. Using models of the type described in Section 17.5 which contain a reflecting middle layer, the double refraction produced in each half of the thickness of the shell can be determined by observing the same point from opposite sides of the shell. If these values correspond to virtual stresses σ_{R1} and σ_{R2} represented by OA and OB respectively, in Figure 8.11, the maximum bending stress σ_b is represented by half the length of the line connecting the end points A and B of the vectors, i.e. by CA or CB while the membrane stress σ_n is represented by OC. The validity of this construction is obvious when considered in reverse. The membrane stress

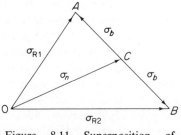

Figure   8.11.   Superposition   of membrane and bending stresses in a shell

σ_n is constant throughout the thickness and can therefore be represented by the single vector OC in each half of the shell. The maximum bending stress has the same magnitude but the opposite sign at corresponding points on the two surfaces of the shell. The direction of the algebraically greater principal stress in one half therefore differs by 90° from that in the other so that with the process of doubling the angles these stresses correspond to equal but oppositely directed vectors CA and CB. These added to the vector OC obviously produce OA representing the resultant stress at one surface of the shell and OB that at the other.

8.8 Examples of photoelastic investigations of uniaxial and biaxial states of stress

As stated previously, only two of the three values necessary to characterize a biaxial state of stress are given by the isochromatics and the isoclinics; an additional procedure must be applied in order to obtain the third value necessary for a complete analysis. In a uniaxial stress system, however, the isochromatics alone supply all the information required. At points on a load free edge of a plate, the direction of one principal stress concides with the edge while the other is perpendicular to it. The principal stress (say σ_2) perpendicular to the edge vanishes, so that the state of stress at such points is uniaxial. Here, σ_1 is identical with $(\sigma_1 - \sigma_2)$ and is thus given directly by the parameters of the isochromatics.

In most problems of two dimensional stress distribution, the maximum stress occurs at some point on a free edge. A knowledge of the edge stresses will therefore often be all that is required. The photoelastic solution of such problems is extremely simple.

By separating the principal stresses, it can be shown that the stresses acting normal to the axis of a thin beam or rod are not only zero at the free edges but are everywhere small. In such problems, the difference of principal stresses $(\sigma_1 - \sigma_2)$ is thus practically identical with the normal stress acting parallel to the axis.

8.8.1   *Application to beam problems*

In Figure 8.12, two beams loaded in different ways are compared. The upper beam is simply supported at its ends and is loaded by a central concentrated force while the lower beam is subjected to pure bending.

The stress distribution of the lower beam fully corresponds to the predictions of the elementary theory of bending; the stress is zero at the neutral surface and varies linearly with the distance from this surface. Except in the outer parts of the beam where the bending moment varies, the stress distribution is the same over each cross section. Thus, stresses calculated from the elementary theory will obviously be in very close agreement with the actual

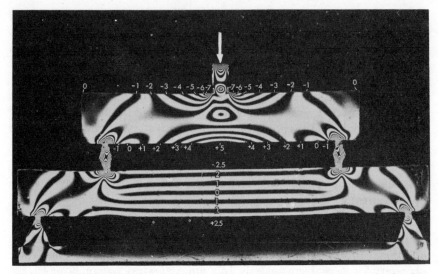

Figure 8.12. Isochromatic patterns for beams under three- and four-point loading

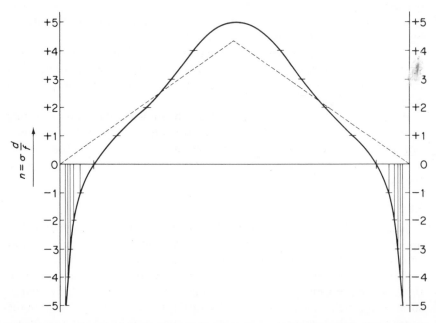

Figure 8.13. Comparison between the stresses determined photoelastically and those calculated from the elementary theory of bending along the lower edge of the beam under three-point loading shown in Figure 8.13

stresses. In accordance with St Venant's principle, deviations from this simple stress distribution are confined to regions near the points of application of the external forces.

In the upper beam, the stress distribution is much different. The edge stresses here are no longer constant, of course, since the bending moment varies along the beam. It is also found, however, that the stresses in this case differ substantially from those predicted by the elementary theory. Figure 8.13 compares the actual stresses along the lower edge with the predicted values, from which the following differences can be observed.

(*a*) The actual stress is not proportional to the bending moment, which varies linearly along the beam.

(*b*) Near the centre of the beam, the actual stresses are higher than the predicted ones.

(*c*) The actual stresses do not exhibit the discontinuity of slope at the centre indicated by the elementary theory. (In the exact solution of this problem by the theory of elasticity, this discontinuity does not exist.)

(*d*) Instead of diminishing to zero at the ends of the beam, the actual stresses change sign near the points of support and again increase. This effect is due to the high radial compression produced by the external forces acting at the supports.

The elementary theory of bending requires the external forces to be distributed over the cross section in the same manner as shear forces, i.e. parabolically. The effects produced when the external forces are actually distributed in this way are demonstrated by means of a model having the form given in Figure 8.14. The isochromatic fringe pattern is shown in Figure

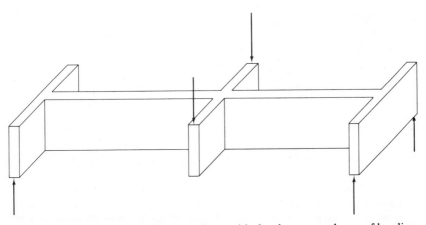

Figure 8.14.   A beam loaded in accordance with the elementary theory of bending

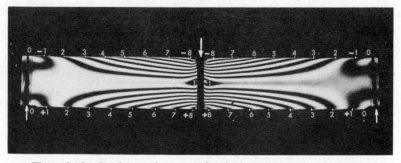

Figure 8.15.   Isochromatic pattern for the beam shown in Figure 8.14

8.15. In this case, the edge stresses practically coincide with those of the elementary theory as shown by Figure 8.16.

The series of experiments illustrated by Figure 8.17 was carried out to check the influence of differently reinforced central parts of a beam loaded by a central concentrated force. As can be seen, the stresses in the outer parts remain practically unchanged; the differences occurring in the central parts are obviously due to the differences of section modulus only.

The stresses produced in two cantilevers having different depths of section are compared in Figure 8.18 under different loading conditions. The section modulus of the lower cantilever was twice that of the upper. For the case

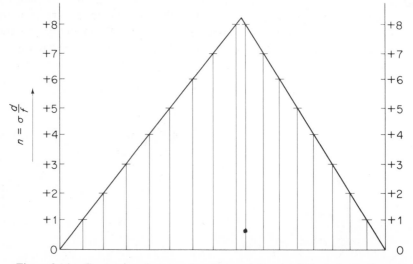

Figure 8.16.   Comparison between the stresses determined photoelastically and those calculated from the elementary theory of bending along the lower edge of the beam under three-point loading shown in Figure 8.14

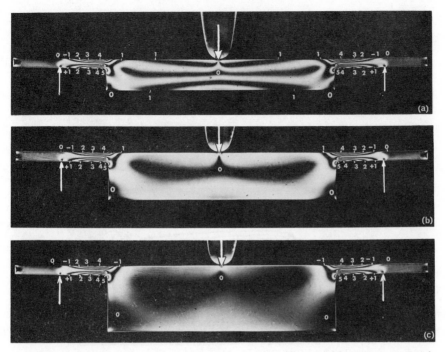

Figure 8.17. Isochromatic patterns for simply supported beams with a constant central load showing the influence of the depth of the central section

illustrated in Figure 8.18*a*, the cantilevers were loaded by equal forces while for that in Figure 8.18*b*, they were loaded by forces chosen to produce equal deflections. As indicated by the isochromatics, the maximum stress in the lower cantilever is half that in the upper one with equal loads; with equal deflections, it is $\sqrt{2}$ times higher.

8.8.2 *Problems of optimum shape*

Photoelasticity is particularly suitable for determining the optimum shape of machine or constructional parts. Ideally, these should be of a minimum weight or volume. This means that the stress everywhere, or at least at every point of the boundary, should be the same. Since the fringe orders of the isochromatics are proportional to the stresses at a load free edge, it follows that, in a part having optimum shape, the edge will coincide with an isochromatic. Since the stress distribution depends on the position and direction of the external forces acting, the optimum shape will of course be different if the loading system is altered.

A beam subjected to pure bending obviously has optimum shape when its edges are parallel; the edge stresses are then constant as shown by the lower

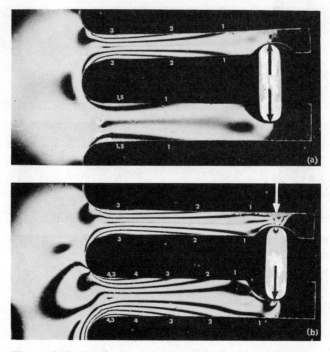

Figure 8.18. Isochromatics in cantilevers having different depths of section with (*a*) equal loads (*b*) equal deflections

beam in Figure 8.12. In the case of a beam loaded by a single concentrated force, however, the bending moment is linearly proportional to the distance from the point of support and the section modulus should vary in the same way. This means that, if the thickness is constant, the depth of section of the beam should be proportional to the square root of the distance from the points of support.

If a simply supported beam is traversed by a moving concentrated load, the maximum bending moment occurring under the load varies parabolically.

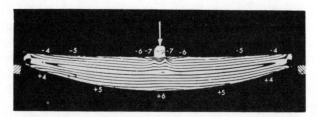

Figure 8.19. Isochromatics in a beam of parabolically varying depth under three-point loading

A beam having such a shape that the maximum stress is approximately constant for all positions of the load is shown in Figure 8.19. With the load in any given position such as at the centre of the span as shown, the edge stresses of course vary along the beam; they vary less, however, than in a beam of constant depth (cf. upper beam, Figure 8.12).

In curved beams, the magnitude as well as the signs of the stresses are different at the inner and outer edges respectively, Figure 8.20. Constant stress can

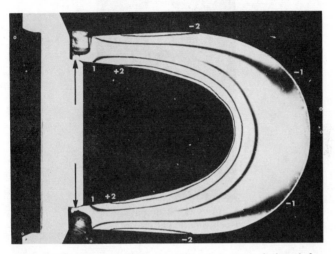

Figure 8.20. Isochromatics in part of a press designed for approximately constant tensile stress at the inner edge

therefore only be obtained on one side. In general, the tensile stress will be critical as in the part of a press illustrated.

In many other problems where practical considerations exclude the possibility of ideal conditions being fully realized, photoelastic investigations can lead to improvements of design by allowing the comparison of one shape with another or by indicating where a change of shape of a part may be desirable, for example to reduce the stresses in a region of high stress concentration.

Figure 8.21 shows a comparison of two hooks having different shapes.

The distribution of edge stresses in a part of a press can be read from Figure 8.22.

8.8.3 *Reinforced concrete*

A special field of linear photoelasticity is its application to problems of reinforced concrete. The stress distribution in reinforced concrete differs essentially from that in homogeneous material only if transverse cracks are

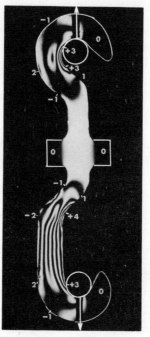

Figure 8.21.   Comparison of
isochromatic patterns of two
hooks of different shape

formed in the concrete on the tension side so that it is unable to support tensile stresses. Such cracks can be simulated in a photoelastic model by cutting, e.g. with a razor blade, after the material has been heated until it is soft. The steel reinforcement can be simulated by cementing a strip of ordinary photoelastic material along the cracked edge. In order to effect the same relative rigidity as the actual reinforcement, the cross-section of the strip is increased in the ratio of the elastic moduli of the steel and concrete.

A model of a reinforced concrete beam having both tension and compression reinforcement and supporting a non-central concentrated load is illustrated in Figure 8.23. Near the ends of the beam, cracks were omitted in place of diagonal reinforcement.

It is possible, of course, to use other materials such as metal or glass wires as reinforcement. The stresses in these cannot be obtained photoelastically but may be measured, for example, by strain gauges.

8.8.4   *Redundant frameworks*

Linear photoelasticity can be applied to the analysis of statically indeterminate frameworks. The method consists of locating points of zero bending

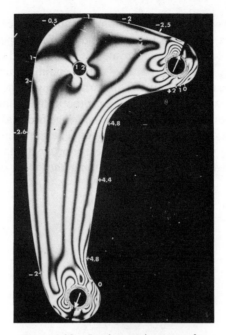

Figure 8.22. Isochromatic pattern for part of a press loaded by forces as indicated, from which the distribution of edge stresses can be read

moment in the members; assuming hinges to exist at these points, the framework is transformed into a statically determinate one.

The position of points of zero bending moment can easily be read from the isochromatic pattern. Such a point obviously occurs where the stress is constant or is symmetrically distributed over the cross-section. Since the stress perpendicular to the axis of a member is negligible, constant or symmetrically distributed fringe orders of the isochromatics means the absence of bending stresses.

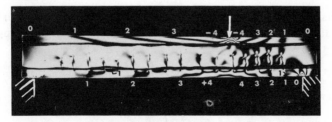

Figure 8.23. Isochromatic pattern for a simulated reinforced concrete beam

In practice, points of zero bending moment can easily be located by drawing a line to connect the points of zero fringe order on opposite edges of the member; the point of zero moment coincides with the mid point of this line.

The forces acting through a point of zero bending moment can also be determined from the isochromatics. The normal component N of the normal stress acting in the axial direction (which is practically uniformly distributed over the cross section) can be read from the fringe order at the edge points abreast of the point of zero bending moment. The transverse component T produces a bending moment which is linearly proportional to the distance from the section AA, Figure 8.24, containing the point of zero moment.

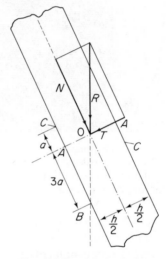

Figure 8.24. Method of determining the direction of the resultant force R acting through a point of zero bending moment

The effects of these two components obviously cancel one another at the edge points of zero fringe order. The stress σ_N due to the normal component N is

$$\sigma_N = N/h,$$

and the bending stress σ_T caused by the transverse component T is

$$\sigma_T = 6Tx/h^2,$$

where h is the depth of section. The distance $x = a$ from the section AA of the

Figure 8.25. Isochromatic pattern
of a part showing a section of zero
bending moment with known direction of the resultant force

points where $\sigma_N = -\sigma_T$ can be deduced from these equations:

$$\frac{N}{T} = -\frac{3a}{h/2}.$$

This relation is the basis of the method of plotting the direction of the resultant R of N and T as shown in Figure 8.24. The distance AB, equal to three

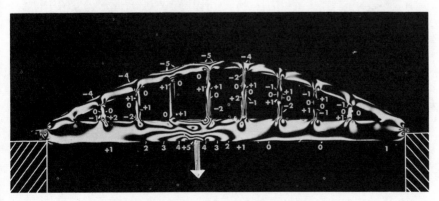

Figure 8.26. Isochromatic pattern for an arch from which points of zero bending
moment may be located

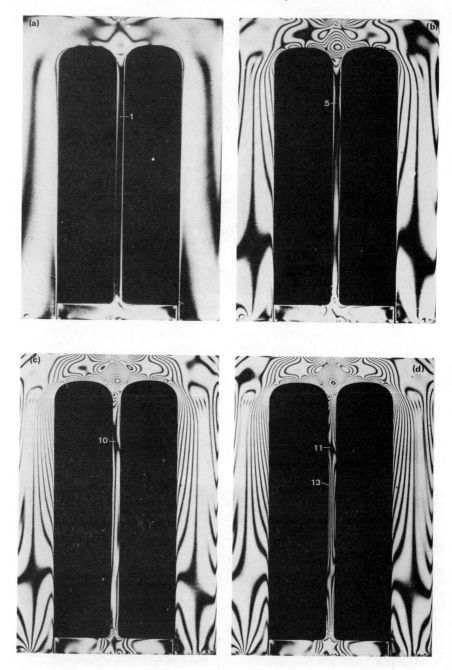

Figure 8.27. Isochromatics in an axially loaded column showing the transition from pure compression to a combination of compression and bending

times the distance AC of the point of zero stress from AA is marked off on the opposite side of AA. The resultant force R acts along the line OB. With this direction and the value of N known, the value of T can be derived.

Figure 8.25 demonstrates the position of the zero points with known direction of the resultant force, which, in this case, obviously coincides with the external force.

Figure 8.26 shows an example of a framework to which the above method could be applied. In addition to the position of points of zero bending moment, all other stress data can of course also be obtained photoelastically.

8.8.5 *Problems of stability*

The application of photoelasticity to the investigation of stability problems produces interesting results and allows the limits of applicability of analytical methods to be demonstrated.

In the example described here, the influence of the shape of the fixed ends of a column was investigated. The column was formed integrally within a relatively rigid frame to which it was connected through large fillets at one end and with a flat base at the other end as shown in Figure 8.27. A concentrated load was applied at the centre of the cross beam of the frame. A proportion of this load, which could be read from the isochromatics, was transmitted to the column. By this means, the column was loaded under

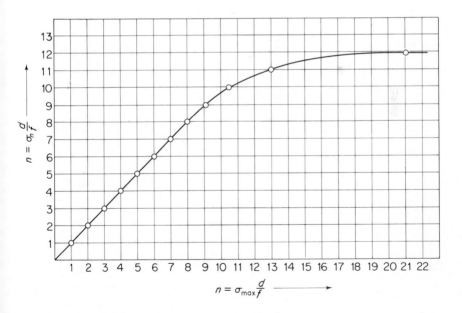

Figure 8.28.   Curve of maximum stress versus direct stress in fringes for the column shown in Figure 8.27

approximately constant deformation conditions, thus avoiding the dangers of sudden buckling and failure.

The isochromatics, given in Figure 8.27, show that, with low values of the load, the column is in a state of pure compression; when the load reaches a certain value, which can be read from the isochromatics, bending stresses begin to appear. When bending is present, the direct normal stress and hence the magnitude of the load can be read from the fringe orders at the points of inflexion.

The curve of Figure 8.28 shows the maximum stress σ_{max} plotted against the direct stress σ_n. From this curve, the safe load which may be applied can easily be read.

9

THE SEPARATION
OF PRINCIPAL STRESSES

9.1 Separation of principal stresses by the shear difference method

9.1.1 *General basis*

The shear difference method for separating the principal stresses, which can be applied to plates, three-dimensional and other types of problem, is based on the equations of equilibrium in Cartesian co-ordinates. A number of different versions of the method have been developed, some employing graphical and other numerical procedures. In some of these, the equations of equilibrium are applied to infinitesimal elements and in others to elements of finite dimensions. The physical meaning of the different forms in which these equations appear is, of course, the same in every case, namely, the conditions of equilibrium of the forces acting on a body or on any part of it. The various versions thus differ merely in the manner in which the equations are manipulated in order to achieve their integration and in the different forms of path, such as straight or curved lines and stress trajectories, along which they are integrated. The results obtained and their accuracy can therefore be expected to be the same whichever variation of the method is employed. The accuracy depends, of course, upon that of the measured optical data and the values derived from them. In the graphical procedure which will now be described, however, the latter can be checked before the final integration is performed so that the results obtained are more reliable.

9.1.2 *Graphical procedures*

In order to determine the required values graphically, lines of constant shear stress τ_{xy} must first be plotted. From these, the slopes $\partial \tau_{xy}/\partial x$ or $\partial \tau_{xy}/\partial y$ which are required for the integration can be derived in different ways. Alternatively, curves of τ_{xy} over certain chosen sections can be read from them.

The lines of constant shear stress τ_{xy} are derived from the isochromatics and the isoclinics using equation (1.7), i.e.

$$\tau_{xy} = \tfrac{1}{2}(\sigma_1 - \sigma_2)\sin 2\varphi. \tag{1.7}$$

The required values of τ_{xy} can be read from Mohr's circle (see Section 1.4). In practice, it is convenient to use a nomogram consisting of Mohr's circles of different diameter corresponding to isochromatics of different parameter or order and radial lines of different inclination corresponding to isoclinics of different parameter as shown in Figure 9.1. As can readily be seen, horizontal lines on this diagram are lines of constant shear stress τ_{xy} while vertical lines are lines of constant $(\sigma_x - \sigma_y)$. Since one quadrant of a circle contains all possible values of τ_{xy} (their sign can here be neglected) only one quarter of the complete diagram need be drawn.

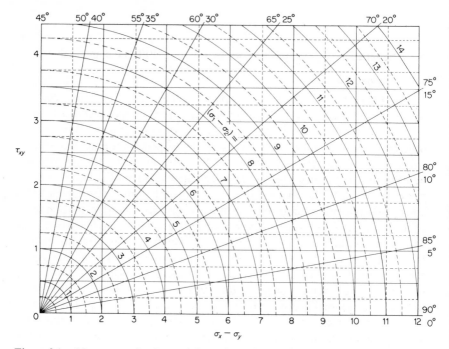

Figure 9.1.   Nomogram for determining values of τ_{xy} and $\sigma_x - \sigma_y$ from those of φ and $(\sigma_1 - \sigma_2)$

In order to plot a system of lines τ_{xy} = constant, the isochromatics with crossed and parallel polaroids and the isoclinics are projected on to a sheet of white paper and drawn in, preferably in different colours. Alternatively, they can be traced from photographs. It is convenient to draw the isochromatics of integral parameter in full lines and those of intermediate parameter in dotted lines. Similarly, the isoclinics of parameters 0°, 15°, 30°, 45°, 60° and 75° can be drawn in full lines and those of parameters 5°, 10°, 20°, 25°, 35°,..., 85° (or 7·5°, 22·5°, 37·5°,..., 82·5°) in dotted lines. Both the isochromatics and the isoclinics in dotted lines can be omitted in regions where those in full lines are sufficiently close together.

A sheet of tracing paper is next fixed over the system of isochromatics and isoclinics. The lines τ_{xy} = constant are now drawn on this paper commencing with the line τ_{xy} = 0. This line is identical with the isoclinic φ = 0°. The points of intersection between the lines τ_{xy} = constant and the 45° isoclinic are then marked. This is done noting, as can be seen from equation (1.7), that the parameters of the former are one half of those of the isochromatics. This is followed by marking the points of intersection between the lines τ_{xy} = 0·25, 0·5, 0·75, 1·0, 1·25, etc., and the 15° and 75° isoclinics. Here the parameters of the former are one quarter those of the isochromatics. Finally, additional points are plotted as found necessary using the nomogram. It is often found that the lines τ_{xy} = constant can be drawn with good accuracy even when only a few points on the curves are known.

The values listed in Table 9.1 can be used as an alternative or in addition to the nomogram of Figure 9.1 for determining points on the curves τ_{xy} = constant. In this table are listed the parameters of the isoclinics which pass through the points of intersection between lines of constant shear stress of different parameter and the whole and half order isochromatics obtained by the normal procedure of using crossed and parallel polaroids. Thus, for example, the line τ_{xy} = 0·75 intersects the isochromatic of parameter 4 at the points of intersection between the latter and the 11° and 79° isoclinics. The isoclinics listed with irregular parameters can be found by interpolation or if necessary can be determined experimentally. The table given can of course be expanded if desired, the necessary values being calculated from equation (1.7). It can be used following either the rows or columns to mark the points of intersection on the system of isochromatics and isoclinics. The required lines are obtained by drawing smooth curves through corresponding points.

Errors and inaccuracies in drawing the lines τ_{xy} = constant, can be avoided if the following rules are observed.

1. The lines τ_{xy} = constant correspond to the contours of a smooth continuous hill which in general is free from irregularities.

2. At its intersection with the 45° isoclinic, the direction of a line τ_{xy} = constant is identical with that of the isochromatic.

3. A line τ_{xy} = constant does not enter regions in which the parameters of the isochromatics are less than twice its own.

4. In regions where the isoclinics and the isochromatics have the same direction, the lines τ_{xy} = constant also have the same direction.

A further check on the reliability of the lines determined can be obtained using the conditions of equilibrium. The integral of the shear stresses acting

Table 9.1 List of parameters of isoclinics intersecting isochromatics having parameters of integral number of half wave lengths, where $\tau_{xy} = 0; 0\cdot25; 0\cdot5; \ldots$

parameter of isochromatic	$\tau_{xy}=$ 0	0·25	0·5	0·75	1·0	1·25	1·5	1·75	2·0	2·25	2·5
0·5	0°	45°									
1·0	0°	15°, 75°	45°								
1·5	0°	9·75°, 80·25°	20·9°, 69·1°	45°							
2·0	0°	7·75°, 82·25°	15°, 75°	24·3°, 65·7°	45°						
2·5	0°	5·75°, 84·25°	11·8°, 78·2°	19·4°, 70·6°	26·6°, 63·4°	45°					
3·0	0°	4·8°, 85·2°	9·75°, 80·25°	15°, 75°	20·9°, 69·1°	28·3°, 61·7°	45°				
3·5	0°	4·1°, 85·9°	8·4°, 81·6°	12·7°, 77·3°	17·4°, 72·6°	22·8°, 67·2°	29·5°, 60·5°	45°			
4·0	0°	3·7°, 86·3°	7·2°, 82·8°	11·0°, 79·0°	15°, 75°	19·3°, 70·7°	24·3°, 65·7°	30·5°, 59·5°	45°		
4·5	0°	3·2°, 86·8°	6·4°, 83·6°	9·75°, 80·25°	13·2°, 76·8°	16·4°, 73·6°	20·9°, 69·1°	25·5°, 64·5°	31·5°, 58·5°	45°	
5·0	0°	2·9°, 87·1°	5·8°, 84·2°	8·7°, 81·3°	11·8°, 78·2°	15°, 75°	18·4°, 71·6°	22·2°, 67·8°	26·6°, 63·4°	32·1°, 57·9°	45°
5·5	0°	2·65°, 87·35°	5·25°, 84·75°	7·9°, 82·1°	10·7°, 79·3°	13·6°, 76·4°	16·4°, 73·6°	19·8°, 70·2°	23·4°, 66·6°	27·5°, 62·5°	32·6°, 57·4°
6·0	0°	2·4°, 87·6°	4·8°, 85·2°	7·2°, 82·8°	9·75°, 80·25°	12·3°, 77·7°	15°, 75°	17·8°, 72·2°	20·9°, 69·1°	24·3°, 65·7°	28·2°, 61·8°

on any cross-section of a plate must equal the algebraic sum of the components of the external forces acting parallel to the section on either side of it. This integral can be determined by plotting the variations of the shear stress over the section considered from its points of intersection with the lines τ_{xy} = constant. In this way a cross-section of the hill of shear stresses is obtained. The required integral is equal to the area of this cross section.

The sign of τ_{xy} is taken to be positive if the shear stresses acting on the edges of a rectangular element have the directions shown in Figure 9.2a. If the shear stresses act in the opposite directions (Figure 9.2b) the sign is reversed. These

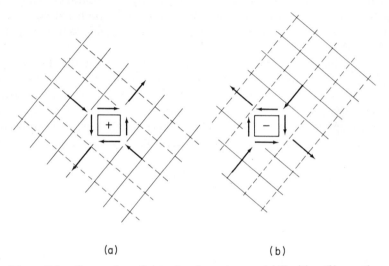

(a)   (b)

Figure 9.2. Convention of signs for shear stresses. (a) Positive, (b) negative

directions can readily be determined from the sign of $(\sigma_1 - \sigma_2)$ and the directions of the stress trajectories. This is most simple in the case of uniaxial stress systems in which the single principal stress is resolved into a normal stress and a shear stress, the directions of which can be determined by inspection. Biaxial stress systems can be reduced to uniaxial ones by superposing a hydrostatic stress, a process which does not affect the shear stresses. The sign of τ_{xy} within a plate can obviously change only at lines $\tau_{xy} = 0$.

The system of lines τ_{xy} = constant which is obtained in any particular case depends upon the direction chosen for the axes of co-ordinates. Rotation of the axes through 90° produces the same system of lines but with the signs reversed.

Integration of the equations of equilibrium in order to separate the principal stresses can be accomplished in different ways. One method consists of determining the values of $\partial \tau_{xy}/\partial x$ or $\partial \tau_{xy}/\partial y$ and plotting a curve of these values over the section considered. Integration of the equations of equilibrium (1.13) gives

$$\sigma_x = \sigma_{x0} - \int_{x_0}^{x} \frac{\partial \tau_{xy}}{\partial y}\, dx,$$

$$\sigma_y = \sigma_{y0} - \int_{y_0}^{y} \frac{\partial \tau_{xy}}{\partial x}\, dy.$$

Thus, if $\partial \tau_{xy}/\partial y$ is integrated step by step along a line parallel to the x axis, the result gives the values of σ_x at the corresponding points along that line. Similarly, values of σ_y are obtained by integrating $\partial \tau_{xy}/\partial x$ along a line parallel to the y axis.

Several different methods can be applied to determine the slopes $\partial \tau_{xy}/\partial x$ or $\partial \tau_{xy}/\partial y$, the choice depending on the conditions. The simplest method consists of measuring the distances between the lines $\tau_{xy} = $ constant parallel to the x or y axes. These distances are inversely proportional to $\partial \tau_{xy}/\partial x$ and $\partial \tau_{xy}/\partial y$ respectively. Thus, if Δx is the distance parallel to the x axis between two lines $\tau_{xy} = $ constant whose parameters differ by $\Delta \tau_{xy}$, Figure 9.3, then

$$\frac{\partial \tau_{xy}}{\partial x} \simeq \frac{\Delta \tau_{xy}}{\Delta x}.$$

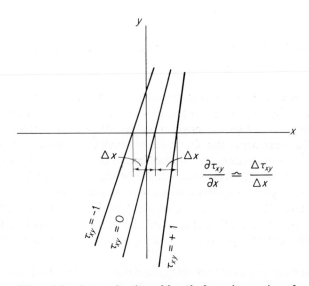

Figure 9.3.   Determination of $\partial \tau_{xy}/\partial x$ from the spacing of the lines $\tau_{xy} = $ constant

Similarly,

$$\frac{\partial \tau_{xy}}{\partial y} \simeq \frac{\Delta \tau_{xy}}{\Delta y}.$$

The above method is unsuitable for determining the required slopes of τ_{xy} at points on a line parallel to one axis when the lines $\tau_{xy} = $ constant are inclined at small angles to the other axis. This difficulty can be overcome in the following way. If, for example, the value of $\partial \tau_{xy}/\partial x$ is required at a point on a section AB parallel to the y axis, Figure 9.4, where the lines $\tau_{xy} = $ constant are

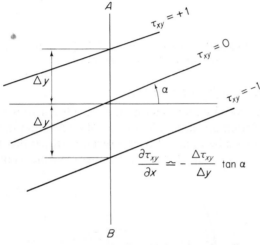

Figure 9.4. Determination of $\partial \tau_{xy}/\partial x$ when lines $\tau_{xy} = $ constant are inclined at small angles to the x-axis

approximately parallel to the x axis, the value of $\partial \tau_{xy}/\partial y$ is first determined in the manner previously described. If we denote by α the angle between the lines $\tau_{xy} = $ constant and the x axis then obviously

$$\frac{\partial \tau_{xy}/\partial x}{\partial \tau_{xy}/\partial y} = -\tan \alpha.$$

The required value of $\partial \tau_{xy}/\partial x$ is therefore obtained by multiplying $\partial \tau_{xy}/\partial y$ by $-\tan \alpha$.

When reliable results cannot be obtained by the above method due to irregular shape of the lines $\tau_{xy} = $ constant, several cross sectional curves of the hill of shear stresses parallel to the x axis can be drawn. The tangents of the angles of inclination of these curves to the x axis at the points of intersection with the line AB give the required values of $\partial \tau_{xy}/\partial x$, Figure 9.5.

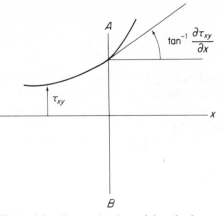

Figure 9.5. Determination of $\partial\tau_{xy}/\partial x$ from
the slope of the hill of shear stresses

9.1.3 *Equilibrium of a finite part of a plate*

An alternative procedure for determining the normal stresses is based on the conditions of equilibrium of a finite part of a plate. Let the distribution of the normal stresses be required over a section AB parallel to the y axis, Figure 9.6. The section is divided into a number of elements of finite length

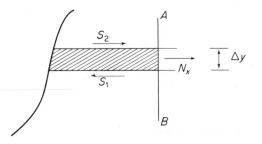

Figure 9.6. Determination of normal stresses
from the condition of equilibrium of forces acting
on a finite strip

by a set of other sections parallel to the x axis. The mean normal stress σ_{xm} acting on any of the elements of the section AB can be obtained from the condition of equilibrium of the forces acting in the x-direction on the corresponding elementary strip of the plate. If the edge of this strip which forms part of the external boundary of the plate is load-free, the corresponding force vanishes. The condition of equilibrium of the forces is then expressed by

$$N_x + S_2 - S_1 = 0,$$

where $S_{1,2}$ are the shear forces and N_x is the normal force acting on the edges of the strip as indicated in Figure 9.6. $S_{1,2}$ are equal to the cross sectional areas of the hill of shear stresses, i.e. the areas under the shear stress curves for the corresponding edges of the strip and N_x is equal to the difference of these areas. If Δy is the length of the element on AB then obviously the mean normal stress σ_{xm} acting on this element is $N_x/\Delta y$. The required curve of σ_x over the section is drawn using the values of the mean normal stresses acting on all the elements of the section obtained in the above manner and considering the boundary conditions at the edges of the plate. The latter, in general, enable the two end points of the curve to be determined from the isochromatics.

The lengths of the elements into which AB is divided by the transverse sections may be either uniform or variable and should be chosen such that it would be expected that irregularities in the shape of the curve of normal stresses would be detected. Thus, while longer elements produce more reliable values of the mean stresses, sudden variations might be lost if they are made too long. On the other hand, the results obtained are usually no more reliable if the elements are too short than if they are too long.

The results obtained by the above procedure can be compared with those by the preceding one if desired.

A method by means of which the directional angles of the isopachics can be determined from those of the lines $\tau_{xy} = $ constant is described in Section 9.3.2.

9.1.4 Numerical procedure

The problem of separating the principal stresses can be solved numerically by the shear difference method. This form of solution can be expedited by the use of an electronic computer.

The integration is performed using steps of finite length and the differential equations of equilibrium are accordingly transformed into equations of finite differences, e.g.

$$\frac{\Delta \tau_{xy}}{\Delta x} = -\frac{\Delta \sigma_y}{\Delta y}.$$

If Δx and Δy are chosen of equal length, this equation simplifies to

$$\Delta \tau_{xy} = -\Delta \sigma_y$$

and hence

$$\sigma_{y1} = \sigma_{y0} - \sum_0^1 \Delta \tau_{xy}. \tag{9.1a}$$

Similarly,

$$\sigma_{x1} = \sigma_{x0} - \sum_0^1 \Delta \tau_{xy}. \tag{9.1b}$$

From these equations the stresses σ_{y1} or σ_{x1} can be determined at any point provided the value of σ_{y0} or σ_{x0} is known at one point on the line of integration. The required value can usually be determined from the parameter of the isochromatic at the point in which the section considered intersects the boundary of the plate.

In order to determine the slope of τ_{xy} with sufficient accuracy, the values of $\Delta\tau_{xy}$ are determined as the difference of τ_{xy} at corresponding points along two neighbouring lines parallel to AB lying on opposite sides and equidistant from it. In Figure 9.7, these lines are those denoted by a and c while the

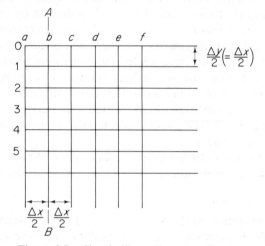

Figure 9.7. Sketch illustrating the numerical application of the shear difference method

section AB corresponds to the line denoted by b. These lines are divided into elements of equal length by a perpendicular system of lines which are denoted by 0, 1, 2, etc.

The shear stresses τ_a and τ_c along the lines a and c at their points of intersection with the lines 0, 1, 2, etc., can be determined from the equation

$$\tau = \tfrac{1}{2}(\sigma_1 - \sigma_2)\sin 2\varphi, \tag{1.7}$$

or from the nomogram in Figure (9.1).

For convenience, the values of $(\sigma_1 - \sigma_2)$ can be expressed in terms of the parameters of the isochromatics from which they are obtained. If the stresses are required in kp/cm² or lb/in², the results must finally be multiplied by f/d in accordance with equation (5.20). The angle φ is measured clockwise between the normal to the section of integration and the direction of the greater principal stress σ_1 and can be obtained from the parameters of the isoclinics.

Having determined the values of $\Delta\tau_{xy}$ at the points of intersection on the line b, the normal stress σ_y along AB can be calculated from equation (9.1a), starting at the point bo. This point is chosen to coincide with the essential point of known stress and will therefore generally lie on the boundary of the plate.

The normal stress σ_x can be calculated from equation (1.8) which may be written:

$$\sigma_x = \sigma_y \pm \sqrt{[(\sigma_1 - \sigma_2)^2 - 4\tau_{xy}^2]}. \tag{9.2}$$

In equation (9.2) the positive sign before the root is taken if $\sigma_x > \sigma_y$ as is the case when the algebraically greater principal stress σ_1 is inclined to σ_x at an angle of less than 45°. In the opposite case, when $\sigma_x < \sigma_y$, the negative sign is taken.

If the principal stresses are required, they can be obtained using the following relation which is evident from Mohr's circle:

$$\sigma_1, \sigma_2 = \tfrac{1}{2}(\sigma_x + \sigma_y) \pm \tfrac{1}{2}(\sigma_1 - \sigma_2).$$

In practice it is convenient to tabulate the various values as shown in Table 9.2. When the normal stresses are required over a field, the method can be applied simultaneously to a series of lines parallel to AB as shown in Figure 9.7. The lines a, b, c are then replaced by b, c, d and so on.

Table 9.2   Calculation of stresses by the shear difference method

point	τ_a	τ_b	$\Delta\tau_{xy}$	$\Sigma\Delta\tau_{xy}$	σ_y	$(\sigma_1 - \sigma_2)$	σ_x

9.1.5   *Example of separation of principal stresses by the shear difference method*

The practical application of the different procedures described in the preceding sections is demonstrated in the following analysis of a short-span beam with a hole under three-point loading.

The dark and bright field isochromatic patterns for the beam are given in Figure 9.8. The isoclinics are shown in Figure 9.9 and the stress trajectories

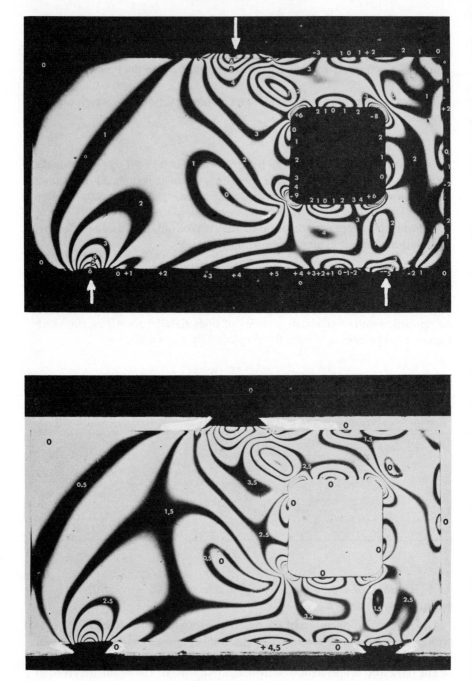

Figure 9.8.   Isochromatic pattern of a short span beam with a hole

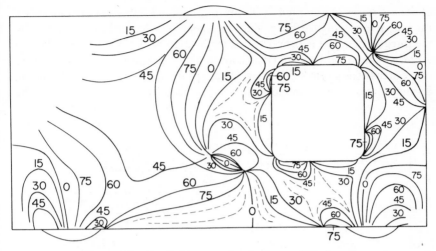

Figure 9.9.   Isoclinic pattern for the beam of Figure 9.8

derived from them are drawn in Figure 9.10. The lines τ_{xy} = constant plotted from the data provided by the isochromatics and the isoclinics with the aid of the nomogram of Figure 9.1 are shown in Figure 9.11. In order to determine the normal stresses on the vertical line AB of Figure 9.11, sections of the hill

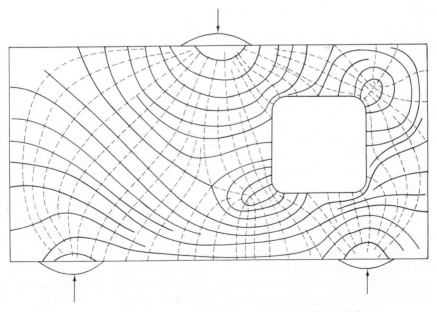

Figure 9.10.   Stress trajectories for the beam of Figure 9.8

of shear stresses represented by its contour lines in Figure 9.11 are plotted on the lines a to i in Figure 9.12a. Since the shear force S along a line is given by the integral of τ_{xy}, it is proportional to the area below the curves of Figure 9.12a, which can be determined by counting the small squares. The differences of the areas of adjacent sections, which are proportional to the corresponding normal forces N, are plotted as equivalent rectangles in Figure 9.12b. The

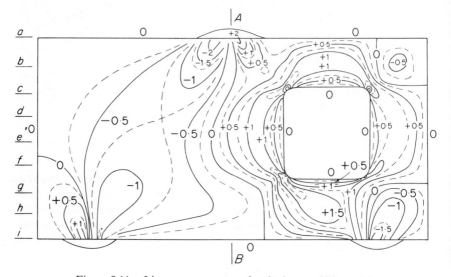

Figure 9.11.   Lines $\tau_{xy} = $ constant for the beam of Figure 9.8

horizontal dimension of each rectangle represents the mean value σ_{xm} of the normal stress over the distance Δy between the corresponding sections. Observing these mean values, the curve of σ_x can then be estimated as shown. The value of σ_x at the lower edge coincides of course with that corresponding to the isochromatic fringe order. In Figure 9.12, the stress is plotted to units of wavelength (λ) corresponding to the order or parameter of the isochromatics.

The method of integration is indicated in Figure 9.12c. The values of $\partial\tau_{xy}/\partial x$ given by the inclination of the curves of τ_{xy} on the lines a to i at their points of intersection with the line AB in Figure 9.12a are recorded in Figure 9.12c (to units of wavelength per unit length, i.e. λ/cm). Since σ_y vanishes at the lower edge, the integral obtained by summing the small squares below the curve of $\partial\tau_{xy}/\partial x$ is identical with σ_y, which can then be plotted as shown. Finally, σ_x is obtained by adding the values of $(\sigma_x - \sigma_y)$ and σ_y. As can be seen, practically the same curve for σ_x is obtained by both methods.

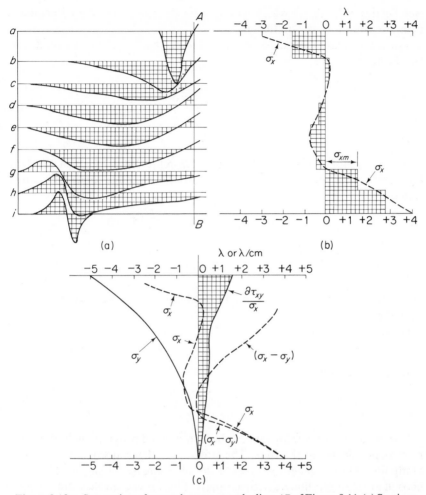

Figure 9.12. Separation of normal stresses on the line *AB* of Figure 9.11. (*a*) Sections of the hill of shear stresses. (*b*) Method based on equilibrium of forces acting on finite strips. (*c*) Method of integration of the equations of equilibrium

9.2 Separation of principal stresses in plates loaded by coplanar forces without using the isoclinics

In cases where the isoclinics have not been determined or are not sufficiently reliable, several methods of separating the principal stresses can be applied. These methods can, of course, also be used to check the results obtained by other procedures involving the use of both the isoclinics and the isochromatics. One group of methods of this type is based on the equations of compatibility (see Section 9.3). Those which will now be described are derived primarily from the equations of equilibrium.

For a thin plate loaded by coplanar forces, the equations of equilibrium (see Section 1.6) are

$$\frac{\partial \sigma_x}{\partial x} + \frac{\partial \tau_{xy}}{\partial y} = 0, \qquad (1.13a)$$

$$\frac{\partial \sigma_y}{\partial y} + \frac{\partial \tau_{xy}}{\partial x} = 0. \qquad (1.13b)$$

At every point on a straight load-free edge represented by $y = $ a constant, the stress components σ_y and τ_{xy} are zero so that the edge conditions can be stated:

$$\sigma_y = 0; \qquad \frac{\partial \sigma_y}{\partial x} = 0; \qquad \frac{\partial^2 \sigma_y}{\partial x^2} = 0;$$

$$\tau_{xy} = 0; \qquad \frac{\partial \tau_{xy}}{\partial x} = 0; \qquad \frac{\partial^2 \tau_{xy}}{\partial x^2} = 0.$$

Further, since τ_{xy} is zero, σ_x and σ_y are identical with the principal stresses σ_1 and σ_2.

Hence

$$\sigma_x = (\sigma_x - \sigma_y) = (\sigma_1 - \sigma_2) = (\sigma_1 + \sigma_2) = (\sigma_x + \sigma_y)$$

and from equation (1.13a)

$$\frac{\partial \sigma_x}{\partial x} = -\frac{\partial \tau_{xy}}{\partial y} = \frac{\partial(\sigma_1 - \sigma_2)}{\partial x}. \qquad (9.3)$$

The equations of equilibrium are here integrated progressing in the direction perpendicular to the edge. The stresses are determined simultaneously along lines of finite length parallel to the edge, Figure 9.13a. These lines are chosen at small but finite distances apart. The stresses on these lines will be represented in order by the subscripts 0, 1, 2, etc., where 0 refers to the edge itself. The distance between the edge and line 1 will be denoted by Δy_{01}, that between lines 1 and 2 by Δy_{12} and so on.

The curve for $(\sigma_1 - \sigma_2)_0$, which can be obtained from the isochromatics, is first drawn. Equation (9.3) shows that the slope of this curve is equal to that of $(\tau_{xy})_0$ perpendicular to the edge, i.e. $(\partial \tau_{xy}/\partial y)_0$, but is of opposite sign. This slope is used to obtain approximate values of $(\tau_{xy})_1$ from

$$(\tau_{xy})_1 = -\left[\frac{\partial(\sigma_1 - \sigma_2)}{\partial x} \right]_0 (\Delta y)_{01}.$$

In order to find $(\sigma_y)_1$, the mean value of the slopes $(\partial \sigma_y/\partial y)_0$ and $(\partial \sigma_y/\partial y)_1$ should be used. Equation (1.13b) shows however that since $(\partial \tau_{xy}/\partial x)_0$ is zero,

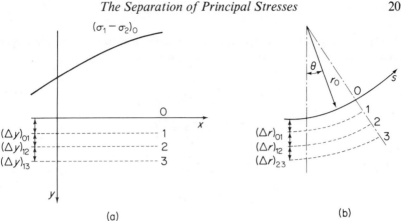

(a)　　　　　　　　　　　　　(b)

Figure 9.13. Method of separation of principal stresses without using the isoclinics at (a) a straight edge, and (b) a curved edge

$(\partial\sigma_y/\partial y)_0$ must be zero and that $(\partial\sigma_y/\partial y)_1 = -(\partial\tau_{xy}/\partial x)_1$. With these values $(\sigma_y)_1$ can be determined from the equation

$$(\sigma_y)_1 = \frac{1}{2}\left[\left(\frac{\partial\sigma_y}{\partial y}\right)_0 + \left(\frac{\partial\sigma_y}{\partial y}\right)_1\right](\Delta y)_{01} = \frac{1}{2}\left[\left(\frac{\partial\sigma_y}{\partial y}\right)_1\right](\Delta y)_{01}.$$

Having determined $(\tau_{xy})_1$ and $(\sigma_y)_1$, $(\sigma_x)_1$ can now be determined using the relation

$$(\sigma_{x1} - \sigma_{y1})^2 = (\sigma_1 - \sigma_2)_1^2 - 4(\tau_{xy})_1$$

or

$$\sigma_{x1} = \sqrt{[(\sigma_1 - \sigma_2)_1^2 - 4(\tau_{xy})_1]} + \sigma_{y1},$$

where, as before the values of $(\sigma_1 - \sigma_2)$ are obtained from the isochromatics and the sign of the root is dictated by the continuity of the stresses.

In this manner, the components σ_{x1}, σ_{y1} and τ_{xy1} along the line 1 are obtained and using these values we may now proceed to determine the components along line 2. Thus, the slope $(\partial\sigma_x/\partial x)_1$ is identical with $-(\partial\tau_{xy}/\partial y)_1$ and is used to obtain $(\tau_{xy})_2$ from

$$(\tau_{xy})_2 = (\tau_{xy})_1 + \left(\frac{\partial\tau_{xy}}{\partial y}\right)_1 (\Delta y)_{12}.$$

The slope $(\partial\tau_{xy}/\partial x)_2$ is identical with $-(\partial\sigma_y/\partial y)_2$ and thus permits the evaluation of $(\sigma_y)_2$ from the relationship

$$(\sigma_y)_2 = (\sigma_y)_1 + \frac{1}{2}\left[\left(\frac{\partial\sigma_y}{\partial y}\right)_1 + \left(\frac{\partial\sigma_y}{\partial y}\right)_2\right](\Delta y)_{12}.$$

The next step is the determination of $(\sigma_x)_2$. This procedure is continued progressively over the whole plate. A check on the accuracy is provided by the results obtained at the other edges of the plate.

If the integration is started at a curved edge, it is more convenient to use polar co-ordinates. The equations of equilibrium are then

$$(\sigma_r - \sigma_\theta) + r\frac{\partial \sigma_r}{\partial r} + \frac{\partial \tau_{r\theta}}{\partial \theta} = 0, \qquad (1.14a)$$

$$\frac{1}{r}\frac{\partial \sigma_\theta}{\partial \theta} + 2\frac{\tau_{r\theta}}{r} + \frac{\partial \tau_{r\theta}}{\partial r} = 0. \qquad (1.14b)$$

The conditions at a load-free edge $r = r_0$, a constant, are represented by

$$\sigma_r = 0; \qquad \frac{\partial \sigma_r}{\partial \theta} = 0; \qquad \frac{\partial^2 \sigma_r}{\partial \theta^2} = 0, \qquad (9.4)$$

$$\tau_{r\theta} = 0; \qquad \frac{\partial \tau_{r\theta}}{\partial \theta} = 0; \qquad \frac{\partial^2 \tau_{r\theta}}{\partial \theta^2} = 0, \qquad (9.5)$$

$$\sigma_\theta = -(\sigma_r - \sigma_\theta) = (\sigma_1 - \sigma_2) = (\sigma_1 + \sigma_2) = (\sigma_r + \sigma_\theta), \qquad (9.6)$$

$$\frac{1}{r_0}\frac{\partial \sigma_\theta}{\partial \theta} = -\frac{\partial \tau_{r\theta}}{\partial r} \qquad (9.7)$$

$$\frac{\partial \sigma_r}{\partial r} = -\frac{\sigma_r - \sigma_\theta}{r_0} = \frac{\sigma_1 - \sigma_2}{r_0} \qquad (9.8)$$

These equations are applied to short sections of the edge over which the curvature may be regarded as constant. For such sections the parallel lines are replaced by circular arcs, described about the centre of curvature, Figure 9.13*b*. As before, these lines are denoted in order commencing with the edge by the subscripts 0, 1, 2, etc., and the distances between them by Δr_{01}, Δr_{12}, etc.

The curve of $(\sigma_1 - \sigma_2)$ along the edge, obtained from the fringe pattern or otherwise, is first plotted. The slope of this curve with respect to the tangent to the edge enables $(\tau_{r\theta})$ to be evaluated. It is given, approximately, by

$$\tau_{r\theta} = \frac{1}{r_0}\left[\frac{\partial(\sigma_1 - \sigma_2)}{\partial \theta}\right]_0 (\Delta r)_{01} = \frac{\partial(\sigma_1 - \sigma_2)}{\partial s}(\Delta r)_{01},$$

where s represents distance measured along the edge. The values of $(\sigma_r - \sigma_\theta)$ are then computed from

$$(\sigma_r - \sigma_\theta)^2 = (\sigma_1 - \sigma_2)^2 - 4\tau_{r\theta}^2, \qquad (9.9)$$

the values of $(\sigma_1 - \sigma_2)$ for line 1 being obtained from the fringe pattern.

The values for $(\partial \sigma_r/\partial r)_1$, obtained from equation (1.14a) after inserting the values of $(\sigma_r - \sigma_\theta)_1$ and $(\partial \tau_{r\theta}/\partial \theta)_1$, together with $(\partial \sigma_r/\partial r)_0$ from equation (9.8) furnish the data necessary to calculate $(\sigma_r)_1$.

Thus,

$$(\sigma_r)_1 = \frac{1}{2}\left[\left(\frac{\partial \sigma_r}{\partial r}\right)_0 + \left(\frac{\partial \sigma_r}{\partial r}\right)_1\right](\Delta r)_{01}.$$

σ_r is now inserted in equation (9.9) to obtain $\sigma_{\theta 1}$. The curve for $\sigma_{\theta 1}$ is then plotted in order to find the slope $(\partial \sigma_\theta/\partial \theta)_1 = r_1 \, \partial \sigma_1/\partial s$.

Finally, values of $\partial \tau_{r\theta}/\partial r$ for use in the next step are determined from equation (1.14b) after substituting for $(\partial \sigma_\theta/\partial \theta)_1$, $(\tau_{r\theta})_1$ and $r = r_1$.

In practice it is convenient to use polar co-ordinates for a few steps only and then to convert into Cartesian co-ordinates as the operations with these are easier.

The results obtained by the method described, which is based on the equations of equilibrium of the first order, can be checked and improved by applying the following equations which are derived from the equations of equilibrium of the second order and the equation of compatibility. In Cartesian co-ordinates, these are

$$\frac{\partial^2 \sigma_x}{\partial x^2} = \frac{\partial^2 \sigma_y}{\partial y^2}, \tag{9.10}$$

$$\frac{\partial^2(\sigma_x - \sigma_y)}{\partial x \, \partial y} = \frac{\partial^2 \tau_{xy}}{\partial x^2} - \frac{\partial^2 \tau_{xy}}{\partial y^2}, \tag{9.11}$$

$$\frac{\partial^2(\sigma_x - \sigma_y)}{\partial x^2} - \frac{\partial^2(\sigma_x - \sigma_y)}{\partial y^2} = -4\frac{\partial^2 \tau_{xy}}{\partial x \, \partial y}, \tag{9.12}$$

$$\frac{\partial^2(\sigma_x + \sigma_y)}{\partial x \, \partial y} = -\left(\frac{\partial^2 \tau_{xy}}{\partial x^2} + \frac{\partial^2 \tau_{xy}}{\partial y^2}\right), \tag{9.13}$$

$$\frac{\partial^2(\sigma_x + \sigma_y)}{\partial y^2} = \frac{\partial^2(\sigma_x + \sigma_y)}{\partial x^2} = \frac{\partial^2(\sigma_x - \sigma_y)}{\partial y^2} - 2\frac{\partial^2 \tau_{xy}}{\partial x \, \partial y}$$

$$= \frac{\partial^2(\sigma_x - \sigma_y)}{\partial x^2} + 2\frac{\partial^2 \tau_{xy}}{\partial x \, \partial y}. \tag{9.14}$$

These equations are derived in the following manner:
If the first order equations of equilibrium

$$\frac{\partial \sigma_x}{\partial x} + \frac{\partial \tau_{xy}}{\partial y} = 0,$$

$$\frac{\partial \sigma_y}{\partial y} + \frac{\partial \tau_{xy}}{\partial x} = 0,$$

are differentiated with respect to x and y we obtain

$$\frac{\partial^2 \sigma_x}{\partial x^2} + \frac{\partial^2 \tau_{xy}}{\partial x\, \partial y} = 0, \tag{9.15}$$

$$\frac{\partial^2 \sigma_y}{\partial y^2} + \frac{\partial^2 \tau_{xy}}{\partial x\, \partial y} = 0, \tag{9.16}$$

$$\frac{\partial^2 \sigma_x}{\partial x\, \partial y} + \frac{\partial^2 \tau_{xy}}{\partial y^2} = 0, \tag{9.17}$$

$$\frac{\partial^2 \sigma_x}{\partial x\, \partial y} + \frac{\partial^2 \tau_{xy}}{\partial x^2} = 0. \tag{9.18}$$

Subtracting equation (9.16) from equation (9.15) and equation (9.18) from equation (9.17) we obtain equations (9.10) and (9.11) respectively. The left hand side of equation (9.12) can be written in the form

$$\frac{\partial^2 \sigma_x}{\partial x^2} + \frac{\partial^2 \sigma_y}{\partial y^2} - \left(\frac{\partial^2 \sigma_y}{\partial x^2} + \frac{\partial^2 \sigma_x}{\partial y^2} \right).$$

The equation of compatibility is

$$\left(\frac{\partial^2}{\partial x^2} + \frac{\partial^2}{\partial y^2} \right)(\sigma_x + \sigma_y) = 0, \tag{2.30}$$

which gives

$$\frac{\partial^2 \sigma_x}{\partial x^2} + \frac{\partial^2 \sigma_y}{\partial y^2} = -\left(\frac{\partial^2 \sigma_y}{\partial x^2} + \frac{\partial^2 \sigma_x}{\partial y^2} \right),$$

so that

$$\frac{\partial^2}{\partial x^2}(\sigma_x - \sigma_y) - \frac{\partial^2}{\partial y^2}(\sigma_x - \sigma_y) = 2\left(\frac{\partial^2 \sigma_x}{\partial x^2} + \frac{\partial^2 \sigma_y}{\partial y^2} \right) = -4\frac{\partial^2 \tau_{xy}}{\partial x\, \partial y}$$

using equations (9.15) and (9.16). Equation (9.13) is obtained by adding equations (9.16) and (9.17). Equation (9.14) is obtained by writing the equation of compatibility in the form

$$\frac{\partial^2 (\sigma_x + \sigma_y)}{\partial y^2} = -\frac{\partial^2 (\sigma_x + \sigma_y)}{\partial x^2},$$

the left hand side of which can be written

$$\frac{\partial^2}{\partial y^2}[(\sigma_x - \sigma_y) + 2\sigma_y] = \frac{\partial^2(\sigma_x - \sigma_y)}{\partial y^2} + 2\frac{\partial^2\sigma_y}{\partial y^2}$$

$$= \frac{\partial^2(\sigma_x - \sigma_y)}{\partial y^2} - 2\frac{\partial_2\tau_{xy}}{\partial x\,\partial y} \qquad \text{from equation (9.16)}$$

$$= \frac{\partial^2(\sigma_x - \sigma_y)}{\partial x^2} + 2\frac{\partial^2\tau_{xy}}{\partial x\,\partial y} \qquad \text{from equation (9.12).}$$

In polar co-ordinates, the equations of equilibrium produce the following equation

$$\frac{\partial^2\sigma_r}{\partial r^2} = \frac{2}{r^2}(\sigma_r - \sigma_\theta) + \frac{1}{r}\frac{\partial\sigma_\theta}{\partial r} + \frac{1}{r^2}\frac{\partial^2\sigma_\theta}{\partial\theta^2}, \qquad (9.19)$$

where $(\partial\sigma_\theta/\partial r)$ can be determined from

$$\frac{\partial\sigma_\theta}{\partial r} = \frac{\partial(\sigma_\theta - \sigma_r)}{\partial r} + \frac{\partial\sigma_r}{\partial r}. \qquad (9.20)$$

The equation of compatibility in polar co-ordinates is

$$\left(\frac{\partial^2}{\partial r^2} + \frac{1}{r}\frac{\partial}{\partial r} + \frac{1}{r^2}\frac{\partial^2}{\partial\theta^2}\right)(\sigma_r + \sigma_\theta) = 0. \qquad (9.21)$$

The results should be checked after each step by means of equations (9.10) to (9.14) or equations (9.19) to (9.21). The complete result can be checked in several ways, e.g. by drawing the isopachics, i.e. lines along which $(\sigma_x + \sigma_y) = (\sigma_1 + \sigma_2) = $ constant. These should obey the rules for isopachics given in Section 9.3.2. Further, the shear force produced by the external loads on any section can be calculated and this should equal the integrated shear stresses over the section. Similarly, the mean normal stresses σ_x and σ_y over any section can be calculated from the geometry of the plate and the loads.

The mean normal stresses σ_{xm} and σ_{ym} within the whole plate can be found from the components of the loads in the x- and y-directions. Consider a plate in equilibrium under the action of two equal and opposite loads acting at the points A and B, Figure 9.14. The loads are represented by their components X and Y parallel to the x and y axes respectively.

The total stress over any cross-section parallel to the y axis lying between the points A and B is equal to X. The mean stress σ_{xm} over the whole plate is given by $\int \sigma_x \, dA/At$, where A is the surface area and t the thickness of the plate and the integration is carried out between sections passing through the

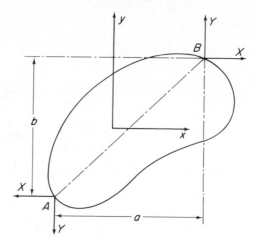

Figure 9.14.   Determination of the mean normal
stresses in a plate

points A and B. Thus,

$$\sigma_{xm} = \frac{aX}{At}.$$

Similarly,

$$\sigma_{ym} = \frac{bY}{At}.$$

If there are more than two forces acting on the plate, the components acting parallel to the x and y axes are split into pairs of equal and opposite forces. Each of these pairs produce a certain mean stress. The total mean normal stress σ_{xm} or σ_{ym} is the sum of these. The procedure is illustrated in Figure 9.15 where, for simplicity, the X-components only of the applied forces are shown. Then,

$$\sigma_{xm} = [a_{13}X_3 + a_{24}X_2 + a_{14}(X_1 - X_3)]/At.$$

The value of σ_{ym} is obtained in a similar manner.

Another procedure for verifying the results consists of plotting the isoclinics from the isochromatics and the stresses derived from them. The parameter of an isoclinic is given by the equation

$$\tan 2\varphi = \frac{2\tau_{xy}}{\sigma_x - \sigma_y}.$$

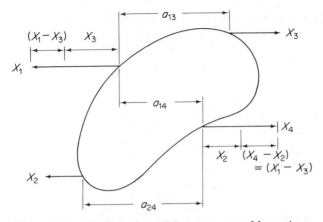

Figure 9.15. Division of parallel components of forces into equal and opposite pairs

From Figure 9.16 the angle γ' between a stress trajectory s_1 and the isoclinic is given by

$$\tan O\widehat{A}B = \tan(180 - \gamma') = OB/OA = \delta s_2/\delta s_1,$$

i.e.

$$\tan \gamma' = -\delta s_2/\delta s_1 = -\rho_2/\rho_1, \tag{9.22}$$

where ρ_1 and ρ_2 are the radii of curvature of the stress trajectories s_1 and s_2 respectively.

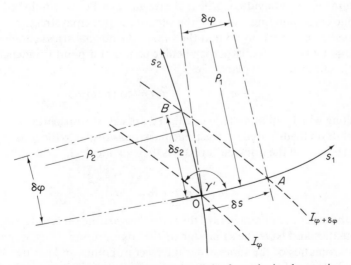

Figure 9.16. Method of plotting isoclinics from the isochromatics and the stress trajectories

From the Lamé–Maxwell equations of equilibrium (1.16),

$$\frac{\partial \sigma_1}{\partial s_1} + \frac{\sigma_1 - \sigma_2}{\rho_2} = 0,$$

which gives

$$\rho_2 = -\frac{\sigma_1 - \sigma_2}{\partial \sigma_1 / \partial s_1}.$$

Hence, inserting this value in equation (9.22)

$$\tan \gamma' = \frac{(\sigma_1 - \sigma_2)/\rho_1}{\partial \sigma_1 / \partial s_1}.$$

At a load-free edge of radius $r = r_0$, $\partial \sigma_1 / \partial s_1 = \partial(\sigma_1 - \sigma_2)/\partial s_1$, and $\delta s_1 = r_0 \delta \theta$ so that

$$(\tan \gamma')_0 = \frac{(\sigma_1 - \sigma_2)_0/r_0}{1/r_0[\partial(\sigma_1 - \sigma_2)/\partial\theta]_0} = \frac{(\sigma_1 - \sigma_2)_0}{[\partial(\sigma_1 - \sigma_2)/\partial\theta]_0}.$$

9.3   Isopachics. Graphical methods of construction

9.3.1   *Introduction*

As shown in Section 5.4, the isochromatic pattern in a plate provides values of the difference in the two principal stresses. One method by which the separation of the individual principal stresses can be accomplished is to determine corresponding values of the sum of the principal stresses.

It was shown in Section 1.3 that the sum of the normal stress components acting on any two mutually perpendicular planes at a point is independent of the directions of those planes, i.e.

$$(\sigma_x + \sigma_y) = (\sigma_1 + \sigma_2) = \text{constant.} \tag{1.5}$$

Lines along which equation (1.5) is satisfied are called isopachics.

At points on load-free edges, the principal stress normal to the edge, say σ_2, is zero. The sum of the principal stresses is then identical with their difference, i.e.

$$\sigma_1 = (\sigma_1 - \sigma_2) = (\sigma_1 + \sigma_2).$$

Here the orders or parameters of the isopachics are the same as those of the isochromatics and have the same sign as the edge stresses. Thus, isopachics and isochromatics of the same order intersect at points on load free boundaries. The sum of the principal stresses can also be determined from the isochromatics at points on a boundary subjected to normal loading of known distribution. The intensity p of the loading is then identical with σ_2 so that

$$(\sigma_1 + \sigma_2) = (\sigma_1 - \sigma_2) + 2p.$$

At points on a loaded boundary where the distribution of the loading is unknown or at points within the field of the plate, the sum of the principal stresses can be obtained by several different methods, e.g. from the isopachics. These can be determined by graphical or numerical procedures or experimentally.

We consider a plate in the xy plane in a biaxial state of stress and let σ_x, σ_y be the normal stresses in any two mutually perpendicular directions in the plane of the plate at any point. The stress σ_z normal to the plane of the plate is assumed to be zero so that, from equation (2.7c) we have for the transverse or lateral strain

$$\epsilon_z = -\frac{v}{E}(\sigma_x + \sigma_y) = \frac{\delta d}{d},$$

where δd represents the change in thickness of the plate at the point considered. The change in thickness is therefore proportional to the sum of the normal stresses. If the thickness d of the plate before loading is uniform, then lines of constant thickness in the loaded plate are identical with lines along which the sum of the normal stresses is constant, i.e. with the isopachics. The name isopachic is derived from the Greek words meaning 'constant thickness'. Measurement of the lateral strain can be accomplished using a lateral extensometer (Section 9.5.1).

Analytical methods of determining the isopachics are based either on the equations of equilibrium or the equations of compatibility or on both. The equation of compatibility in two dimensions, i.e.

$$\nabla^2(\sigma_1 + \sigma_2) = \nabla^2(\sigma_x + \sigma_y) = \nabla^2\Sigma = 0, \tag{2.32}$$

is identical in form with the equation governing the distribution of potential in a plane electric field. The well known methods used in this connection can therefore be applied to determine the isopachics.

The laws of plane potentials are readily understood when considered in terms of a practical example. A plane potential field can be induced in a thin sheet of metal by applying electric sources of different voltage to its edges. The distribution of voltage and the flow of electric current within the field of the sheet are governed by the laws of potentials. Lines within the field along which the voltage is constant are called equipotentials. An equipotential connects one point on an edge of the plate with another such point and cannot take the form of a closed loop if the sources are applied to the edges of the sheet only. Lines which are everywhere tangential to the direction of the flow of electric current are called flow lines or streamlines. Since the direction of the current flow is everywhere normal to the direction of constant voltage, it follows that the streamlines and equipotentials together form an orthogonal net. This net is also isometric. By this, we mean that, with curved equipotentials, the distance between two adjacent streamlines varies in the same way as that between adjacent equipotentials which they intersect if the parameters of successive equipotentials differ by equal amounts. If, for example, two adjacent streamlines and two adjacent equipotentials intersect to form a near or curvilinear square, the intersection of these streamlines with two other adjacent equipotentials will also form a square, Figure 9.17. If the squares are

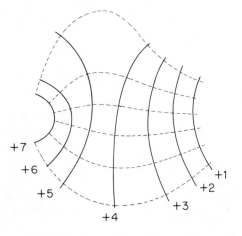

9.17. Equipotentials and streamlines

sufficiently small, their diagonals are approximately equal in length and perpendicular to each other.

If the parameters of the isopachics show more than one maximum and one minimum at the boundary of a plate, where, it will be recalled, they are identical with the parameters of the isochromatics, the equipotentials will show one or more saddle points in the interior of the field. These saddle points represent points of stagnation of the current flow. In a normal saddle point two equipotentials of the same parameter intersect at right-angles as shown in Figure 9.18. In exceptional cases such as in a plate having multiple symmetry, two or more saddle points may coincide. More than two equipotentials then intersect at the saddle point.

The above general rules allow the isopachics to be drawn by trial and error. When a system of equipotentials satisfying the aforementioned boundary conditions has been sketched in, it can be checked by means of the streamlines. After gaining some experience, it will usually be found unneces-

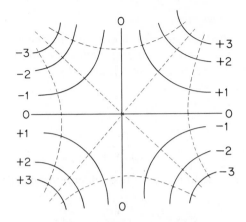

Figure 9.18.   Equipotentials and streamlines
about a saddle point

sary to draw the streamlines in every case. Obvious deviations from the isometry can be recognized from the system of isopachics alone.

Figure 9.19 shows the isopachic pattern for a short span beam with a hole, the isochromatics for which are given in Figure 9.8. These curves were plotted from the sum of the values of the normal stresses determined by the shear difference method in Section 9.1.5. Five saddle points can be distinguished in this pattern.

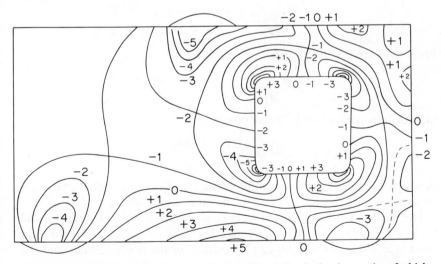

Figure 9.19.   Isopachics for a short span beam with a hole, the isochromatics of which
are shown in Figure 9.8

9.3.2   *Properties of isopachics*

Additional rules which are helpful in determining the isopachics by the above method can be derived from the equations of equilibrium. The slope of an isopachic with respect to the x axis is given by

$$
\tan \vartheta = -\frac{\partial \Sigma / \partial x}{\partial \Sigma / \partial y} = -\frac{\partial (\sigma_x + \sigma_y) / \partial x}{\partial (\sigma_x + \sigma_y) / \partial y}
$$

$$
= \frac{\partial (\sigma_x - \sigma_y) / \partial x - 2\partial \sigma_x / \partial x}{\partial (\sigma_x - \sigma_y) / \partial y + 2\partial \sigma_y / \partial y}.
\tag{9.23}
$$

From the equations of equilibrium in two dimensions, equations (1.13), we have

$$
\frac{\partial \sigma_x}{\partial x} = -\frac{\partial \tau_{xy}}{\partial y}; \qquad \frac{\partial \sigma_y}{\partial y} = -\frac{\partial \tau_{xy}}{\partial x}.
$$

Inserting these values in equation (9.23) we obtain

$$
\tan \vartheta = \frac{\partial (\sigma_x - \sigma_y) / \partial x + 2\partial \tau_{xy} / \partial y}{\partial (\sigma_x - \sigma_y) / \partial y - 2\partial \tau_{xy} / \partial x}.
\tag{9.24}
$$

Similarly, the angle ϑ' between the isopachic and the stress trajectory s_1 is given by

$$
\tan \vartheta' = -\frac{\partial \Sigma / \partial s_1}{\partial \Sigma / \partial s_2} = -\frac{\partial (\sigma_1 + \sigma_2) / \partial s_1}{\partial (\sigma_1 + \sigma_2) / \partial s_2}
$$

$$
= \frac{\partial (\sigma_1 - \sigma_2) / \partial s_1 - 2\partial \sigma_1 / \partial s_1}{\partial (\sigma_1 - \sigma_2) / \partial s_2 + 2\partial \sigma_2 / \partial s_2}.
\tag{9.25}
$$

From the Lamé–Maxwell equations (1.16) we have

$$
\frac{\partial \sigma_1}{\partial s_1} = -\frac{(\sigma_1 - \sigma_2)}{\rho_2}; \qquad \frac{\partial \sigma_2}{\partial s_2} = -\frac{(\sigma_1 - \sigma_2)}{\rho_1}.
$$

Substituting in equation (9.25) we obtain

$$
\tan \vartheta' = \frac{\partial (\sigma_1 - \sigma_2) / \partial s_1 + 2(\sigma_1 - \sigma_2) / \rho_2}{\partial (\sigma_1 - \sigma_2) / \partial s_2 - 2(\sigma_1 - \sigma_2) / \rho_1},
\tag{9.26}
$$

in which ρ_1 and ρ_2 are the radii of curvature of the orthogonal stress trajectories s_1 and s_2 which intersect at the point considered.

The angle ϑ can be evaluated from equation (9.24) after the lines τ_{xy} and $(\sigma_x - \sigma_y) = $ constant have been plotted. The values to be inserted in the right hand side of equation (9.26) can be read from the isochromatics and the isoclinics.

The parameter gradients of the isopachics in the directions of the x, y axes and in the direction normal to the isopachics can be determined from the equations

$$\partial(\sigma_x + \sigma_y)/\partial x = -\partial(\sigma_x - \sigma_y)/\partial x - 2\,\partial\tau_{xy}/\partial y,$$

$$\partial(\sigma_x + \sigma_y)/\partial y = \partial(\sigma_x - \sigma_y)/\partial y + 2\,\partial\tau_{xy}/\partial x,$$

$$\partial(\sigma_x + \sigma_y)/\partial s_3 = \frac{\partial(\sigma_x + \sigma_y)/\partial x}{\sin\vartheta} = -\frac{\partial(\sigma_x + \sigma_y)/\partial y}{\cos\vartheta},$$

respectively, where ϑ is the directional angle of the isopachic and s_3 denotes the streamline passing through the point. These equations can also be used to determine the distance between two isopachics having unit difference of parameter:

$$\Delta s_3 = \frac{\sin\vartheta}{\partial(\sigma_x + \sigma_y)/\partial x} = -\frac{\cos\vartheta}{\partial(\sigma_x + \sigma_y)/\partial y}.$$

Both equations (9.24) and (9.26) can be used to determine the directions of the isopachics at any point in a field. Their greatest value, however, is that they enable us to establish rules which are applicable in particular cases. The following rules are derived from equation (9.24):

1. At points where $\partial(\sigma_x - \sigma_y)/\partial x = 0$ and $\partial(\sigma_x - \sigma_y)/\partial y = 0$, i.e. where the system of lines of constant $(\sigma_x - \sigma_y)$ form a maximum, a minimum, or a saddle point,

$$\tan\vartheta = -\frac{\partial\tau_{xy}/\partial y}{\partial\tau_{xy}/\partial x} = -\frac{1}{\dfrac{\partial\tau_{xy}/\partial x}{\partial\tau_{xy}/\partial y}}.$$

The slope of the isopachic is thus equal to the inverse slope of the line $\tau_{xy} =$ constant, i.e. the angle between the isopachic and the x axis is equal to the angle between the line $\tau_{xy} =$ constant and the y axis. The line bisecting the co-ordinate axes therefore also bisects the angle between the isopachic and the line $\tau_{xy} =$ constant (Figure 9.20).

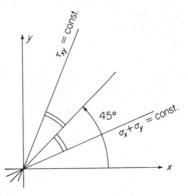

Figure 9.20. Relation between the directions of the isopachic and the line $\tau_{xy} =$ constant at a point where $\partial(\sigma_x - \sigma_y)/\partial x = \partial(\sigma_x - \sigma_y)/\partial y = 0$

2. At points where $\partial\tau_{xy}/\partial x = \partial\tau_{xy}/\partial y = 0$, i.e. where the system of lines τ_{xy} form a maximum, minimum or saddle point,

$$\tan \vartheta = \frac{\partial(\sigma_x - \sigma_y)/\partial x}{\partial(\sigma_x - \sigma_y)/\partial y}.$$

This equation shows that the isopachic and the line $(\sigma_x - \sigma_y) = $ constant are equally inclined to the x axis but lie on opposite sides of it. The x axis therefore bisects the angle between the isopachic and the line $(\sigma_x - \sigma_y) = $ constant (Figure 9.21).

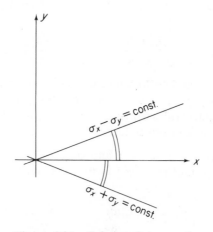

Figure 9.21. Relation between the directions of the isopachic and the line $(\sigma_x - \sigma_y) = $ constant at a point where $\partial\tau_{xy}/\partial x = \partial\tau_{xy}/\partial y = 0$

3. Where the system of isopachics form a saddle point, $\tan \vartheta = 0/0$. If the x axis is chosen to coincide with the direction of the isochromatic through the point then $\partial(\sigma_x - \sigma_y)/\partial x = \partial(\sigma_1 - \sigma_2)/\partial x = 0$. Then, as can be seen from equation (9.24), $\partial\tau_{xy}/\partial y$ must also vanish. The line $\tau_{xy} = $ constant is therefore perpendicular to the isochromatic. It was shown in Section 8.6.3 that a necessary though not sufficient condition for the existence of a saddle point is that the isoclinic and one of the stress trajectories are also perpendicular to the isochromatic. This occurs, for example, at axes of symmetry. The other condition to be satisfied is

$$\partial(\sigma_x - \sigma_y)/\partial y - 2\,\partial\tau_{xy}/\partial x = 0,$$

or

$$\partial(\sigma_1 - \sigma_2)/\partial y = 2\,\partial\tau_{xy}/\partial x.$$

This means that the slope of the lines $\tau_{xy} = $ constant parameter gradient is one half that of the isochromatics and has the same sign.

The following rules are derived from equation (9.26):

1. At a point where $\partial(\sigma_1 - \sigma_2)/\partial s_1 = \partial(\sigma_1 - \sigma_2)/\partial s_2 = 0$, i.e. where the isochromatics form a maximum, minimum or saddle point (but not an isotropic point), $\tan \vartheta' = -\rho_1/\rho_2$.

Since $1/\rho = \partial\varphi/\partial s$, this can be written

$$\tan \vartheta' = -\frac{\partial\varphi/\partial s_2}{\partial\varphi/\partial s_1} = \cot \gamma',$$

where γ' is the angle between the isoclinic $\varphi = $ constant and the direction of s_1. Thus the angle between the isoclinic and the stress trajectory s_2 is equal to the angle between the isopachic and the stress trajectory s_1. The bisector of the angle between the stress trajectories is therefore also the bisector of the angle between the isopachic and the isoclinic (see Figure 9.22).

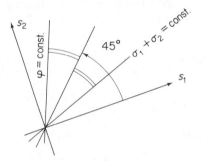

Figure 9.22. Relation between the directions of the isopachic and the isoclinic at a point where $\partial(\sigma_1 - \sigma_2)/\partial s_1 = \partial(\sigma_1 - \sigma_2)/\partial s_2 = 0$

2. At a point where $(\sigma_1 - \sigma_2)/\rho_1 = (\sigma_1 - \sigma_2)/\rho_2 = 0$ or $\partial\varphi/\partial s_1 = \partial\varphi/\partial s_2 = 0$, i.e. at a maximum, minimum or saddle point of the parameters of the isoclinics,

$$\tan \vartheta' = \frac{\partial(\sigma_1 - \sigma_2)/\partial s_1}{\partial(\sigma_1 - \sigma_2)/\partial s_2}.$$

This shows that the slopes of the isopachic and the isochromatic with respect to the stress trajectory s_1 are equal but of opposite sign. Hence one of the two stress trajectories bisects the angle between the isochromatic and the isopachic (Figure 9.23).

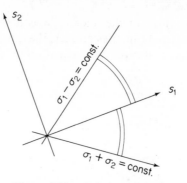

Figure 9.23. Relation between the directions of the isopachic and the isochromatic at a point where $\partial \varphi / \partial s_1 = \partial \varphi / \partial s_2 = 0$

3. Along a load-free boundary which, as we have seen, must coincide with a stress trajectory, say s_1, the stress normal to the boundary is everywhere zero. We then have $\sigma_2 = 0$ and $\partial \sigma_2 / \partial s_1 = 0$. Hence, from equations (9.25) and (9.26),

$$\tan \vartheta' = \frac{-\partial \sigma_1 / \partial s_1}{\partial(\sigma_1 - \sigma_2)/\partial s_2 - 2(\sigma_1 - \sigma_2)/\rho_1}$$

$$= \frac{-\partial(\sigma_1 - \sigma_2)/\partial s_1}{\partial(\sigma_1 - \sigma_2)/\partial s_2 - 2(\sigma_1 - \sigma_2)/\rho_1}. \tag{9.27}$$

At a straight load-free boundary, $\rho_1 = \infty$ so that equation (9.27) reduces to

$$\tan \vartheta' = -\frac{\partial(\sigma_1 - \sigma_2)/\partial s_1}{\partial(\sigma_1 - \sigma_2)/\partial s_2},$$

which shows that the directions of the isopachic and the isochromatic coincide (Figure 9.24).

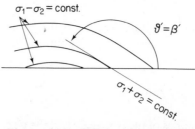

Figure 9.24. Relation between the directions of the isopachic and the isochromatic at a point on a straight load-free edge

At points on a load-free concave boundary, the stresses parallel and perpendicular to the boundary have the same sign. At points close to such a boundary, the absolute value of the sum of the principal stresses is therefore greater than that of their difference. Thus, an isopachic which originates at a point on the boundary with an isochromatic of the same order will deviate from it within the plate in the direction of isochromatics of lower order (see Figure 9.25). At points on a load-free convex boundary the reverse will apply.

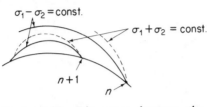

Figure 9.25. Divergence between the isopachics and the isochromatics at points on a load-free concave edge

9.3.3 Determination of the isopachics by means of the s- and t-lines

In the following method, proposed by Föppl,[16] two families of lines are used, one of which is perpendicular to the isoclinics. The points of intersection of the isopachics with the other family are determined.

The conditions of equilibrium in the directions of the stress trajectories s_1, s_2, are given by the equations

$$\frac{\partial(\sigma_1 + \sigma_2)}{\partial s_1} + \frac{\partial(\sigma_1 - \sigma_2)}{\partial s_1} + 2(\sigma_1 - \sigma_2)\frac{\partial\varphi}{\partial s_2} = 0, \qquad (9.28a)$$

$$\frac{\partial(\sigma_1 + \sigma_2)}{\partial s_2} - \frac{\partial(\sigma_1 - \sigma_2)}{\partial s_2} + 2(\sigma_1 - \sigma_2)\frac{\partial\varphi}{\partial s_1} = 0. \qquad (9.28b)$$

The angle between the isoclinic and the stress trajectory s_1 passing through a point will be denoted by γ'. Multiplying equation (9.28a) by $\sin \gamma'$ and equation (9.28b) by $\cos \gamma'$ and adding, we obtain

$$\frac{\partial(\sigma_1 + \sigma_2)}{\partial s_1}\sin \gamma' + \frac{\partial(\sigma_1 + \sigma_2)}{\partial s_2}\cos \gamma' + \frac{\partial(\sigma_1 - \sigma_2)}{\partial s_1}\sin \gamma'$$

$$-\frac{\partial(\sigma_1 - \sigma_2)}{\partial s_2}\cos \gamma' + 2(\sigma_1 - \sigma_2)\left(\frac{\partial\varphi}{\partial s_2}\sin \gamma' + \frac{\partial\varphi}{\partial s_1}\cos \gamma'\right) = 0. \qquad (9.29)$$

The last term of this equation obviously vanishes; it represents the change of the angle φ along the isoclinics which is zero.

A pair of lines is now drawn through the point, Figure 9.26. One of these lines is perpendicular to the isoclinic and thus includes an angle equal to γ' with the stress trajectory s_2. Such lines are called t-lines. The second family

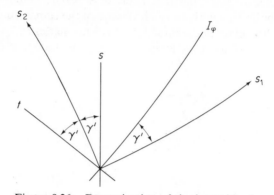

Figure 9.26.   Determination of the isopachics by
means of the s- and t-lines

of lines is the reflection of the t-lines at the stress trajectories, i.e. they are drawn at an angle equal to γ' on the other side of the stress trajectory s_2. These lines are called s-lines (not to be confused with the stress trajectories). In terms of this system, equation (9.29) reads

$$\frac{\partial(\sigma_1 + \sigma_2)}{\partial s} = \frac{\partial(\sigma_1 - \sigma_2)}{\partial t}.$$

In the application of the method, the system of s- and t-lines is first drawn. Then, beginning at a load free edge, the distance between the points of intersection of two neighbouring isochromatics on the t-line is measured. This distance is equal to that between the points of intersection of the isopachics along the s-line. While the above procedure is quite simply performed, it does not produce satisfactory results in certain cases, e.g. at a line of symmetry. This is because the s- and t-lines, both of which are in a direction perpendicular to the axis of symmetry, indicate a vanishing slope of the isochromatics and the isopachics in this direction. This difficulty can be overcome by means of a variation of the above method in which the stress trajectories s_1, s_2 are replaced by a system of co-ordinates x, y. The following system of equations, derived from equations (1.13), is applied.

$$\frac{\partial(\sigma_x + \sigma_y)}{\partial x} + \frac{\partial(\sigma_x - \sigma_y)}{\partial x} + 2\frac{\partial\tau_{xy}}{\partial y} = 0,$$

$$\frac{\partial(\sigma_x + \sigma_y)}{\partial y} - \frac{\partial(\sigma_x - \sigma_y)}{\partial y} + 2\frac{\partial\tau_{xy}}{\partial x} = 0.$$

The angle between the line τ_{xy} = constant and the x axis will be denoted by α. Multiplying the first of the above equations by sin α and the second by cos α and adding produces the result

$$\frac{\partial(\sigma_x + \sigma_y)}{\partial s'} = \frac{\partial(\sigma_x - \sigma_y)}{\partial t'},$$

where t' is the line perpendicular to the line τ_{xy} = constant (which includes an angle equal to α with the y axis) while the y axis bisects the angle between the t' and s' lines. With this procedure, the x-, y-directions can be chosen in such a way that the corresponding s'- and t'-lines are in directions which avoid the difficulty mentioned above.

9.3.4 *Determination of parameters of streamlines at the edges of a plate*

The accuracy of construction of the streamlines and hence of the isopachics can be improved if the parameters of the streamlines at the edges of a plate are known. Each streamline connects two points on the edges where this parameter is the same. These parameters can be determined by means of the isochromatics. Some constant which has the same value over the whole field must be chosen; for convenience, the parameter of the streamline at the starting point on the edge is chosen as zero.

On a straight load free edge where, it will be recalled, the directions of the isopachics and the isochromatics coincide, the points of intersection of the streamlines can be found directly from the directions of the isochromatics. Let s_1, Figure 9.27, be the stress trajectory coinciding with the edge. Neighbouring isochromatics of parameters $n + 1$ and n intersect s_1 in the points

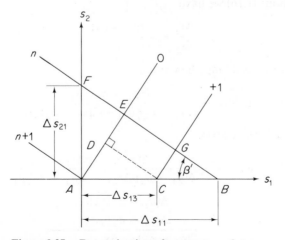

Figure 9.27.   Determination of parameters of streamlines along a straight load-free edge

A and *B*, distant Δs_{11} apart. Choosing *A* as the starting point, the parameter of the streamline intersecting the edge in this point is taken as zero. Let Δs_{13} be the distance from *A* of the required point *C* in which the streamline of parameter $+1$ intersects the edge. The streamlines through *A* and *C* intersect the isochromatic (or isopachic) through *B* in the points *E* and *G* respectively. Obviously,

$$\Delta s_{13} = \frac{CD}{\cos \beta'},$$

where β' is the inclination of the isochromatic to the edge. Since the points *A*, *E* and *G* lie at the corners of a near square, we have:

$$CD = GE = AE = \Delta s_{21} \cos \beta',$$

where $\Delta s_{21} = AF$ is the distance in the direction of the normal stress trajectory s_2 between the neighbouring isochromatics. Hence,

$$\Delta s_{13} = \Delta s_{21}.$$

At a curved load free edge, the directional angle ϑ' between the isopachic and the edge, again represented by s_1, is given by

$$\tan \vartheta' = -\frac{\partial(\sigma_1 + \sigma_2)/\partial s_1}{\partial(\sigma_1 + \sigma_2)/\partial s_2},$$

or, since here $(\sigma_1 + \sigma_2) = (\sigma_1 - \sigma_2)$,

$$\tan \vartheta' = -\frac{\partial(\sigma_1 - \sigma_2)/\partial s_1}{\partial(\sigma_1 - \sigma_2)/\partial s_2}. \tag{9.30}$$

From equations (1.16) we have

$$\frac{\partial \sigma_2}{\partial s_2} = -\frac{(\sigma_1 - \sigma_2)}{\rho_1}$$

so that equation (9.30) may be written

$$\tan \vartheta' = -\frac{\partial(\sigma_1 - \sigma_2)/\partial s_1}{\partial(\sigma_1 - \sigma_2)/\partial s_2 - 2(\sigma_1 - \sigma_2)/\rho_1},$$

or, in finite difference form,

$$\tan \vartheta' = -\frac{\Delta(\sigma_1 - \sigma_2)/\Delta s_{11}}{\Delta(\sigma_1 - \sigma_2)/\Delta s_{21} - 2(\sigma_1 - \sigma_2)/\rho_1}$$

$$= \frac{\dfrac{\Delta(\sigma_1 - \sigma_2)/\Delta s_{11}}{\Delta(\sigma_1 - \sigma_2)/\Delta s_{21}}}{\dfrac{2(\sigma_1 - \sigma_2)}{\rho_1} \dfrac{\Delta s_{21}}{\Delta(\sigma_1 - \sigma_2)} - 1}.$$

Substituting $\Delta(\sigma_1 - \sigma_2) = 1$ and $(\sigma_1 - \sigma_2) = n$, this equation reduces to

$$\tan \vartheta' = \frac{\Delta s_{21}/\Delta s_{11}}{(2n/\rho_1)\Delta s_{21} - 1} = \frac{\Delta s_{21}}{(2n/\rho_1)\Delta s_{11}\Delta s_{21} - \Delta s_{11}}. \qquad (9.31)$$

The procedure is as follows : A straight line is drawn to represent the development of the curved edge corresponding to the stress trajectory s_1, Figure 9.28.

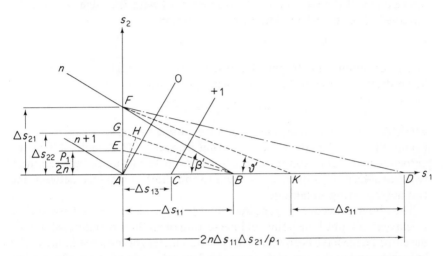

Figure 9.28.   Determination of parameters of streamlines along a curved load-free edge

On this line a length AB is marked off corresponding to the distance along the curved edge between the starting point A and the point of intersection B of the neighbouring isochromatic. This isochromatic intersects the normal stress trajectory s_2 through A in the point F. Thus,

$$AB = \Delta s_{11}, \qquad AF = \Delta s_{21}, \qquad A\widehat{B}F = \beta'.$$

A distance $AE = \rho_1/2n$ is next marked off on AF. Then obviously,

$$\tan A\widehat{B}E = \rho_1/2n\,\Delta s_{11}.$$

Through F, a line is then drawn parallel to EB to intersect s_1 in D. This gives

$$AD = AF \cot A\widehat{B}E = 2n\,\Delta s_{11}\,\Delta s_{21}/\rho_1.$$

The length $DK = \Delta s_{11}$ is now set off from D directed towards A if the edge is convex and in the opposite direction if it is concave. This produces

$$AK = 2n\,\Delta s_{11}\,\Delta s_{21}/\rho_1 - \Delta s_{11},$$

and

$$\tan A\widehat{K}F = \frac{\Delta s_{21}}{(2n\,\Delta s_{11}\,\Delta s_{21}/\rho_1) - \Delta s_{11}} = \tan \vartheta'$$

as can be seen from equation (9.31).

The required distance Δs_{13} between the starting point and the point of intersection of the streamline of parameter $+1$ with the edge can now be obtained either by calculation from the equation

$$\Delta s_{13} = \Delta s_{11} \tan \vartheta'$$

or by continuing the graphical construction. If, through B, Figure 9.28 a line is drawn parallel to FK to intersect s_2 in G, then

$$AG = \Delta s_{22} = \Delta s_{13}$$

where Δs_{22} is the distance measured in the direction of s_2 between the isopachic through A and the neighbouring isopachic. The length AH of the perpendicular from A on the line BG is the shortest distance between these two isopachics. This distance is equal to the shortest distance between the two neighbouring streamlines.

The above procedure is repeated, interpolating as necessary between the isochromatics, until the points of intersection of all the streamlines along the developed edge have been obtained. The parameters of the streamlines at the beginning and at the end of this line are identical, a fact which can be used to check the accuracy of the results. If the parameter gradient is plotted along the developed edge, the total area under the resulting curve must be zero, i.e. the positive and negative areas must be equal.

9.4 Numerical methods of determining isopachics

9.4.1 *Iteration method*

Equation 2.31 leads to the following approximate formula established by Liebmann:

$$\Sigma_0 = \frac{\Sigma_1 + \Sigma_2 + \Sigma_3 + \Sigma_4}{4} \tag{9.32}$$

where Σ_0 is the sum of the normal stresses at the centre of a square and $\Sigma_{1,2,3,4}$ are the corresponding values at the four corners. This formula can be applied to determine values of $(\sigma_1 + \sigma_2)$ within the field of a plate by an iterative process.

A coarse network of large squares is first drawn over the field of the plate, Figure 9.29a. Initial values are then assigned to the corners of the squares. Where the latter coincide with a free boundary, the actual values indicated by

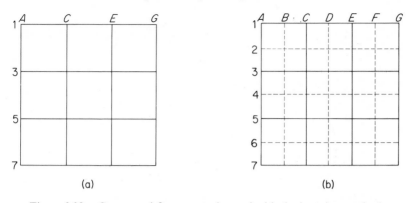

Figure 9.29.   Coarse and finer networks used with the iteration method

the orders of the isochromatics are used; these and any other known values remain unchanged throughout the procedure while the values at all other points are progressively improved. The initial values at points where they are unknown can be chosen arbitrarily since any set of such values will eventually lead to the correct solution. The labour and time involved can be substantially reduced, however, by estimating values which will approximate to the final ones.

In the next step, improved values of Σ at the nodes are determined by applying equation (9.32) to each in turn. For example, an improved value for Σ_{C3} is given by the mean of the values at the points $C1$, $E3$, $C5$, $A3$ (or $A1$, $E1$, $E5$, $A5$). When the values at all the nodes have been improved once, the cycle is repeated in the same order and this process is continued until the differences between the results of successive cycles are within the limits of error of the known boundary values. The limit of accuracy possible with squares of this size has now been reached. A more accurate solution within the field may now be obtained by subdividing the squares into a finer network, Figure 9.29b. The final values of the initial coarse network become the initial values of the finer network. These will be further improved as the iteration process is applied to determine the values of $(\sigma_1 + \sigma_2)$ at the additional nodes introduced by the finer net.

As mentioned previously, the rapidity with which a satisfactory solution is obtained depends on the accuracy of the initial values chosen at points where they are unknown. A first approximation to the value of Σ at such a point can be determined by linearly interpolating between the known values at two points on the boundary lying on a straight line passing through the point. Thus, considering only the line 1–2 in Figure 9.30, an approximate value of Σ at the point O is

$$\Sigma_0 = \frac{1}{a_1 + a_2}(a_2\Sigma_1 + a_1\Sigma_2).$$

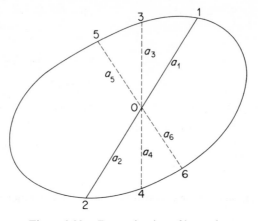

Figure 9.30. Determination of key values

A better value can be obtained by evaluating Σ_0 in the above manner as the line is rotated in turn into different positions 3–4, 5–6, etc. The average of all these values is a close approximation to the correct value. This procedure need be applied to only a few specially selected or key points such, for example, as point $C3$ in Figure 9.29. By applying equation (9.32) in turn to each internal point and four other neighbouring points disposed either diagonally or normally about the first, the value of Σ_{C3} will lead to approximate values at all other points of the first set of squares.

If the corners of a square do not coincide with the edge as will occur, for example, at a curved boundary, equation (9.32) is replaced by

$$\Sigma_0 = \frac{st}{s+t}\left[\frac{\Sigma_1}{t(1+t)} + \frac{\Sigma_2}{s(1+s)} + \frac{\Sigma_3}{(1+t)} + \frac{\Sigma_4}{(1+s)}\right], \qquad (9.33)$$

where s and t have the meaning indicated in Figure 9.31. Equation (9.33) is known as the four point influence equation.

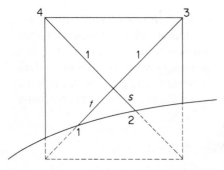

Figure 9.31. Notation for the application of Liebmann's equation to an incomplete square

A simple example of the iteration procedure is given in Figure 9.32 where the numbers at the points of intersection of the network with the boundary

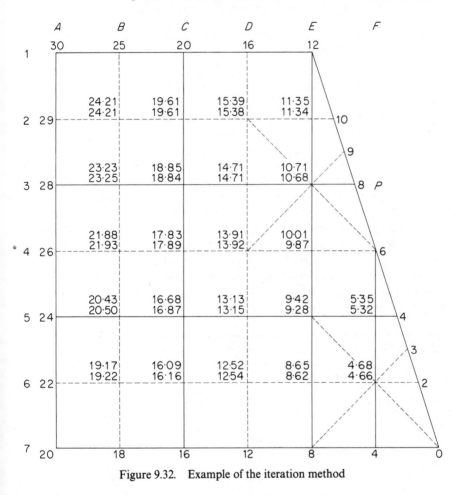

Figure 9.32. Example of the iteration method

are taken to represent the edge values of Σ as determined from the iso-chromatics. A key value for the point $E5$ is first calculated from the values at the normally and diagonally disposed boundary points using equation (9.33):

$$\Sigma_{E5} = \frac{1}{4}\left[\frac{1}{1\cdot5}(\tfrac{1}{2} \times 16 + 1 \times 6) + \tfrac{1}{3}(1 \times 30 + 2 \times 0)\right.$$

$$\left. + \tfrac{1}{3}(1 \times 12 + 2 \times 8) + \frac{1}{2\cdot67}(2 \times 4 + \tfrac{2}{3} \times 24)\right] = 9\cdot42.$$

An initial value for the point $C3$ is given by the mean of the values at the points $A1, E1, E5, A5$:

$$\Sigma_{C3} = \tfrac{1}{4}(30 + 12 + 9\cdot42 + 24) = 18\cdot85.$$

An initial value for the point $E3$ can then be calculated from the values at the points $P, E5, C3, E1$ using equation (9.33):

$$\Sigma_{E3} = \frac{\tfrac{1}{3}}{1 + \tfrac{1}{3}}\left[\frac{8}{\tfrac{1}{3}(1 + \tfrac{1}{3})} + \frac{9\cdot42}{2} + \frac{18\cdot85}{1 + \tfrac{1}{3}} + \frac{12}{2}\right] = 10\cdot71.$$

An initial value for the point $C5$ may then be calculated from the values at the points $A3, E3, E7, A7$:

$$\Sigma_{C5} = \tfrac{1}{4}[28 + 10\cdot71 + 8 + 20] = 16\cdot68.$$

Using the approximate values of Σ at the nodes of the coarse network, first approximations to the values at the nodes of the finer network can now be calculated. The values at the points $B2, D2, B4$, etc., are determined from the values at diagonally disposed points and those at the points $C2, E2, B3$, etc., from the values at normally disposed points using equation (9.32) for complete squares and equation (9.33) for incomplete squares.

After the initial values have been determined at all the nodes, they are improved by repeating the whole cycle. In the given example, this produces only minor changes in the values. For comparison, Table 9.3 gives the results

Table 9.3   Values of $(\sigma_1 + \sigma_2)$ for the problem of Figure 9.32 obtained from solution by computer using (a) 4 nodes, (b) 22 nodes.

row	nodes	B	C	column D	E	F
2	22	24·22	19·64	15·43	11·35	
3	4		18·94		10·71	
3	22	23·26	18·89	14·74	10·68	
4	22	21·92	17·92	13·94	9·97	
5	4		17·05		9·28	
5	22	20·50	16·93	13·15	9·26	5·32
6	22	19·16	16·14	12·47	8·59	4·65

(corrected to the second decimal place) when the same problem was solved by means of an electronic computer. Here, the 4 and 22 equations of the types (9.32) or (9.33) relating to the nodes of the coarse and fine networks respectively were solved simultaneously. It can be seen that the values obtained by the iteration procedure nowhere differ by more than about 0·5 % from those of the computer solution. The two sets of values obtained by the computer indicate that only a minor increase in accuracy is obtained when the coarse net containing 4 nodes is replaced by the fine net with 22 nodes.

9.4.2 Influence method

Another method of determining the sum of the normal stresses at a point within the field of a plate from known values at the edges or on some other closed curve within the plate has been suggested by F. Vadovic.[17] This method is based on the fact that the influence of an edge value on the value of Σ_0 at an internal point depends on the reciprocal of the distance between the two points. The value of Σ_0 is calculated from

$$\Sigma_0 = \frac{\dfrac{\Sigma_1/a_1 + \Sigma_2/a_2}{1/a_1 + 1/a_2}\dfrac{1}{a_2 a_2} + \dfrac{\Sigma_3/a_3 + \Sigma_4/a_4}{1/a_3 + a/a_4}\dfrac{1}{a_3 a_4} + \cdots}{\dfrac{1}{a_1 a_2} + \dfrac{1}{a_3 a_4} + \cdots},$$

in which the terms have the meaning indicated in Figure 9.30.

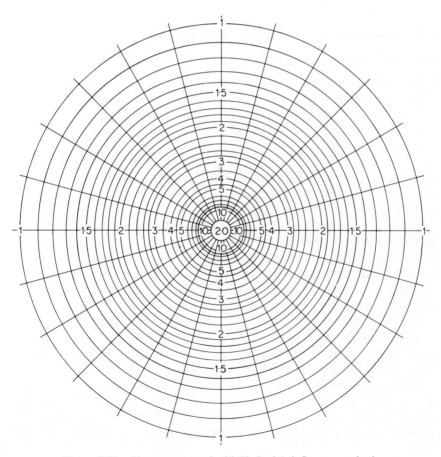

Figure 9.33. Nomogram used with Vadovic's influence method

The application of the method is simplified by using a nomogram of the form shown in Figure 9.33 consisting of a system of concentric circles and radial lines. The numbers relating to the circles are proportional to the reciprocals of their radii. The nomogram is reproduced on transparent paper and placed over a drawing of the profile of the plate in such a way that its centre coincides with the point of interest. Since Σ_0 is independent of the unit of a, the factor $1/a$ relating to each point on the boundary can be obtained by reading from the nomogram the number of the circle passing through the point.

The points from which the values are taken should be distributed along the edge so that they include equal angles at the point of interest; in general, those lying on the radial lines of the nomogram will be sufficient. If greater accuracy is required, additional points can be chosen, e.g. midway between these lines.

9.5   Experimental methods of determining isopachics

9.5.1   *Lateral extensometers*

As shown in Section 9.3.1, the sum of the principal stresses at any point of a plate in a two dimensional state of stress is proportional to the lateral strain at the point. This may be determined by direct measurement of the change of thickness of the plate by means of a lateral extensometer. Since the changes of thickness involved are extremely small, such instruments must be very sensitive. In a plate of 10 mm thickness, the maximum change of thickness will generally lie within the range 0·001 to 0·01 mm. In order to obtain accurate results, the extensometer must be capable of measuring to within a small fraction of such changes.

One of the simplest and most successful forms of lateral extensometer is that developed by Hiltscher,[18] shown in Figure 9.34. This consists of a U-shaped frame carrying a precision measuring instrument on one side and a centre which can be adjusted by a micrometer screw at the other. The dial indicator has a range of 20×10^{-4} mm which is divided into 200 divisions and allows readings to be estimated to 10^{-5} mm.

The extensometer is suspended by a system of chords in such a way that its axis is normal to the plate while the measuring points coincide with the point of interest. The weight of the extensometer is balanced by means of a counterweight suspended over a free running pulley.

The system of isopachics can be determined from point by point measurement over the field of the plate of the thickness in the unloaded and loaded states. Alternatively, a lateral extensometer may be used to determine key values of $(\sigma_1 + \sigma_2)$ to expedite the solution by numerical methods (Section 9.4.1).

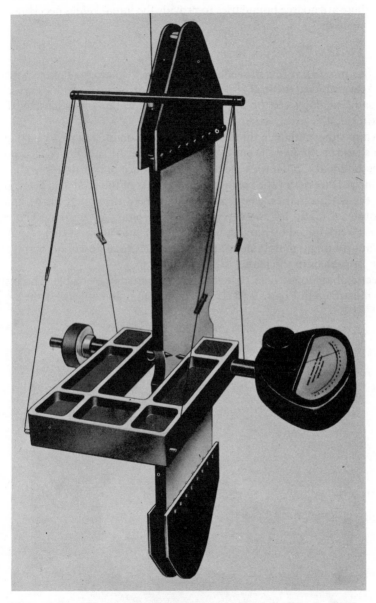

Figure 9.34.   Hiltscher lateral extensometer. Courtesy Sharples Photomechanics Ltd

The lateral extensometer method has the advantage that it can be applied when there are no free boundaries to provide the initial values necessary with other methods.

9.5.2 *Electrical analogy method*

In Section 9.3.1 it was shown how the electrical analogy of plane potential leads to a graphical method for the determination of the isopachics in a plate. The same analogy can be applied to determine the isopachics by direct experiment.

To apply the method, a thin sheet of metal or conducting paper having the same shape as the plate is cut to a suitable scale. At suitably spaced points on the boundary, potentials are applied which are proportional to the sum of principal stresses at the corresponding points of the model as determined from the isochromatics. The circuit is completed through a probe and incorporates a meter for measurement of the voltage or current. The equipotentials are traced by moving the probe over the field in such a way that a constant reading is maintained at the meter. These equipotentials correspond to the isopachics in the plate.

Electrical analogue field plotters are commercially available. The instrument illustrated in Figure 9.35 is a 40 channel unit and allows for the supply

Figure 9.35.   Sharples–Fylde 40 channel analogue field plotter.
Courtesy Sharples Photomechanics Ltd

voltage at each electrode on the boundary of the model to be individually controlled by a rheostat.

9.5.3 Membrane analogy method

This method is based on the identity of form of equation (2.32) with that of the lateral ordinate of a membrane loaded by forces acting at its edges. A frame is erected vertically such that its contour is geometrically similar to that of the plate while its height at any point is proportional to the sum of the normal stresses at the corresponding point of the edge of the plate. A thin rubber membrane or soap film is then stretched over the upper end of the frame. Lines of constant height of the membrane correspond to the isopachics in the plate.

Since this analogy is valid only when the slope of the membrane is small, care must be taken to ensure that significant errors are not introduced in regions of high stress concentration.

9.5.4 Interferometric methods

Optical methods of determining the isopachic patterns in plates by means of interferometers are described in Chapter 10.

10

INTERFEROMETRIC METHODS

10.1 Introduction

Optical data additional to those measured in the polariscope can be obtained experimentally by methods based on the phenomenon of interference.

Some of these methods provide the sum of the principal stresses $(\sigma_1 + \sigma_2)$ which, together with the difference $(\sigma_1 - \sigma_2)$ derived from the isochromatics, allows the separation of the principal stresses. Whole field methods of this type depend on the production of an interference pattern in which the fringes correspond to the isopachics. With models which are initially plane parallel in the optical sense, interference patterns of this type can be obtained directly by means of an interferometer. With models which are not plane parallel, the required interference patterns can be obtained by applying the moiré method. Moiré patterns are produced by the superposition of the interference patterns before and after loading the model. With most procedures, a clear moiré pattern is obtained only when the birefringence of the model is negligible, otherwise two systems of fringes appear. This means that it is often necessary to manufacture a second model from a material of low optical sensitivity. By one of the methods which will be described, however, the isochromatic and isopachic fringe patterns are obtained simultaneously.

By other methods, the principal stresses are determined individually. Such methods involve point by point measurement of the optical path differences produced by the stresses with an interferometer, or of the corresponding variations of intensity of the light with a photomultiplier. With classical interferometers such as the Mach–Zehnder and Michelson types, the path differences are obtained by comparing two optical paths, one of which includes the model. In photoelasticity, simple optical arrangements can often be used although these produce less perfect interference phenomena than the more delicate and expensive instruments.

10.2 Optical basis of photoelastic interferometry

Let us consider the effects produced when a birefringent plate is inserted in the path of a beam of light and let L denote the geometrical distance between two reference points in air on opposite sides of the plate. The index of refraction for air may be taken as unity so that, in the absence of the model, the optical path between the reference points is equal to the length L.

If an unstressed plate of refractive index n_0 and uniform thickness d is placed normally in the path of the beam, Figure 10.1a, the length of the optical path between the reference points becomes

$$(L - d) + n_0 d = L + (n_0 - 1)d.$$

The absolute path difference or retardation produced by the plate is therefore equal to $(n_0 - 1)d$.

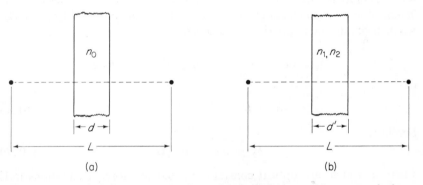

Figure 10.1. Birefringent plate in the path of a light beam (a) unstressed plate, (b) stressed plate

If the plate is now stressed in its own plane, both the index of refraction and the thickness are altered. Let n_1, n_2 denote the refractive indices of the material for light vibrating parallel to the principal stresses σ_1, σ_2 respectively and let the thickness of the stressed plate be d', Figure 10.1(b). The change of thickness is:

$$\Delta d = -\frac{v}{E}(\sigma_1 + \sigma_2)d, \tag{10.1}$$

so that

$$d' = \left[1 - \frac{v}{E}(\sigma_1 + \sigma_2)\right]d.$$

The total optical paths of the components of light vibrating parallel to σ_1, σ_2 are then

$$L + (n_{1,2} - 1)d',$$

so that the absolute retardations produced by the stressed plate are

$$R_{1,2} = (n_{1,2} - 1)d'.$$

The net variations of the absolute retardations due to the application of the stresses are therefore

$$\delta_{1,2} = (n_{1,2} - 1)d' - (n_0 - 1)d$$

$$= (n_{1,2} - n_0)d + (n_{1,2} - 1)\Delta d. \tag{10.2}$$

The retardations $\delta_{1,2}$ can be expressed in terms of the stresses. For example, from equation (10.1),

$$\delta_1 = (n_1 - n_0)d + (n_1 - n_0)\Delta d + (n_0 - 1)\Delta d.$$

The magnitude of the second term on the right hand side of this equation is small compared with that of the others and may be neglected. Inserting the values of $(n_1 - n_0)$ and Δd given by equations (5.1a) and (10.1) respectively, we then obtain the approximate equation:

$$\delta_1 = (a\sigma_1 + b\sigma_2)d - (n_0 - 1)\frac{v}{E}(\sigma_1 + \sigma_2)d,$$

or

$$\delta_1 = (a'\sigma_1 + b'\sigma_2)d. \tag{10.3a}$$

Similarly

$$\delta_2 = (a'\sigma_2 + b'\sigma_1)d, \tag{10.3b}$$

where a' and b' are optical constants related to those of equations (5.1) through

$$a' = a - \frac{(n_0 - 1)v}{E} \tag{10.4a}$$

$$b' = b - \frac{(n_0 - 1)v}{E}. \tag{10.4b}$$

From equations (10.3) we obtain

$$\delta_1 + \delta_2 = (a' + b')(\sigma_1 + \sigma_2)d, \tag{10.5a}$$

$$\delta_1 - \delta_2 = (a' - b')(\sigma_1 - \sigma_2)d. \tag{10.5b}$$

Solving equations (10.5) for the stresses produces

$$\sigma_1 = \frac{a'\delta_1 - b'\delta_2}{(a'^2 - b'^2)d};$$

$$\sigma_2 = \frac{a'\delta_2 - b'\delta_1}{(a'^2 - b'^2)d}.$$

In the usual photoelastic procedures, the relative retardation $R = \delta_1 - \delta_2$ is measured. Comparing equations (10.4), (10.5) and (5.4) it can be seen that the stress optic coefficient C is related to the above optical constants through

$$C = (a' - b') = (a - b) \qquad (10.6)$$

with the approximation mentioned above.

If the birefringence of the material is negligible so that $n_{1,2} \simeq n_0$, equation (10.2) reduces to

$$\delta = (n_0 - 1)\Delta d = -\frac{v}{E}(n_0 - 1)(\sigma_1 + \sigma_2)d,$$

i.e. the absolute retardation is directly proportional to the sum of the principal stresses.

10.3 Practical application

Since interference phenomena are best observed when the light has a high degree of coherence, small retardations are usually desirable when a normal lamp is used as the source. Certain methods however, such as those based on the production of fringes of superposition (Section 3.27), may be applied with retardations up to about 10 mm. If a laser is used as the source, much greater retardations are permissible since the coherence of the light is much greater than that of ordinary lamps.

10.4 Application of the Mach–Zehnder interferometer

One arrangement allowing point-by-point determination of absolute retardations using the Mach–Zehnder interferometer is shown in Figure 10.2. After passing through the colour filter F and rotatable polarizer P, light from

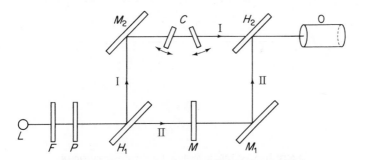

Figure 10.2. Optical system for the observation of interference patterns using the Mach–Zehnder interferometer

the lamp L is split into two beams which are directed round opposite sides of a rectangle and finally partially recombined by the system of half mirrors H_1, H_2 and mirrors M_1, M_2. The model M is inserted in the path of one of these beams while the other includes a compensator C. This compensator consists of a pair of glass plates which can be tilted so that the optical paths within them can be adjusted. By this means the difference of the optical paths of the two beams can be reduced to zero, e.g. before loading the model.

The variations of the absolute retardations produced when the model is loaded can be determined either by counting the number of fringes passing a point of the interference pattern observed through the optical system O or from the further adjustments of the compensator required to reduce the optical path differences to zero. Measurements made after rotating the polarizer P until its axis is parallel to each of the principal stresses σ_1, σ_2 in turn yield the corresponding values of δ_1, δ_2. The results may be checked by measuring the relative retardation $R = \delta_1 - \delta_2$ observed when the model is placed between a polarizer and an analyser. For this purpose, an analyser can be inserted between M and M_1.

10.5 Application of the Michelson interferometer

A whole field fringe pattern of the moiré type representing lines $(\sigma_1 + \sigma_2) =$ constant can be obtained by a procedure employing the Michelson interferometer, Figure 10.3.

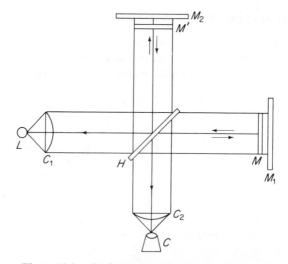

Figure 10.3.   Optical system for the observation of interference patterns using a Michelson interferometer.

The divergent light from the lamp L is made parallel by the condenser lens C_1 and is split by the half mirror H. Part of the light is transmitted through H and the model M is reflected by the mirror M_1 and after passing through the model a second time, returns to H where it is partially reflected. The other part of the light, which is reflected by H, passes through a transparent plate M' having the same thickness as M. This light is reflected by the mirror M_2 and, after passing through M' again, is partially transmitted through H. Behind H the two parts are united and interfere. This interference system is photographed by means of the camera C after passing through the field lens C_2 where the parallel light is made convergent. The film in the camera is exposed twice, once before loading the model and a second time after loading. The superposition of the two images produces the moiré pattern.

A clear moiré pattern is obtained by this method only when the birefringence of the model is negligible, otherwise two systems of fringes appear. Some improvement can be effected by placing a polarizer between C and H with its axis parallel to one of the principal stresses. The difficulty cannot be fully overcome by this means, however, since the light beam will be split in two whatever the direction of polarization may be.

10.6 Application of the series interferometer

In a method developed by D. Post,[19] the isopachic pattern is obtained optically using a series interferometric arrangement based on the production of fringes of superposition (Section 3.27). The model M (Figure 10.4) is placed between two of a system of three partial mirrors. In practice these mirrors are formed by the surfaces of transparent plates, one of which forms

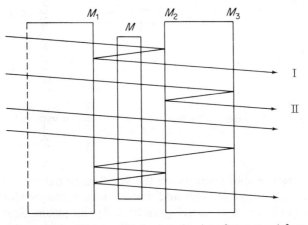

Figure 10.4. Light paths in a series interferometer (after D. Post)

a double mirror; for improved definition of the interference pattern, the reflectivity is increased by the deposition of thin metallic films on the surfaces (Section 3.25).

Interference occurs between rays which have approximately the same lengths of optical path. If the optical path between the mirrors M_1 and M_2 (including that within the model M) is approximately equal to that between M_2 and M_3, interference will occur between rays which follow geometrically similar paths through the system. For example, rays such as I which have suffered one double reflection between M_1 and M_2 but have passed without reflection through M_3 will interfere with rays such as II which have passed M_1 without reflection but have suffered a double reflection between M_2 and M_3.

Interference will also occur between rays which have taken mutually similar paths involving multiple reflections. These, together with light which has passed through the system without being reflected, superimpose an

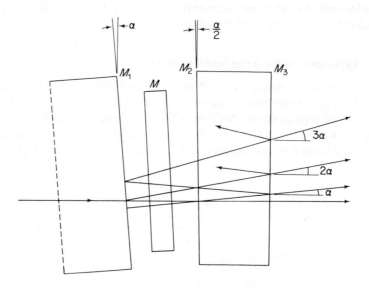

Figure 10.5. Light paths in a series interferometer with inclined mirror system (after D. Post)

undesirable background intensity on the interference pattern produced by rays of the types I and II. This additional intensity can be eliminated by using an inclined mirror system (Figure 10.5). By this means, light following paths of different lengths is projected in different directions. The camera is then placed to intercept the desired rays.

10.7 Composite interference patterns

In a method used by Nisida and Saito,[20,21,22] employing a Mach–Zehnder interferometer, a composite interference pattern of isopachic and isochromatic fringes is obtained from which both the sum and the difference of the principal stresses over the whole field in two-dimensional problems can be determined. The arrangement is similar to that shown in Figure 10.2 except that the polarizer P is omitted. On entering the stressed model the ordinary light following path II is split into two polarized components of equal amplitude vibrating in the directions of the principal stresses σ_1, σ_2. After partial reflection at H_2, each of these components interferes with the corresponding component of the same amplitude of the light which has followed path I. If the optical paths in the interferometer are made equal before loading the model, the path differences produced by the stresses between light following path II and vibrating parallel to σ_1 and light following path I is

$$\delta_1 = (n_1 - 1)d' - (n_0 - 1)d.$$

The interfering components may therefore be represented by

$$y_{\mathrm{I}} = A \sin\left(\omega t + \frac{2\pi}{\lambda}\delta_1\right),$$

$$y_{\mathrm{II}} = A \sin \omega t,$$

respectively. The resultant wave is:

$$y = y_I + y_{II} = A\left[\left(1 + \cos\frac{2\pi}{\lambda}\delta_1\right)\sin \omega t + \sin\frac{2\pi}{\lambda}\delta_1 \cos \omega t\right],$$

of amplitude

$$A\sqrt{\left[\left(1 + \cos\frac{2\pi}{\lambda}\delta_1\right)^2 + \sin^2\frac{2\pi}{\lambda}\delta_1\right]} = A\sqrt{\left[2\left(1 + \cos\frac{2\pi}{\lambda}\delta_1\right)\right]},$$

and intensity

$$I_1 = kA^2\left(1 + \cos\frac{2\pi}{\lambda}\delta_1\right).$$

Similarly, the intensity of the other wave resulting from the interference of light vibrating parallel to σ_2 and light following path I is

$$I_2 = kA^2\left(1 + \cos\frac{2\pi}{\lambda}\delta_2\right).$$

The intensities of these two waves are superimposed so that the combined intensity I is given by

$$I = I_1 + I_2 = kA^2 \left[2 + \cos \frac{2\pi}{\lambda} \delta_1 + \cos \frac{2\pi}{\lambda} \delta_2 \right]$$

$$= \frac{I_0}{2} \left[1 + \cos \frac{2\pi}{\lambda} \frac{(\delta_1 + \delta_2)}{2} \cos \frac{2\pi}{\lambda} \frac{(\delta_1 - \delta_2)}{2} \right] \quad (10.7)$$

where $I_0 = 2kA^2$. Substituting for the quantities $(\delta_1 + \delta_2)$ and $(\delta_1 - \delta_2)$ in terms of the stresses as defined by equations (10.5) and observing equation (10.6), equation (10.7) becomes

$$I = \frac{I_0}{2} \left[1 + \cos \left\{ \frac{2\pi}{\lambda} \frac{(a' + b')}{2} (\sigma_1 + \sigma_2)d \right\} \cos \left\{ \frac{2\pi}{\lambda} \frac{C}{2} (\sigma_1 - \sigma_2)d \right\} \right]. \quad (10.8)$$

Equation (10.8) shows that the distribution of light intensity in the interference pattern depends on both the sum and the difference of the principal stresses. At a point where

$$\frac{2\pi}{\lambda} C(\sigma_1 - \sigma_2)d = 2m\pi,$$

corresponding to a dark point on the normal isochromatic pattern,

$$I = \frac{I_0}{2} \left[1 + \cos \left\{ \frac{2\pi}{\lambda} \frac{(a' + b')}{2} (\sigma_1 + \sigma_2)d \right\} \right].$$

The interference pattern consists of both isopachics and isochromatics with the isochromatics appearing in half tone. The isopachics change from lines of maximum darkness to lines of maximum brightness or the reverse each time they cross an isochromatic. Interference patterns of this type are shown in Figures 10.8 to 10.13.

With the arrangement described above, appreciable errors may be introduced in regions where the deformation of the model is severe. These can be eliminated by placing the model in an immersion cell containing a liquid of matching refractive index. The differences of optical paths in the interferometer produced by the stresses are then

$$\delta_{1,2} = (n_{1,2} - n_0)d'.$$

It can be shown that the intensity at any point of the interference pattern is now given by

$$I = \frac{I_0}{2} \left[1 - \cos \left\{ \frac{2\pi}{\lambda} \frac{(a + b)}{2} (\sigma_1 + \sigma_2)d \right\} \cos \left\{ \frac{2\pi}{\lambda} \frac{C}{2} (\sigma_1 - \sigma_2)d \right\} \right],$$

in which the optical constants a, b are those defined by equations (5.1).

With the immersion technique, the sensitivity of the method with respect to the number of isopachic fringes produced is substantially less than when the model is in air.

10.8 Photometric interferometric method

In a method developed by Favre and Schumann,[23] all of the information necessary to determine the individual values of the principal stresses at a point in a two dimensional state is determined optically without the use of a specialized interferometer. The method depends on the interference of light rays multiply reflected at the two quasi plane parallel faces of the model. By measuring the intensity of the resultant reflected waves vibrating parallel to the directions of the two principal stresses, the variations of absolute retardation and hence the values of the stresses can be determined.

As shown in Section 3.22 only the first two reflections contribute significantly towards the total intensity of reflected light. The intensities of these reflections, given by equations (3.34) and (3.37), are β and $(1 - \beta)^2\beta$, Figure 10.6. The corresponding amplitudes are $\sqrt{\beta}$ and $(1 - \beta)\sqrt{\beta}$. Due to reflection,

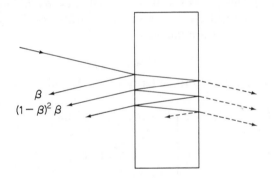

Figure 10.6. Multiple reflection of light at the surfaces of a quasi plane parallel model

the phase of the first reflected ray differs by π from that of the incident ray. Since the second ray is produced after reflection in the medium of higher refractive index there is no change of phase due to reflection. As shown in Section 3.22 however there is a change of phase of $4\pi nd/\lambda$ due to the difference in length of the optical paths. Considering the first two reflections only, the resultant vibration parallel to the direction of the principal stress σ_1 can be expressed by

$$y = \sqrt{(\beta)} \cos(\omega t + \pi) + (1 - \beta)\sqrt{(\beta)} \cos\left(\omega t + \frac{4\pi n_1 d'}{\lambda}\right), \qquad (10.9)$$

where n_1 is the refractive index for waves vibrating parallel to σ_1 and d' is the thickness of the stressed plate.

Equation (10.9) may be expressed in the usual way in the form $y = C \cos(\omega t + D)$. The intensity $J_1 = 2C^2$ is found to be

$$J_1 = 2\beta - \beta^3 - 2\beta(1 - \beta)\cos\frac{4\pi n_1 d'}{\lambda}.$$

Since β is a small quantity (approximately 0·05 for $n = 1·6$), this approximates to

$$J_1 = 2\beta\left[1 - \cos\frac{4\pi n_1 d'}{\lambda}\right]. \tag{10.10a}$$

Similarly, the intensity of components vibrating parallel to σ_2 is

$$J_2 = 2\beta\left[1 - \cos\frac{4\pi n_2 d'}{\lambda}\right]. \tag{10.10b}$$

The optical path difference between the first and second reflections in the unstressed plate is $2n_0 d$ and will be denoted by R_0. The corresponding path differences R_1, R_2 for waves vibrating parallel to σ_1, σ_2 in the stressed plate are $2n_1 d'$, $2n_2 d'$ respectively. The absolute retardations δ_1, δ_2 between the faces due to the stresses are therefore

$$\delta_{1,2} = 2(n_{1,2} d' - n_0 d).$$

Expressing equations (10.10) in terms of R_0, δ_1 and δ_2 produces

$$J_1 = 2\beta\left[1 - \cos\frac{2\pi}{\lambda}(\delta_1 + R_0)\right], \tag{10.11a}$$

$$J_2 = 2\beta\left[1 - \cos\frac{2\pi}{\lambda}(\delta_2 + R_0)\right]. \tag{10.11b}$$

As shown by Favre,[24] the absolute retardations δ_1, δ_2 are related to the stresses through

$$\delta_1 = (a''\sigma_1 + b''\sigma_2)d, \tag{10.12a}$$

$$\delta_2 = (a''\sigma_2 + b''\sigma_1)d, \tag{10.12b}$$

where a'', b'' are optical constants. These constants are related to those of equations (5.1) and (10.3) through the approximate equations:

$$a'' = 2\left(a - \frac{n_0 v}{E}\right) = 2\left(a' - \frac{v}{E}\right), \tag{10.13a}$$

$$b'' = 2\left(b - \frac{n_0 v}{E}\right) = 2\left(b' - \frac{v}{E}\right). \tag{10.13b}$$

Solving equations (10.12) for the stresses yields

$$\sigma_1 = \frac{a''\delta_1 - b''\delta_2}{(a''^2 - b''^2)d},$$ (10.14a)

$$\sigma_2 = \frac{a''\delta_2 - b''\delta_1}{(a''^2 - b''^2)d}.$$ (10.14b)

These equations may be used to evaluate the individual principal stresses. Since the difference between a'' and b'' is small compared with their absolute values, however, the accuracy of the results is doubtful. Favre and Schumann point out that the accuracy can be greatly improved if the relative retardation in transmitted light, which is capable of fairly precise measurement, is also determined. The relative retardation R is given by

$$R = C(\sigma_1 - \sigma_2)d = (\delta_1 - \delta_2)/2,$$ (10.15)

where $C = (a'' - b'')/2$.

Adding equations (10.14) and combining with equation (10.15) produces

$$\sigma_1, \sigma_2 = \frac{\delta_1 + \delta_2}{2(a'' + b'')d} \pm \frac{R}{2Cd},$$

which may be expressed in the form

$$\sigma_1, \sigma_2 = \frac{1}{2Cd}(\gamma R_m \pm R),$$ (10.16)

where $\gamma = 2C/(a'' + b'') = 1/(1 + b''/C)$ and $R_m = \frac{1}{2}(\delta_1 + \delta_2)$ is the mean absolute retardation.

The basic optical system is represented in Figure 10.7. Light from the source falls on the half mirror H where it is partially reflected to pass through the polarizer P on to the model M. That part of the light which is reflected at the

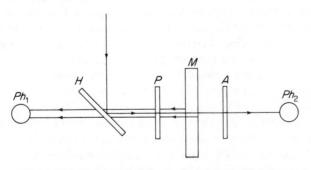

Figure 10.7. Schematic of optical system for the determination of absolute and relative retardations (after H. Favre and W. Schumann)

front and rear surfaces of the model returns through P to H where it is partially transmitted to the photomultiplier Ph_1. The light which is transmitted through the model passes through the analyser A to the photomultiplier Ph_2.

The variations of voltage resulting from the changes of intensity of the light entering the photomultipliers as the load is applied to the model are displayed on the screen of an oscilloscope against the load as represented by the potential difference over a resistance strain gauge attached to a calibration specimen loaded together with the model. The absolute retardations R_1, R_2 are determined from the oscillograms representing the intensities J_1, J_2 of the reflected light measured by the photomultiplier Ph_1 with the polarizer P set parallel to each of the principal stresses σ_1, σ_2 in turn. The relative retardation R is determined from the intensity of the transmitted light detected by Ph_2 with the polarizers set at 45° to the directions of the principal stresses when, from equation (5.14),

$$I = I_0 \sin^2 \frac{\alpha}{2} = I_0 \sin^2 \frac{\pi R}{\lambda}.$$

Other systems used by Schumann and his associates of a more elaborate nature intended primarily for the determination of dynamic states of stress by the photometric-interferometric method are described in Section 19.12.

10.9   Examples of interferometric method

The authors are indebted to M. Nisida for the following examples illustrating composite interference patterns or interfero-stress patterns consisting of both isopachics and isochromatics obtained by means of a Mach–Zehnder interferometer.

The first example illustrates the application of the method to a centrally loaded beam of short span. Figure 10.8 shows the interfero-stress pattern for a model prepared from a plane parallel plate. For comparison, the isochromatic pattern for the same stress distribution obtained with a bright background in an ordinary circular polariscope is shown in Figure 10.9. Figure 10.10 shows the corresponding interference pattern of a similar model which was not truly plane parallel obtained by the moiré method.

The second example given involved the determination of the stress distribution in the vicinity of a hole in a plate. The plate was loaded in its own plane by a pin passing through the hole in an annular disc fitted in the hole in the plate. The load was applied through stirrups in the direction normal to one of the straight edges while the plate was supported by a row of small pins along the opposite edge. Figures 10.11 and 10.12 show the interfero-stress pattern and bright field isochromatic pattern respectively for the case when the hole was sufficiently remote from the edges that the plate

Figure 10.8. Interfero-stress pattern of a plane parallel beam with
a central load. Courtesy Nisida and Saito

Figure 10.9. Bright field isochromatic pattern of the beam shown in Figure
10.8. Courtesy Nisida and Saito

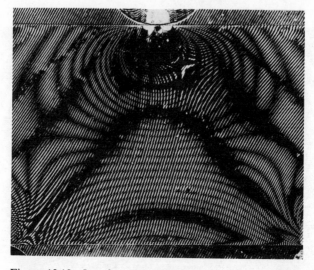

Figure 10.10   Interfero-stress pattern of a centrally loaded beam obtained by the moiré method. Courtesy Nisida and Saito

Figure 10.11.   Interfero-stress pattern of an infinite plate loaded by a circular pin. Courtesy Nisida and Saito

Figure 10.12. Isochromatic pattern of the plate shown in Figure 10.11. Courtesy Nisida and Saito

could be regarded as infinite in extent. Figure 10.13 shows the interfero-stress pattern obtained after the same model had been cut in such a way that the centre of the hole was situated at a distance equal to the diameter from the edge of the plate.

Figure 10.13.   Interfero-stress pattern of a semi-infinite plate loaded by a circular pin. Courtesy Nisida and Saito

10.10   Holography

By means of the holographic method, an image of a body may be reproduced in which its three-dimensional characteristics are preserved. Proposed originally by Gabor[25] for application to electron microscopy, the method has been developed and refined by later investigators, notably Leith and Upatnieks.[26]

The process consists of two parts, namely, (i) the production of the hologram and (ii) the reconstruction from the hologram of the image of the body. In two beam holography, the hologram is produced by the interference of two beams of monochromatic light possessing high spatial coherency, usually obtained by splitting the beam of a suitable laser. One of the beams, which will be referred to as the signal beam, is transmitted or scattered by the body while the other, known as the reference beam, avoids the body, Figure 10.14(a). A photographic plate is placed to intercept the light of both beams.

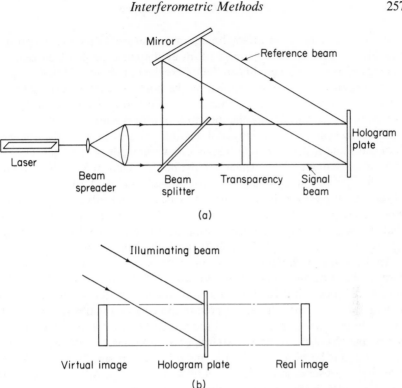

Figure 10.14. Typical arrangement for holographic reconstruction of a transparency, (*a*) recording, (*b*) image reconstruction

Each point of the photographic plate receives light from every point of the body. Since the length of optical path varies according to the position of the point on the body, however, the rays arrive at the plate with phase differences between them. Provided that the optical path differences are within the coherence length of the light, the two beams interfere at the plate and the variations of phases are converted into variations of intensity. These are recorded as a diffraction pattern on the plate which is developed in the usual way.

To reconstruct the image, the photographic plate or hologram is illuminated by a single beam of coherent monochromatic light, Figure 10.14(*b*). Since, as already observed, each point of the hologram receives light from every point of the body, the complete image may be reconstructed by illuminating any small part of the plate. In general, the first order diffracted waves produce two images, one being a virtual image which can be observed by eye or photographed with an ordinary camera while the other is a real image which may be displayed on a screen or photographed directly on a plate without using a lens. If the hologram is replaced in its original position and illuminated by the reference beam at the same angle of incidence as during exposure (the signal beam being blocked), the virtual image appears in the position previously occupied by the body. The image exhibits three-dimensional characteristics, i.e. the parallax effect, and a change of perspective can be observed by varying the direction of observation.

Since the spacing of the fringes of the diffraction pattern on the hologram is of the same order of magnitude as the wavelength of the light used and relative movement of the elements must be restricted to a small fraction of this, the system must possess high mechanical rigidity. This can be accomplished by mounting the elements on a heavy steel table placed on anti-vibration supports. It is also necessary that the photographic emulsion used shall possess very high resolution. Suitable materials are Kodak 649F and Agfa-Gevaert Scientia 14C70, 10E70 and 8E70 emulsions.

Applications of holography to photoelastic investigations have been described by Fourney,[27] Hosp and Wutzke,[28,29] Hovanesian[30–33] and others who have shown that isochromatic and isoclinic patterns obtained in this way are identical with those observed in the conventional polariscope. Due to the three-dimensional nature of the image, however, the method has the advantage that isochromatic patterns corresponding to different angles of incidence can be observed from a single hologram simply by altering the direction of observation. A further advantage is the possibility of recording a number of different images on the same hologram.

The optical arrangement is basically that normally used for the production of a hologram from a transparency modified by the insertion of additional optical elements according to the nature of the data to be recorded. Since the laser beam is already plane polarized, it is unnecessary to add a polarizer in some arrangements. Further, since the hologram is produced by the interference of the two beams of coherent light, analysers also are not usually required. The contrast of the interference pattern can be improved by inserting a diffuser screen in front of the model. Since this introduces partial depolarization of the light, a polarizer should then be inserted between the diffuser and the model.

As in conventional photoelasticity, isoclinics of different parameters may be recorded varying the direction of polarization of the two beams; these may be obtained against a bright or a dark background depending on whether the directions of polarization of the two beams are kept parallel or perpendicular to each other. The directions of polarization may be rotated conveniently by means of rotatable half-wave plates inserted in front of the model in the signal beam and in the path of the reference beam. Fourney has found that the definition of the isoclinics is improved if an analyser is placed in the signal beam between the model and the hologram with its principal axis oriented in the direction of the isoclinic being recorded.

If both beams are circularly polarized by inserting a quarter-wave plate oriented in the usual manner in the path of each, isochromatic patterns free from isoclinics may be recorded; these may be of either the bright or dark field types depending on whether the beams are circularly polarized in the same or the opposite sense.

Figure 10.15 shows an arrangement used by E. Hosp and H. Wutzke for the production of holograms of plane photoelastic models. The beam from

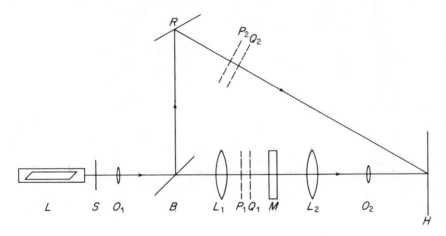

Figure 10.15. Optical system for holography of plane photoelastic models (after Hosp and Wutzke)

the laser L passes through the shutter S, is spread by the objective lens O_1 and then split into a signal beam and a reference beam by the beam splitter B. The model M is inserted in the parallel beam of light produced by the field lens L_1 in the signal beam which is then projected by the second field lens L_2 and the objective O_2 on to the hologram plate H. The reference beam is reflected by the mirror R on to the hologram plate where it interferes with the signal beam. Polarizers P_1, P_2 and quarter-wave plates Q_1, Q_2 are inserted into the beams when a particular state of polarization is required.

Figure 10.16 illustrates holographic reconstructions obtained by Hosp and Wutzke of a circular ring with a frozen axially symmetrical state of stress showing isochromatics and isoclinics in (a) a bright field and (b) a dark field. For these, the directions of vibration of the two linearly polarized beams were inclined at about 45° to the vertical.

By means of procedures known as hologram interferometry, it is possible to observe or record isopachic patterns. In one procedure, a hologram is first prepared of the model before it is loaded using circularly polarized light. After is has been developed, the hologram is replaced in exactly the same

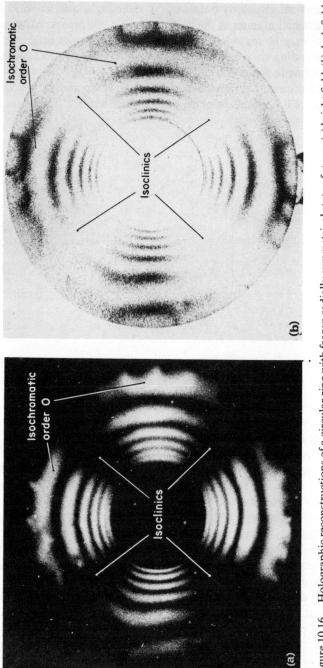

Figure 10.16.   Holographic reconstructions of a circular ring with frozen radially symmetrical state of stress, (*a*) bright field, (*b*) dark field. Courtesy Hosp and Wutzke

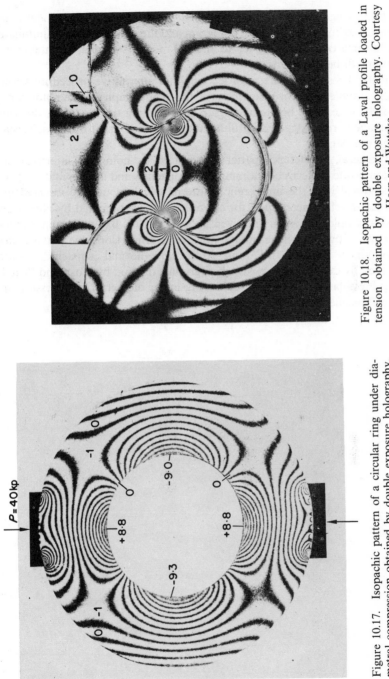

Figure 10.18. Isopachic pattern of a Laval profile loaded in tension obtained by double exposure holography. Courtesy Hosp and Wutzke

Figure 10.17. Isopachic pattern of a circular ring under diametral compression obtained by double exposure holography. Courtesy Hosp and Wutzke

position as it occupied during exposure. The position of the model is left undisturbed throughout. If the hologram and the model are now illuminated by the same two beams as before and the model is loaded, an interference pattern will be observed.

Identical results can be recorded using an alternative procedure known as the double exposure method. As the name implies, the hologram is exposed twice, once before and again after loading the model. In this case, the image is reconstructed by illuminating the hologram by a single beam in the normal way.

In general, interference patterns produced by hologram interferometry are composite patterns containing both isopachics and isochromatics. These are identical with the interference patterns which would be observed in a classical interferometer such, for example, as those obtained by Nisida and Saito, examples of which are given in Section 10.9. By employing a model material of low optical sensitivity, however, normal isopachic patterns may be obtained. Figures 10.17 and 10.18 show normal isopachic patterns obtained by the double exposure holographic method by Hosp and Wutzke using Plexiglas models of a circular ring under diametral compression and a Laval profile loaded in tension respectively.

11

BIREFRINGENT COATING METHOD

11.1 Introduction

In the application of the birefringent coating method, it is not necessary to make a model of the part to be investigated. Instead, a thin layer of birefringent material is bonded to the surface of the part itself. Assuming that the adhesion between the contacting surfaces is complete, the deformation under load of the part will be communicated to the coating. Since the part will usually be of an opaque material, a reflection technique is required for the observation of the photoelastic effects produced. In order to obtain high reflectivity, the surface of the part may be polished. Alternatively, a reflective type of adhesive containing aluminium powder may be used for bonding the coating to the part.

11.2 Reflection polariscope

The optical effects produced in the coating are observed by means of a reflection type polariscope. In the reflection polariscope, the polarizer and first quarter-wave plate are mounted in parallel with the second quarter-wave plate and the analyser, in contrast with the conventional transmission polariscope in which the optical components are in line. Light from the lamp L passes through the polarizer P (Figure 11.1) and falls at a small angle of

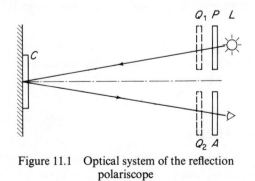

Figure 11.1   Optical system of the reflection
polariscope

incidence on the birefringent coating C. After passing through the coating, the light is reflected at the rear surface and passes through the coating in the reverse direction. The optical effects produced are observed through the analyser A. The quarter-wave plates Q_1, Q_2 can be rotated out of or into the field of view depending on whether the observations are to be made with plane or circularly polarized light. For increased portability, the elements of the reflection polariscope are usually mounted on a tripod.

Figure 11.2 illustrates a reflection polariscope manufactured by the Photostress Corporation. Observations in both circular and plane polarized white

Figure 11.2.   A reflection polariscope. Courtesy Automation Industries U.K.

light may be made, the effect of the quarter-wave plates being eliminated when necessary by rotating them into line with the polaroids. The instrument is designed to allow synchronous rotation of the polaroids for the observation of isoclinics.

11.3   Application of coating

While the conventional photoelastic model materials are suitable for use as birefringent coating, others specifically developed for the purpose are commercially available.* A suitable cold curing liquid epoxy resin with hardener is used as the adhesive.

Special procedures are required when birefringent coatings have to be applied to curved or irregular surfaces. The coating may be applied in liquid form by brushing, spraying or dipping. These methods suffer from the disadvantage that uniformity of thickness of the coating is extremely difficult to achieve. An alternative procedure and the one which is by far the most popular is the contoured sheet method in which a thin sheet of partially polymerized plastic is formed to the surface to which it is to be bonded.

11.4 Contoured sheet method

A silicone rubber or metal mould suitable for casting a thin flat sheet of plastic of uniform thickness must first be prepared. This can consist of a frame having internal dimensions somewhat greater than those of the sheet of plastic required placed on a flat plate which has been carefully levelled. If the mould is of metal, a thin coating of release agent should be applied to prevent adhesion of the plastic.

The resin and hardener are heated separately to a temperature of about 40 °C to improve the flow and are then mixed and thoroughly stirred. Due to the exothermic reaction which now takes place, the temperature rises. The mixture is poured into the mould at a temperature between 50 °C and 65 °C depending on the material, care being taken to avoid the introduction of air bubbles. The mould is then covered to exclude dust.

After a time of several hours depending on the material, its thickness and the ambient temperature, the plastic reaches a semi-polymerized state in which it is solid but extremely flexible like soft rubber. After coating the hands and the previously cleaned surface of the part with mineral oil such as liquid paraffin to prevent sticking, the plastic is removed from the mould and placed over the surface. At this stage the plastic may be trimmed if necessary with scissors. The plastic sheet is then carefully formed with the hands to the exact contour of the surface, care being taken to prevent the formation of air pockets between the contacting surfaces. The plastic is left in position on the part for approximately 24 hours until polymerization is complete. It is then removed and after thorough degreasing and cleaning, is cemented back in position on the surface of the part. When the curing cycle of the adhesive has been completed, the part is ready for test.

* Materials produced by Photostress Corporation available from Photostress Co., 101 Geiger Road, Philadelphia, Pa. 19115, U.S.A.; Sharples Photomechanics Ltd., Europa Works, Wesley Street, Bamber Bridge, Preston PR5 4PB, England; Deltalab, Le Chevallon, 38 Voreppe, France. Materials produced by Photolastic Inc. available from Photolastic Inc., 67 Lincoln Highway, Malvern, Pa, U.S.A.; Welwyn Electric Ltd, 70 High Street, Teddington, Middlesex, England.

11.5   Recording and analysis of data

As in the transmission polariscope, the directions of the principal stresses or strains are indicated by the isoclinics and their differences by the order of the isochromatics. In order to minimize errors, it is essential that the thickness of the coating be kept as small as possible. The optical effects produced are consequently small and the ordinary isochromatic pattern may contain insufficient fringes to form a satisfactory basis of analysis. For this reason, white light is commonly used and the isochromatic orders determined by colour matching as described in Section 6.2. Alternatively, fractional fringe orders may be measured by the Tardy or Senarmont methods.

From equations (5.4) and (5.20), the difference of principal stresses $(\sigma_1 - \sigma_2)_C$ in the coating, allowing for double passage of the light, is given by

$$(\sigma_1 - \sigma_2)_C = \frac{R}{2Cd} = \frac{nf}{2d}. \tag{11.1}$$

The principal strains are related to the principal stresses through Hooke's equations:

$$\epsilon_1 = \frac{1}{E_C}(\sigma_1 - v\sigma_2)_C; \qquad \epsilon_2 = \frac{1}{E_C}(\sigma_2 - v\sigma_1)_C, \tag{11.2}$$

from which

$$\varepsilon_1 - \varepsilon_2 = \frac{1 + v_C}{E_C}(\sigma_1 - \sigma_2)_C. \tag{11.3a}$$

Similarly, for the strains at the surface of the part we have

$$\varepsilon_1 - \varepsilon_2 = \frac{1 + v_S}{E_S}(\sigma_1 - \sigma_2)_S. \tag{11.3b}$$

Hence, assuming that the strains in the coating and at the surface are the same, we have

$$\frac{1 + v_S}{E_S}(\sigma_1 - \sigma_2)_S = \frac{1 + v_C}{E_C}(\sigma_1 - \sigma_2)_C,$$

or

$$(\sigma_1 - \sigma_2)_S = \frac{E_S}{E_C}\frac{1 + v_C}{1 + v_S}(\sigma_1 - \sigma_2)_C. \tag{11.4}$$

Equations (11.3) and (11.4) allow the difference of principal strains and the difference of principal stresses respectively at the surface of the part to be evaluated.

Separation of the principal stresses in the coating may be accomplished by the usual two-dimensional methods such as the shear difference and oblique incidence methods. For observations under oblique incidence, the arrangement shown in Figure 11.3 employing two mirrors M_1, M_2 is commonly used. Light from the polarizer is reflected by the first mirror to fall obliquely on the coating; after reflection at the rear surface of the coating it is viewed through the analyser by means of the second mirror.

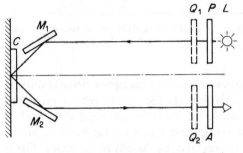

Figure 11.3. Arrangement of the reflection polariscope for observation under oblique incidence

After the individual principal stresses in the coating have been determined, the principal strains ϵ_1, ϵ_2 may be evaluated from equations (11.2). Finally, the individual principal stresses at the surface of the part may be calculated from the equations

$$\sigma_{1S} = \frac{E_S}{1 - v_S^2}(\epsilon_1 + v_S\epsilon_2),$$

$$\sigma_{2S} = \frac{E_S}{1 - v_S^2}(\epsilon_2 + v_S\epsilon_1),$$

derived from Hooke's equations for the surface.

When applying the above analysis to the results of an investigation carried out by the birefringent coating technique, it is essential that careful consideration be given to the possibility of errors arising from the various sources discussed in Section 11.7.

11.6   Birefringent strip methods

The complete state of strain at plane surfaces free from high strain gradients may be determined by the 'two-strip coating' method proposed by E. Mönch.[34] In this method a series of parallel strips of rectangular section is cemented over a lower sheet which is cemented to the surface of the part. In this way, a coating consisting of alternating strips of different thickness is formed, Figure 11.4. Reflective surfaces are provided at the surface of the

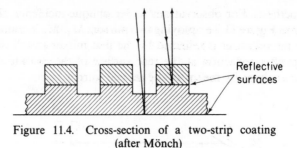

Figure 11.4.   Cross-section of a two-strip coating
(after Mönch)

part and at the junction between the upper strips and the lower sheet so that
the light passes only through the lower sheet or the upper strips as shown.
The reflecting surfaces in the strips are located a small distance above the
level of the sheet to minimize errors arising from variations of the directions
of the principal strains over the height of the strips. The isochromatics and
the system of isoclinics are recorded on colour film using white light. Isochromatic orders are determined by matching the colours with those of a
similar photograph of a calibration specimen.

   The photoelastic effects produced by the lower sheet are approximately the
same as with a uniform sheet; assuming the state of strain to be the same as
that of the surface to which it is attached, the isochromatic fringe order gives
the difference of the principal strains ($\epsilon_1 - \epsilon_2$) while the isoclinics indicate
the angle α between the directions of the principal strains and the direction
of the strips, Figure 11.5. The strain of the outer strips is approximately

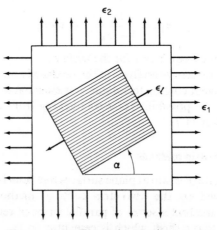

Figure 11.5.   Two-strip coating on a
surface in a uniform state of strain (after
Mönch)

uniaxial and the isochromatic order here gives the strain ϵ_l in the direction of the strip. The three measurements of $(\epsilon_1 - \epsilon_2)$, ϵ_l and α, completely define the state of strain.

A somewhat similar technique in which a series of narrow strips is used as a coating has been suggested by R. O'Regan.[35] The complete state of strain is determined either from the axial strain ϵ_l of the strips together with the values of $(\epsilon_1 - \epsilon_2)$ and α measured using an ordinary coating or from values of ϵ_l obtained with the strips arranged in three different directions.

A technique in which narrow strips of birefringent material are cemented to the test surface and viewed normally in a direction parallel to the surface has been proposed by J. Duffy and T. C. Lee.[36]

11.7   Sources of error

In order to avoid errors when interpreting the results of investigations by the birefringent coating method, it is important that careful consideration be given to the questions of (1) the extent to which the stresses and strains in the test part are modified by the action of the coating, and (2) the extent to which the strains in the coating as indicated by the birefringence correspond with those of the test surface. In general, inaccuracies increase with the thickness of the coating so that it would appear desirable that the coating should be as thin as possible. On the other hand, the thickness must be great enough to produce an optical effect giving satisfactory resolution. In practice, it is usually necessary to compromise between these conflicting requirements although both can be satisfied to a considerable extent by combining the use of a thin coating with the fringe multiplication technique (Section 6.9) as has been done by E. E. Day and others.[37]

Modification of the strains in the part result from the fact that the coating carries part of the load so that the strains are less than they would be in the absence of the coating. This reinforcing effect will obviously depend on the ratio of the thickness of the coating to that of the part, the ratio of the elastic moduli of the two materials, and the nature of the loading. In many problems the effect is negligible but with thin parts, particularly when these are subjected to bending, serious errors can be introduced if it is not taken into account and the appropriate corrections made. The reinforcing effect has been studied by F. Zandman, S. Redman and E. Riegner[38] and by G. S. Holister[39] for some simple cases of loading. The results of these investigations are presented in the form of curves from which correction factors for different ratios of coating thickness/metal thickness for several common metals can be read; the measured strain is divided by the appropriate correction factor to obtain the true strain at the test surface.

The measured strain in the coating will correspond accurately with that of the test surface only if the coating is in a state of plane stress. The various factors which can cause a departure from this state increase with the coating thickness. These factors are:

(a) A difference in the values of Poisson's ratio for the two materials.

(b) Shear forces at the edges of the coating.

(c) Strain gradients over the test surface.

(d) Variations of curvature of the test surface under load.

Effects (a) and (b) have been studied by D. Post and F. Zandman.[40] Effect (a) is usually sufficiently small that it can be neglected. Effect (b) occurs only when the edge of the coating does not coincide with a free edge of the test part and arises from the fact that the deformation here is transmitted from the test surface to the coating by means of shear stresses. The shear stress is zero at the edge of the coating but increases rapidly to a maximum value and then diminishes to zero in a distance of about three to four times the thickness of the coating. Over a zone of this width bordering the edge of the coating, the analysis given previously is invalid. The effect can be reduced by providing a generous fillet of cement around the edges of the coating. When the edge of the coating coincides with a free edge of the part or of a hole or notch within it, the effect does not arise since the shear stress in the part is then also zero.

The effects of factors (c) and (d) have been investigated by D. Post and F. Zandman,[40] and by J. Duffy and his associates.[41,42,43] In regions of high strain gradient tangential to the surface, a strain gradient is produced through the thickness of the coating. One result is that peaks of surface strain tend to be smoothed out in the coating. The birefringence observed in such cases corresponds to the average strain through the thickness of the coating, which differs from the strain at the surface. A similar effect due to strain gradient through the coating is introduced when the loading causes a change of curvature of the test surface.

Investigations have shown that errors due to strain gradient through the coating can be appreciable, particularly when large plastic deformation of the test part occurs and in problems of impact. Zandman suggests that when high strain gradients are expected, such as when a discontinuity occurs in the part, their influence can be assessed by comparing the results obtained using different thicknesses of coating; if the birefringence is proportional to the thickness, no appreciable error is involved. Otherwise, the results can be extrapolated to zero thickness to obtain the correct surface strain.

11.8   Practical applications

The birefringent coating method has been applied to the investigation of many problems of static stress distributions, thermal stresses and dynamic stresses. An application to the study of the stresses in an anisotropic material is described in the following section while an application to the investigation of transient thermal stresses is described in Section 18.6.3.

11.9 Example of the birefringent coating method: Application to fibre-reinforced composites

Problems of the stresses and strains in bodies of anisotropic materials such as fibre-reinforced composites obviously cannot be investigated by means of models made of the usual isotropic photoelastic materials. Further, reinforced photoelastic materials are generally unsuitable because of their lack of transparency. This is usually due not only to opacity of the fibres or a difference in the refractive indices of the fibre and the resin but also to the presence of a large number of minute air bubbles resulting from incomplete impregnation of the fibres by the resin. This problem has been investigated and partly overcome by Pih and Knight[44] using glass fibre reinforcement in an epoxy resin matrix of matching refractive index, combined with a vacuum impregnation technique. The refractive indices must be very accurately matched, however, since even a small difference between them will render the composite translucent due to the large number of fibre-resin interfaces.

The distribution of surface strains in composites may be conveniently studied by means of the birefringent coating method as illustrated by the following example from an investigation by Kedward and Hindle[45] of carbon-fibre- and glass-fibre-reinforced plastic beams under three-point bending. The reinforcement was unidirectional with the fibres parallel to the axes of the beams.

Theoretical considerations indicate that, in general, the planes of principal stress and the planes of principal strain in fibre-reinforced composites are not identical. Thus, on a plane of principal stress, the shear stress is zero

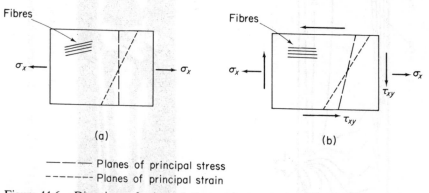

——— Planes of principal stress
------- Planes of principal strain

Figure 11.6. Directions of principal planes of stress and strain in a fibre reinforced material for (a) normal stress inclined to the fibre direction (b) normal and shear stresses parallel and perpendicular to the fibre direction. (After Kedward and Hindle.)

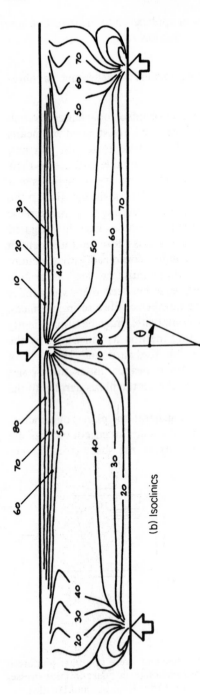

(b) Isoclinics

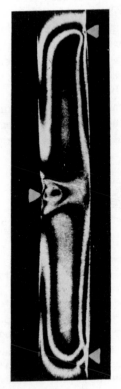

(a) Isochromatics

Figure 11.7. Experimentally derived isochromatics and isoclinics for a carbon-fibre-reinforced plastic beam. Courtesy Kedward and Hindle

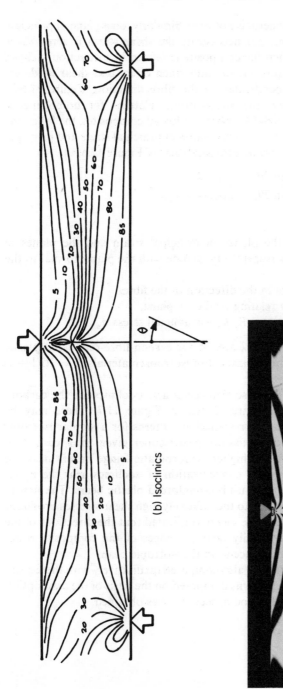

(b) Isoclinics

(a) Isochromatics

Figure 11.8. Experimentally derived isochromatics and isoclinics for an isotropic beam corresponding to that of Figure 11.7. Courtesy Kedward and Hindle

while the shear strain in general is not zero. Similarly, on a plane of principal strain, the shear stress but not necessarily the shear strain is zero. These conditions apply both when the composite is loaded in a direction inclined to the fibre direction, Figure 11.6(a), and when direct and shear loads are applied parallel and perpendicular to the fibre direction, Figure 11.6(b). With the assumption that the stress–strain relations for homogeneous orthotropic materials are valid for fibre-reinforced composites, the following equation relating the directions of the planes of principal stress and principal strain is obtained for the case investigated (that of Figure 11.6(b)):

$$\frac{\tan 2\theta_\epsilon}{\tan 2\theta_\sigma} = \frac{E_x}{2G_{xy}(1 + v_{xy})} = \lambda,$$

where

$\theta_\epsilon, \theta_\sigma$ = angles which the planes of principal strain and the planes of principal stress respectively include with the plane normal to the fibres,

E_x = elastic modulus in the direction of the fibres,

G_{xy} = shear modulus relating to the xy plane,

v_{xy} = Poisson's ratio = $-E_y/E_x$ for uniaxial stress σ_x.

Typical values of the constant λ are 7·5 for carbon-fibre-reinforced plastic and 3 for glass-fibre-reinforced plastic. For isotropic materials, $E = 2G(1 + v)$ and $\lambda = 1$.

The experimentally derived isochromatics and isoclinics for the carbon-fibre-reinforced plastic beam are shown in Figure 11.7. These may be compared visually with the corresponding patterns for a similar isotropic beam (obtained in the transmission polariscope) given in Figure 11.8. Noticeable differences are the higher isochromatic fringe orders near the neutral surface and the greater concentration of isoclinics near the upper and lower surfaces of the carbon-fibre-reinforced plastic beam. The first of these differences was attributed to the relatively high shear stresses produced in the composite beam while the second indicated that the directions of the principal strains changed rapidly near the edges of the composite beam instead of near the neutral surface as in the isotropic beam.

The experimental results were also compared qualitatively with theoretical isochromatic and isoclinic patterns computed on the basis of the assumption referred to above. Generally good agreement was obtained.

12

PLATES UNDER TRANSVERSE BENDING

12.1 Distribution of stresses due to bending

The stresses developed in a plate under transverse bending are distributed in a similar manner to those in a beam which is bent by transverse forces or bending moments. While the stresses in a beam can generally be regarded as uniaxial, however, those in a transversely bent thin plate are approximately biaxial in the plane of the plate.

If a beam extending in the x-direction is bent by a moment M acting in the xz plane, the bending stress σ_x is proportional to M and acts in the plane of the bending moment, i.e. the xz plane. The shear stress τ_{xz} is proportional to the transverse or shear force Q. In the case of a plate parallel to the xy plane, the stresses σ_x, σ_y are proportional to the bending moments m_x, m_y per unit length respectively while τ_{xy} is proportional to the twisting moment m_{xy} per unit length (see Figure 12.1). The shear stresses τ_{xz} and τ_{yz} are pro-

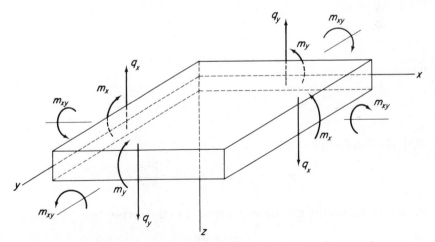

Figure 12.1.   Bending moments, twisting moments and shear forces acting on an element of a plate

portional to the shear forces q_x and q_y per unit length respectively which are components of the total shear force q per unit length.

In thin plates, i.e. plates in which the thickness is small compared with the lateral dimensions, it can be assumed that the middle plane forms a neutral surface and that lines which are originally straight and perpendicular to the middle plane remain so with respect to the neutral surface when the plate is bent provided the deflections are small compared with the thickness. The strains and therefore the stresses σ_x, σ_y and τ_{xy} under elastic conditions can be assumed to vary linearly with the distance z from the middle plane. The shear stresses τ_{xz} and τ_{yz}, which in general are very small, can be assumed to be distributed parabolically over the thickness of the plate (see Figure 12.2).

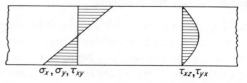

$$\sigma_x, \sigma_y, \tau_{xy} \qquad\qquad \tau_{xz}, \tau_{yx}$$

Figure 12.2.   Distributions of normal and shear
stresses through the thickness of a thin plate

12.2   Relations between moments and displacements in thin plates

We consider a small element cut from a bent plate as shown in Figure 12.3. If we denote by R_x, R_y the radii of curvature in planes parallel to the xz and yz planes respectively, then at distance z from the neutral surface, the strains are

$$\epsilon_x = \frac{z}{R_x}, \qquad \epsilon_y = \frac{z}{R_y}.$$

For small deflections,

$$R_x \simeq -1 \Big/ \frac{\partial^2 w}{\partial x^2}; \qquad R_y \simeq -1 \Big/ \frac{\partial^2 w}{\partial y^2}.$$

With these approximations,

$$\epsilon_x = -z\frac{\partial^2 w}{\partial x^2}, \qquad \epsilon_y = -z\frac{\partial^2 w}{\partial y^2}.$$

For elastic behaviour, the strains are related to the stresses by

$$\epsilon_x = (\sigma_x - v\sigma_y)/E,$$
$$\epsilon_y = (\sigma_y - v\sigma_x)/E,$$

$$(2.6)$$

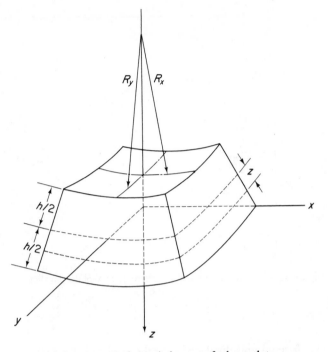

Figure 12.3. Deformed element of a bent plate

from which

$$\sigma_x = \frac{E}{1 - v^2}(\varepsilon_x + v\varepsilon_y) = -\frac{Ez}{1 - v^2}\left(\frac{\partial^2 w}{\partial x^2} + v\frac{\partial^2 w}{\partial y^2}\right),$$

$$\sigma_y = \frac{E}{1 - v^2}(\varepsilon_y + v\varepsilon_x) = -\frac{Ez}{1 - v^2}\left(\frac{\partial^2 w}{\partial y^2} + v\frac{\partial^2 w}{\partial x^2}\right).$$

The moments produced by σ_x, σ_y over unit length of the edges of the element are equal to the bending moments m_x, m_y respectively. Thus,

$$m_x = \int_{-h/2}^{h/2} \sigma_x z \, dz = -\frac{Eh^3}{12(1 - v^2)}\left(\frac{\partial^2 w}{\partial x^2} + v\frac{\partial^2 w}{\partial y^2}\right), \qquad (12.1a)$$

$$m_y = \int_{-h/2}^{h/2} \sigma_y z \, dz = -\frac{Eh^3}{12(1 - v^2)}\left(\frac{\partial^2 w}{\partial y^2} + v\frac{\partial^2 w}{\partial x^2}\right). \qquad (12.1b)$$

To obtain the relation between the twisting moment m_{xy} and the twist $\partial^2 w/\partial x\partial y$ of the plate, we consider an elementary slice *ABCD* of thickness dz situated at distance z from the middle plane of the plate (Figure 12.4). This

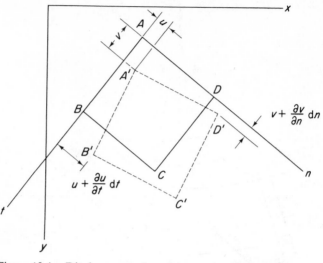

Figure 12.4.   Displacement of an element of a plate parallel to its plane

element is deformed and displaced into the position $A'B'C'D'$. If u, v denote the displacements of A parallel to the axes n, t respectively, the displacement of B parallel to the n axis is $u + (\partial u/\partial t)\,dt$ while that of D parallel to the t axis is $v + (\partial v/\partial n)\,dn$. From equation (2.1) the shear strain is

$$\gamma_{nt} = \frac{\partial u}{\partial t} + \frac{\partial v}{\partial n},$$

while the shear stress is

$$\tau_{nt} = G\gamma_{nt} = G\left(\frac{\partial u}{\partial t} + \frac{\partial v}{\partial n}\right). \tag{12.2}$$

The displacement u is equal to the rotation in the nz plane of the middle surface of the plate at the point A multiplied by the distance z of the slice from the middle surface, i.e.

$$u = -z\frac{\partial w}{\partial n}.$$

Similarly,

$$v = -z\frac{\partial w}{\partial t}.$$

Inserting these values in equation (12.2) gives

$$\tau_{nt} = -2Gz\frac{\partial^2 w}{\partial n\, \partial t}.$$

The twisting moment m_{nt} is obtained by integrating the moments due to τ_{nt} on each elementary slice over the whole thickness of the plate, i.e.

$$m_{nt} = \int_{-h/2}^{h/2} \tau_{nt}z\, dz = -\frac{Gh^3}{6}\frac{\partial^2 w}{\partial n\, \partial t}$$

$$= -\frac{Eh^3}{12(1+v)}\frac{\partial^2 w}{\partial n\, \partial t}. \tag{12.3}$$

The quantity $Eh^3/12(1 - v^2)$ is known as the flexural rigidity of the plate and will be denoted by D. Substituting in equations (12.1) and (12.3) and expressing the latter in terms of an element with its edges parallel to the x, y axes we obtain

$$m_x = -D\left(\frac{\partial^2 w}{\partial x^2} + v\frac{\partial^2 w}{\partial y^2}\right), \tag{12.4a}$$

$$m_y = -D\left(\frac{\partial^2 w}{\partial y^2} + v\frac{\partial^2 w}{\partial x^2}\right), \tag{12.4b}$$

$$m_{xy} = -D(1 - v)\frac{\partial^2 w}{\partial x\, \partial y}. \tag{12.4c}$$

12.3 Relations between stresses and moments

Since the stresses $\sigma_x, \sigma_y, \tau_{xy}$ vary linearly with z, they are related to the maximum stresses at the surface of the plate by

$$\sigma_x = \frac{2z}{h}\sigma_{x\,max}, \qquad \sigma_y = \frac{2z}{h}\sigma_{y\,max}, \qquad \tau_{xy} = \frac{2z}{h}\tau_{xy\,max},$$

Substituting in equations (12.1) and (12.3) produces

$$\sigma_{x\,max} = \frac{6m_x}{h^2},$$

$$\sigma_{y\,max} = \frac{6m_y}{h^2},$$

$$\tau_{xy\,max} = \frac{6m_{xy}}{h^2},$$

or, in general

$$\sigma_x = \frac{12z}{h^3} m_x, \tag{12.5a}$$

$$\sigma_y = \frac{12z}{h^3} m_y, \tag{12.5b}$$

$$\tau_{xy} = \frac{12z}{h^3} m_{xy}. \tag{12.5c}$$

As mentioned earlier, it can be assumed that the shear stresses τ_{xz} and τ_{yz} are distributed parabolically over the thickness and reach their maximum values at the middle surface. The integral of these stresses over the thickness on unit length of the plate is equal to the shear force q_x or q_y.

Thus,

$$q_x = \int_{-h/2}^{h/2} \tau_{xz}\, dz = \tfrac{2}{3} h \tau_{xz\,max},$$

from which

$$\tau_{xz\,max} = \frac{3q_x}{2h}.$$

Similarly,

$$\tau_{yz\,max} = \frac{3q_y}{2h}.$$

Using the equation of the parabola, we obtain for the shear stresses at any point in the plate:

$$\tau_{xz} = \frac{3q_x}{2h}\left(1 - \frac{4z^2}{h^2}\right), \qquad \tau_{yz} = \frac{3q_y}{2h}\left(1 - \frac{4z^2}{h^2}\right).$$

Differentiating these equations and substituting $z = \pm h/2$ we obtain the slopes of τ_{xz} and τ_{yz} at the surfaces of the plate:

$$\left(\frac{\partial \tau_{xz}}{\partial z}\right)_{\pm h/2} = \mp \frac{6q_x}{h^2},$$

$$\left(\frac{\partial \tau_{yz}}{\partial z}\right)_{\pm h/2} = \mp \frac{6q_y}{h^2}.$$

Equations (12.5) show that, at any point in a plate, the stresses σ_x, σ_y, τ_{xy} acting parallel to the plane of the plate are directly proportional to the moments m_x, m_y, m_{xy} respectively. The moments are therefore related to one another in the same way as the stresses. It follows that the relations derived in Section 1.3 between the stresses can also be applied to the bending and twisting moments in a plate, the normal stresses being replaced by the bending moments and the shear stresses by the twisting moments. For example, at any point in a plate there will always exist two mutually perpendicular planes on which the twisting moment is zero. The bending moments on these planes are the principal bending moments m_1 and m_2. By comparison with equation (1.4), these are given by

$$m_1, m_2 = \frac{m_x + m_y}{2} \pm \sqrt{\left[\left(\frac{m_x - m_y}{2}\right)^2 + m_{xy}^2\right]}.$$

The relations between the moments acting on different planes through a point can also be represented by Mohr's circle in which the ordinates correspond to twisting moments and the abscissa to bending moments. The radius of the circle in this case is $(m_1 - m_2)/2$ and its centre is at a distance $(m_1 + m_2)/2 = (m_x + m_y)/2$ from the origin (see Figure 12.5).

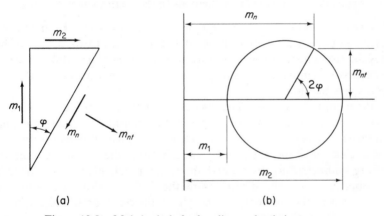

(a)   (b)

Figure 12.5.   Mohr's circle for bending and twisting moments

12.4   Equations of equilibrium

The equations of equilibrium of an element of the plate in terms of the moments and shear forces are

$$\frac{\partial m_x}{\partial x} + \frac{\partial m_{xy}}{\partial y} - q_x = 0, \tag{12.6a}$$

$$\frac{\partial m_y}{\partial y} + \frac{\partial m_{xy}}{\partial x} - q_y = 0, \tag{12.6b}$$

$$p + \frac{\partial q_x}{\partial x} + \frac{\partial q_y}{\partial y} = 0. \tag{12.6c}$$

Equations (12.6a,b) express the conditions of equilibrium of moments acting parallel to the xz, yz planes respectively. The condition of equilibrium of forces acting parallel to the z axis is expressed by equation (12.6c); the transverse load p per unit surface of the plate vanishes in load free regions.

In general, the axial stress σ_z can be neglected in load free regions. The general equations of equilibrium (1.11) then reduce to

$$\frac{\partial \sigma_x}{\partial x} + \frac{\partial \tau_{xy}}{\partial y} + \frac{\partial \tau_{xz}}{\partial z} = 0, \qquad (12.7a)$$

$$\frac{\partial \sigma_y}{\partial y} + \frac{\partial \tau_{xy}}{\partial x} + \frac{\partial \tau_{yz}}{\partial z} = 0, \qquad (12.7b)$$

$$\frac{\partial \tau_{xz}}{\partial x} + \frac{\partial \tau_{yz}}{\partial y} = 0. \qquad (12.7c)$$

12.5   Application of photoelastic methods to thin plates under bending

The stresses and moments in thin plates under transverse bending can be determined photoelastically using models consisting of two layers. In thick plates the stress distribution differs from that in thin plates and can only be determined by employing a general three-dimensional method such as the frozen stress method.

In the methods which will now be described, the model is observed in the direction perpendicular to its plane. With transmitted light, the phase difference at exit is the resultant of the elementary phase differences produced by each element of the plate along its path. Since the stresses in a uniform plate are symmetrically distributed about the middle plane of the plate but have opposite signs the directions of the fast and slow oscillations are interchanged in the two halves. The phase difference produced in the first half of the plate is thus cancelled by that produced in the second half so that the phase difference at exit is zero (Figure 12.6(a)).

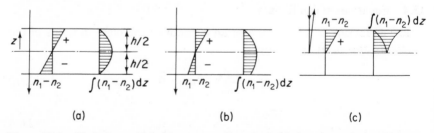

(a)   (b)   (c)

Figure 12.6.   Variations of birefringence and total retardation through the thickness of plates under transverse bending. (a) Transmission through a uniform plate. (b) Transmission through a composite plate of two different materials. (c) Reflection at a central reflecting layer between two sheets of the same material

In order to obtain a finite phase difference, the models can be composed of two layers of different materials having different photoelastic constants. In such models the magnitude of the positive phase difference produced in the tension zone is different from that of the negative phase difference produced in the compression zone so that their sum differs from zero (Figure 12.6(*b*)). Since these models are transparent, the necessary observations can be made in the normal transmission polariscope.

An alternative method is to use models consisting of two layers of the same material with a reflecting surface embedded between them. The light then passes twice through the same zone. If the layers are of equal thickness, the phase difference of the reflected beam at exit is equal to twice that produced in one half of the plate (Figure 12.6(*c*)). This method requires a reflection type polariscope and is preferable in cases where a considerable part of the field of the plate is obscured by the loading device.

12.6 Preparation of models

For models of the transmission type, the materials chosen for the two layers should have a large difference in their photoelastic coefficients so that a satisfactory number of fringes is obtained. The combination of one material having a positive stress optical coefficient C with another having a negative coefficient ensures the best results. One of the methacrylates (Plexiglas, Perspex), which have negative though rather small values of C, is suitable as one of the materials while a polyester or epoxy resin can be used as the other. The cement used must be suitable for both materials. For those mentioned, liquid polyester with hardener has been found satisfactory.

The composite plates are prepared in the following manner. The two faces to be cemented together are first roughened by means of fine sand paper. These faces are then covered with a thin film of the liquid cement. A larger amount of bubble free cement is then poured on one of the faces. The other face is next pressed on it in such a way that any air bubbles present are expelled. The model is then kept under slight pressure (e.g. by placing it under a weight) until the cement is hard. As the upper layer tends to move sideways, some simple device should be provided to keep the layers correctly aligned.

The outer surfaces of the model should be protected against contamination by the liquid cement during the fabricating process. This can be done using adhesive transparent tape or film. The film on the lower surface can then be bent upwards along the edges of the plate so that the liquid cement will cover the slit between the layers and thus prevent the re-entrance of air. The model should be cut to its final shape only after the cement has hardened since slight displacements between the two layers during hardening are hardly avoidable.

Models of the reflection type are usually made of epoxy resin though other materials may also be suitable. After roughening the surfaces to be cemented, one of them is covered with a very thin film of liquid cold curing epoxy resin with hardener. This film is then sprinkled with aluminium powder. In order to obtain a film of good reflectivity, the liquid film of cement should be as thin as possible. Since thin films of epoxy resin tend to become opaque at room temperature, the sheet to which the film has been applied should be kept for a few minutes in an oven at an elevated temperature (40 to 50 °C) before applying the aluminium powder. The surplus powder is then blown away and the sheet is replaced in the oven until the cement has fully hardened. After removing dust or other extraneous material from the reflecting layer, the second sheet is cemented to it in the manner already described.

As an alternative to the above procedure, the two sheets may be cemented together using one of the cold curing reflection type adhesives which are commercially available.

12.7   Interpretation of photoelastic data

As can readily be derived from Figure 12.6(*b*), the relative retardation produced in models of the transmission type composed of two layers having different stress optical coefficients but the same modulus of elasticity is a maximum when the layers are of equal thickness and is then given by

$$R = (C_2 - C_1)\frac{h}{4}(\sigma_1 - \sigma_2)_{max},$$

or, since $\sigma_{max} = 6m/h^2$,

$$R = (C_2 - C_1)\frac{3}{2h}(m_1 - m_2),$$

where R is the relative retardation, C_1 and C_2 the stress optical coefficients of the materials, h the total thickness of the plate, $(\sigma_1 - \sigma_2)_{max}$ the difference of the principal stresses at the surface of the plate and $(m_1 - m_2)$ the difference of the principal bending moments.

If the two layers consist of materials having different moduli of elasticity, the neutral surface no longer coincides with the middle surface but is displaced into the layer having the higher modulus. Similarly, if the layers are of unequal thickness, the neutral surface will not coincide with the common surface between the materials. In such cases, the required relations can be determined from elementary principles provided the necessary data for both layers are known.

The stress distribution in a plate is, of course, dependent on the value of Poisson's ratio for the material. This effect is only secondary, however, and since Poisson's ratio has approximately the same value of about 0·38 for all the model materials here mentioned, errors arising from a difference in the value of Poisson's ratio for the two layers can usually be neglected. The relative retardation produced in models of the reflection type consisting of two layers of equal thickness of the same material is given by

$$R = C\frac{h}{2}(\sigma_1 - \sigma_2)_{max} = C\frac{3}{h}(m_1 - m_2).$$

12.8 Boundary conditions

Since the optical data provide the difference of the principal moments or principal stresses only, one of the various methods of separation must be applied in order to determine their individual values at points where both are unknown. At the edges of a plate, certain relations exist depending on the nature of the support which enable both principal moments to be determined if their difference is known.

(a) Free edge

If, for example, the edge $x = $ constant is perfectly free, the moments m_x and m_{xy} vanish on this edge. Thus, one of the principal moments $m_1 = m_y$ acts parallel to the edge while the other, $m_2 = m_x$, is zero. Hence, $m_y = m_1 = (m_1 - m_2) = (m_1 + m_2)$, Figure 12.7(a). The shear force component q_x

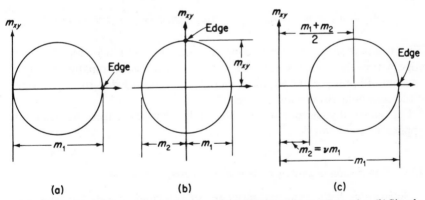

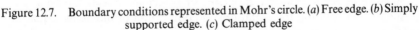

Figure 12.7. Boundary conditions represented in Mohr's circle. (a) Free edge. (b) Simply supported edge. (c) Clamped edge

normal to the edge also vanishes so that the component q_y acting in the direction of the edge is identical with the resultant shear force q, i.e. the edge is a line of shear force. The value of q is inversely proportional to the distance from the edge of an adjacent line of shear force.

(b) Simply supported edge

If the edge x = constant is rigidly supported (but not clamped), then $w = (\partial w/\partial y) = (\partial^2 w/\partial y^2) = 0$. Since also $m_x = 0$, it follows from equation (12.4a) that $\partial^2 w/\partial x^2 = 0$ and therefore $m_y = 0$. Corresponding to equation (1.5) we have $(m_x + m_y) = (m_1 + m_2) =$ constant. Hence $(m_1 + m_2) = 0$ (Navier's edge condition) from which $m_{1,2} = \pm(m_1 - m_2)/2$ or $m_1 = -m_2$, i.e. $m_{xy} \neq 0$ (Figure 12.7(b)). According to these results, the principal moments should act at 45° to the direction of the edge. Due, however, to limitations of the elementary plate theory when applied in the vicinity of an edge (Section 12.10.1), deviations from this condition are usually observed in practice; the actual directions of the principal moments are indicated by the isoclinics.

(c) Rigidly clamped edge

If the edge x = constant is rigidly clamped, then

$$\frac{\partial w}{\partial x} = \frac{\partial w}{\partial y} = \frac{\partial^2 w}{\partial x \, \partial y} = \frac{\partial^2 w}{\partial y^2} = 0.$$

Hence, from equation (12.4c), $m_{xy} = 0$ and $m_y = vm_x$. One of the principal moments again acts parallel to the edge. In this case, Figure 12.7(c),

$$m_1 - m_2 = m_x - m_y = m_x(1 - v),$$

from which

$$m_1 + m_2 = m_x + m_y = \frac{1 + v}{1 - v}(m_1 - m_2) = \frac{1 + v}{1 - v}(m_x - m_y).$$

The resultant shear force q is constant and its direction is normal to the edge within the region where the stresses (i.e. the isochromatic order) are constant. In practice, the edge of a model can be regarded as rigidly clamped if the clamping devices are of steel.

For elastically supported or elastically clamped edges, no generally applicable rules can be established.

12.9 Separation of principal stresses or principal moments

Separation of the principal moments or principal stresses can be accomplished by integrating the equations of equilibrium of moments, equations (12.6), or stresses, equations (12.7), if the values of q_x, q_y or $\partial \tau_{xz}/\partial z, \partial \tau_{yz}/\partial z$ are known. At a free edge, the shear force or shear stress acting perpendicular to the edge vanishes as mentioned above while the component acting parallel to the edge can be determined. For example, along a free edge represented by x = constant, the values of m_y (or σ_y) are indicated by the isochromatics while those of m_{xy} (or τ_{xy}) within the field can be determined from the isochromatics and the isoclinics. From these values, the derivatives $\partial m_y/\partial y$ and $\partial m_{xy}/\partial x$ (or $\partial\sigma_y/\partial y$ and $\partial\tau_{xy}/\partial x$) can be obtained. The values of q_y or $\partial\tau_{yz}/\partial z$ along the edge can then be evaluated from equation (12.6b) or equation (12.7b) while it is also known that $q_x = \partial\tau_{xz}/\partial y = 0$. The derivatives $\partial q_y/\partial y$ or $\partial^2\tau_{yz}/\partial y\,\partial z$ in the direction of the edge can now be obtained and it is also known that $\partial q_x/\partial y = \partial^2\tau_{xz}/\partial y\,\partial z = 0$. The shear forces q_x and q_y (or the corresponding shear stress values) within the field can then be determined by integrating the two equations

$$\frac{\partial q_x}{\partial x} = -\frac{\partial q_y}{\partial y}, \tag{12.8a}$$

$$\frac{\partial q_x}{\partial y} = \frac{\partial q_y}{\partial x}. \tag{12.8b}$$

Equation (12.8a) is simply equation (12.6c) expressed for a load free region where $p = 0$. To derive equation (12.8b), we express the moments in equation (12.6a) in terms of the displacement w in accordance with equations (12.4). Thus,

$$\begin{aligned}
q_x &= \frac{\partial m_x}{\partial x} + \frac{\partial m_{xy}}{\partial y} = -D\left[\frac{\partial^3 w}{\partial x^3} + v\frac{\partial^3 w}{\partial x\,\partial y^2} + (1-v)\frac{\partial^3 w}{\partial x\,\partial y^2}\right] \\
&= -D\left[\frac{\partial^3 w}{\partial x^3} + \frac{\partial^3 w}{\partial x\,\partial y^2}\right] = -D\frac{\partial}{\partial x}\nabla^2 w,
\end{aligned} \tag{12.9a}$$

where $\nabla^2 = \partial^2/\partial x^2 + \partial^2/\partial y^2$. In the same manner we obtain

$$q_y = -D\frac{\partial}{\partial y}\nabla^2 w. \tag{12.9b}$$

It follows from these equations that

$$\frac{\partial q_x}{\partial y} = \frac{\partial q_y}{\partial x} = -D\frac{\partial^2}{\partial x\,\partial y}\nabla^2 w.$$

Determination of the shear forces and separation of the principal moments or principal stresses by the above as well as by other methods are possible only if the values at some point are known, i.e. in practice if there is a free edge. If there is none, a hole may be made in the plate from which separation of the principal stresses, etc., can be started. This hole should be remote from some other region of the plate within which it may be assumed that the stresses are unaffected by the presence of the hole. The stresses in this region determined from the plate containing the hole may then be used as known values in the integration for the plate without the hole.

The shear forces q_x, q_y, q can be determined from the isochromatics and the isoclinics by another method. By inserting the corresponding terms of equation (12.4) it can be shown that the following equations hold true:

$$\frac{\partial m_x}{\partial x} = \frac{1}{1-v}\left[\frac{\partial(m_x - m_y)}{\partial x} + (1+v)\frac{\partial m_{xy}}{\partial y}\right], \tag{12.10a}$$

or

$$\frac{\partial \sigma_x}{\partial x} = \frac{1}{1-v}\left[\frac{\partial(\sigma_x - \sigma_y)}{\partial x} + (1+v)\frac{\partial \tau_{xy}}{\partial y}\right], \tag{12.10b}$$

and

$$\frac{\partial m_y}{\partial y} = \frac{1}{1-v}\left[-\frac{\partial(m_x - m_y)}{\partial y} + (1+v)\frac{\partial m_{xy}}{\partial x}\right], \tag{12.10c}$$

or

$$\frac{\partial \sigma_y}{\partial y} = \frac{1}{1-v}\left[-\frac{\partial(\sigma_x - \sigma_y)}{\partial y} + (1+v)\frac{\partial \tau_{xy}}{\partial x}\right]. \tag{12.10d}$$

Replacing $\partial m_x/\partial x$ in equation (12.6a) in accordance with equation (12.10a) produces

$$q_x = \frac{1}{1-v}\left[\frac{\partial(m_x - m_y)}{\partial x} + 2\frac{\partial m_{xy}}{\partial y}\right]. \tag{12.11a}$$

Similarly,

$$q_y = \frac{1}{1-v}\left[-\frac{\partial(m_x - m_y)}{\partial y} + \frac{2\partial m_{xy}}{\partial x}\right]. \tag{12.11b}$$

As mentioned previously, the values of $(m_x - m_y)$ and m_{xy} can be determined by means of the nomogram, Figure 9.1.

The resultant shear force is given by

$$q = \sqrt{(q_x^2 + q_y^2)}, \tag{12.12}$$

and its directional angle ϑ'' by

$$\tan \vartheta'' = q_y/q_x. \tag{12.13}$$

The directions ϑ'' form a system of lines of shear force independent of the system of stress trajectories, connecting parts of the plate subjected to transverse forces acting in opposite directions such as loaded areas and areas subjected to reactive forces. The spacing of these lines is proportional to the reciprocal of the shear force q. It will be shown that both the directions of the lines of shear force and their spacing can be deduced from the system of lines of constant sum of bending moments. This provides an alternative procedure for the determination of the shear forces in transversely bent plates.

Adding equations (12.4a) and (12.4b) we obtain

$$\frac{m_x + m_y}{1 + v} = \frac{M}{1 + v} = -D\nabla^2 w,$$

where $M = m_x + m_y$. Inserting this result in equations (12.9a) and (12.9b) produces

$$q_x = \frac{1}{1 + v} \frac{\partial M}{\partial x}, \qquad (12.14a)$$

$$q_y = \frac{1}{1 + v} \frac{\partial M}{\partial y}, \qquad (12.14b)$$

i.e. the component of shear force in any direction is proportional to the inclination of the hill of constant sum of bending moments in that direction. The resultant shear force can be determined from equation (12.12).

The directional angle ϑ of a line of constant sum of bending moments (or normal stresses) is given by

$$\tan \vartheta = -\frac{\partial(m_x + m_y)/\partial x}{\partial(m_x + m_y)/\partial y} = -\frac{\partial M/\partial x}{\partial M/\partial y},$$

or, from equations (12.14),

$$\tan \vartheta = -q_x/q_y. \qquad (12.15)$$

Comparing this result with equation (12.13), we see that the lines of shear force intersect the lines of constant sum of bending moments at right angles. The directional angle of q can therefore be determined from that of M:

$$\tan \vartheta'' = -\cot \vartheta.$$

From equations (12.15) and (12.11) we have

$$\tan \vartheta = \frac{\partial(m_x - m_y)/\partial x + 2\partial m_{xy}/\partial y}{\partial(m_x - m_y)/\partial y - 2\partial m_{xy}/\partial x}. \qquad (12.16)$$

Written in terms of the stresses, equation (12.16) becomes

$$\tan \vartheta = \frac{\partial(\sigma_x - \sigma_y)/\partial x + 2\partial\tau_{xy}/\partial y}{\partial(\sigma_x - \sigma_y)/\partial y - 2\partial\tau_{xy}/\partial x}.$$

This equation is identical with equation (9.24) for the directional angle of the isopachics in plates loaded by coplanar forces (Section 9.3.2). Equation (9.26) for the angle ϑ' between the directions of the isopachic and the stress trajectory s, i.e.

$$\tan \vartheta' = \frac{\partial(\sigma_1 - \sigma_2)/\partial s_1 + 2(\sigma_1 - \sigma_2)/\rho_2}{\partial(\sigma_1 - \sigma_2)/\partial s_2 - 2(\sigma_1 - \sigma_2)/\rho_1},$$

is therefore also applicable to transversely bent plates, in which case $\sigma_{1,2}$ can be replaced by $m_{1,2}$. Moreover, the equation of compatibility

$$\nabla^2 M = 0,$$

i.e.

$$\nabla^2(m_x + m_y) = \nabla^2(m_1 + m_2) = 0,$$

or

$$\nabla^2(\sigma_x + \sigma_y) = \nabla^2(\sigma_1 + \sigma_2) = 0, \qquad (12.17)$$

also holds true within load free areas of transversely bent plates. The rules governing isopachics given in Section 9.3.2 are therefore also applicable in the case of plates under transverse bending.

While the above rules are applicable within the body of the plate, deviations are often observed in practice at the edges. For example, since a free edge is a line of shear force, it follows that the lines of constant sum of bending moments should intersect the edge at right angles. Further, it can be deduced from the rules governing isopachics that the directions of the isochromatics and the lines of constant sum of bending moments should coincide at a free edge. The isochromatics should therefore also intersect a straight load free edge of a transversely bent plate at right angles. In practice, this is found to hold true over the whole or major part of the free edges in many cases but not in all.

For the case of a rigidly supported edge, if Navier's edge condition $(m_1 + m_2) = 0$ were exact, such an edge would be a line of constant sum of bending moments, the lines of shear force would intersect the edge at right angles and the stress trajectories would include angles of 45° with the edge. Deviations from these conditions can be observed in most cases.

The reason for the above discrepancies between the predicted results and those observed in practice is that the elementary theory of bending of thin plates, from which many of the equations in this chapter have been derived, applies only approximately to plates of finite width. In the development of the theory, it is assumed that the thickness is small compared with the lateral dimensions and this is clearly not valid near the edges (see also Section 12.10.1).

Since equation (12.17) is not valid for loaded areas of the plate, the initial values required for its application with the methods of the potential theory are those at the borders of the loaded areas as well as at the true boundary of the plate.

The values deduced from the isochromatics and the isoclinics can be used to determine not only the directional angle of the lines of constant sum of bending moments or normal stresses but also their spacing or the slope of the hill of sum of bending moments. If we substitute in the identities

$$\frac{\partial M}{\partial x} = \frac{\partial(m_x + m_y)}{\partial x} = -\frac{\partial(m_x - m_y)}{\partial x} + \frac{2\partial m_x}{\partial x},$$

and

$$\frac{\partial M}{\partial y} = \frac{\partial(m_x + m_y)}{\partial y} = \frac{\partial(m_x - m_y)}{\partial y} + \frac{2\partial m_y}{\partial y},$$

the values of $\partial m_x/\partial x$ and $\partial m_y/\partial y$ given by equations (12.10a) and (12.10c) respectively, we obtain the slopes in the x, y directions:

$$\frac{\partial M}{\partial x} = \frac{1+v}{1-v}\left[\frac{\partial(m_x - m_y)}{\partial x} + 2\frac{\partial m_{xy}}{\partial y}\right] = (1+v)q_x,$$

$$\frac{\partial M}{\partial y} = -\frac{1+v}{1-v}\left[\frac{\partial(m_x - m_y)}{\partial y} - 2\frac{\partial m_{xy}}{\partial x}\right] = (1+v)q_y.$$

The slope of the hill in the direction normal to the lines of constant sum of bending moments, which direction may be designated by s, can then be determined from the expression

$$\frac{\partial M}{\partial s} = \frac{1}{\sin\vartheta}\frac{\partial M}{\partial x} = -\frac{1}{\cos\vartheta}\frac{\partial M}{\partial y} = (1+v)q,$$

while the distance Δs between two lines of constant sum of bending moments, the parameters of which differ by unity, is

$$\Delta s = \frac{\sin\vartheta}{\partial M/\partial x} = \frac{-\cos\vartheta}{\partial M/\partial y}.$$

Separation of the principal stresses can be accomplished in certain cases without the necessity of determining the shear force q by using two equations derived from the equations of equilibrium. Differentiating equations (12.7a, b, c) or (12.6a, b, c) with respect to x, y, z respectively and subtracting the third from the sum of the first two produces

$$\frac{\partial^2\sigma_x}{\partial x^2} + \frac{\partial^2\sigma_y}{\partial y^2} + 2\frac{\partial^2\tau_{xy}}{\partial x\,\partial y} = 0, \tag{12.18a}$$

or, with $p = 0$,

$$\frac{\partial^2 m_x}{\partial x^2} + \frac{\partial^2 m_y}{\partial y^2} + 2\frac{\partial^2 m_{xy}}{\partial x\,\partial y} = 0. \tag{12.18b}$$

In these equations the shear stresses τ_{xz}, τ_{yz} and shear forces q_x, q_y respectively are eliminated so that they can be used for separating principal stresses over the whole plate. Since the second derivatives of the experimental data have to be determined, the accuracy of this method in general is doubtful. In special cases, however such as when $\partial^2 \tau_{xy}/\partial x\, \partial y = 0$, equations (12.18) may be useful.

Separation of bending moments or normal stresses can also be accomplished by integrating equations (12.10). For example, integrating equation (12.10a) produces

$$(m_x)_1 = (m_x)_0 + \frac{1}{1-v}\left[(m_x - m_y)_1 + (1 + v)\int_0^1 \frac{\partial m_{xy}}{\partial y}\, dx\right].$$

As already mentioned, the terms of the right hand sides of equations (12.10), i.e. τ_{xy} or m_{xy} and $(\sigma_x - \sigma_y)$ or $(m_x - m_y)$, can be determined by the same relations as in two-dimensional problems, e.g. by means of the nomogram, Figure 9.1. Lines m_{xy} = constant (identical with lines τ_{xy} = constant) and lines $(m_x - m_y)$ = constant (identical with lines $\sigma_x - \sigma_y$ = constant) may then be drawn. The subsequent procedure is similar to that for two-dimensional problems (Section 9.1). The accuracy of the results of integration of equations (12.10) will obviously be less than in two dimensional cases or by integration of the equations of equilibrium (12.6) or (12.7) since the two systems of lines or two sets of values deduced from the experimental data will introduce greater inaccuracy than one. Moreover, each of these values is multiplied by a factor greater than unity, i.e. by $1/(1 - v)$ or $(1 + v)/(1 - v)$ respectively which further reduces the accuracy.

In addition to the methods given here, some of those described in other chapters, e.g. birefringent coatings (Chapter 11) and stress freezing (Chapter 14) are applicable to plates under transverse bending.

12.10 Rules on the stress distribution in plates under transverse bending

The stresses (or moments and shear forces) at any general point in a load free area of a transversely bent plate or in an isotropic point respectively can be expressed by the following system of equations using general constants:

$$\sigma_x = y(b - vd) + x(va - c) + \sigma_0,$$

$$\sigma_y = y(vb - d) + x(a - vc) + v\sigma_0,$$

$$\tau_{xy} = (ay + bx + \tau_0)(1 - v),$$

$$\partial \tau_{xz}/\partial z = a - c,$$

$$\partial \tau_{yz}/\partial z = b - d.$$

In the case of an isotropic point, σ_0 and τ_0 vanish. It can be shown that these stresses (or the corresponding system of moments and shear forces) satisfy the equations of equilibrium (12.6) and (12.7) and equations (12.10) which are a form of equations of compatibility.

The above system produces the same relations between isochromatics, isoclinics and lines of constant stress sum as those defined by equations (8.2) for plates loaded by coplanar forces. Thus, the isochromatics and isoclinics around an isotropic point form an oblique projection of a spoked wheel and the direction of the line of constant sum of principal moments is determined in the same way as that of the isopachic passing through the isotropic point as described in Section 8.6.2.

The directional angle of the line of constant sum of principal moments at a general point can be determined using the methods described in Section 8.6.3, e.g. from equation (8.20):

$$\tan \vartheta =$$

$$\frac{\tan^2 2\varphi \tan \gamma(\tan \beta - \tan \alpha) + \tan 2\varphi(\tan \beta - \tan \gamma) + \tan \beta(\tan \gamma - \tan \alpha)}{-\tan^2 2\varphi(\tan \beta - \tan \alpha) + \tan 2\varphi \tan \alpha(\tan \beta - \tan \gamma) - (\tan \gamma - \tan \alpha)} ,$$

which is identical with that valid for transversely bent plates. Due to the effect of the transverse shear stresses, isotropic points at load free edges of transversely bent plates may be either positive or negative in contrast to similar points in plates loaded by coplanar forces which can only be of the negative type.

12.11 Practical examples of plates under transverse bending

12.11.1 *Models of the transmission type*

A. Examples demonstrating the effect of torsion.

Figure 12.8 shows the isochromatics in a model of a twisted plate. This model was clamped at its lower edge while a twisting moment was applied

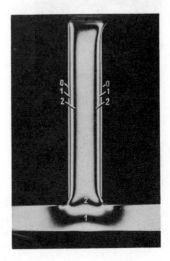

Figure 12.8. Isochromatics in a thin plate subjected to torsion

by means of a couple of forces to its upper edge (which is not fully shown in this photograph).

According to the theory of thin plates, the stresses in this case should be the same over the whole plate. Figure 12.8 shows, however, that the stresses are uniform only within that part of the plate which is at some distance from the edges. Over a region bordering the edges, the stresses diminish to reach zero at the edges. The width of this region is equal to about one half of the thickness of the plate, the latter in this example being about 2/5 of the total width. The reason for this departure from the predicted results is that the thin plate theory applies only approximately to plates of finite width. In the development of the theory it is assumed that the thickness is small compared with the lateral dimensions and this is clearly not valid near the edges. The effect is illustrated in Figure 12.9 which shows the shear flow on a cross

Figure 12.9.   Shear flow in a thin plate

section of a plate. With a plate of finite width, the shear flow is similar to that in a prismatic bar under torsion. In the interior of the plate, the flow lines are parallel to the surface but near the edges the shear stresses change direction through 180°. In this region the shear stresses act normal to the surfaces of

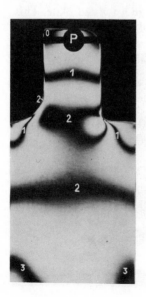

Figure 12.10.   Isochromatic pattern obtained by the two-sheet transmission method for a thin plate clamped at the lower edge and loaded by a concentrated force *P*

the plate and then produce no photoelastic effect when the model is viewed in the direction normal to its plane. This causes the fringe order to drop to zero at the edges.

Figure 12.10 shows the isochromatics in a model of a plate clamped at its lower edge and loaded by a single force concentrated at P. Similar to plates loaded by coplanar forces, stress concentration appears at the fillets. The maximum stress, however, occurs here at some distance from the edge. This is due to an effect similar to that described above.

B. Example of separation of principal stresses or principal bending moments.

The method of separation of principal stresses is illustrated by its application to a model of a rectangular plate supported at the corners and loaded by a single force P applied over a small circular area. Figures 12.11(a) and

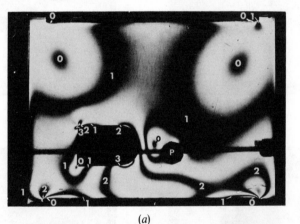

(a)

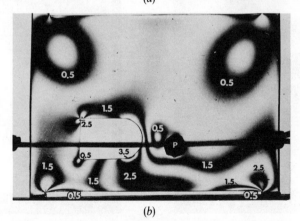

(b)

Figure 12.11. Isochromatics in a plate with a hole loaded by a transverse force P. (a) Dark field. (b) Bright field

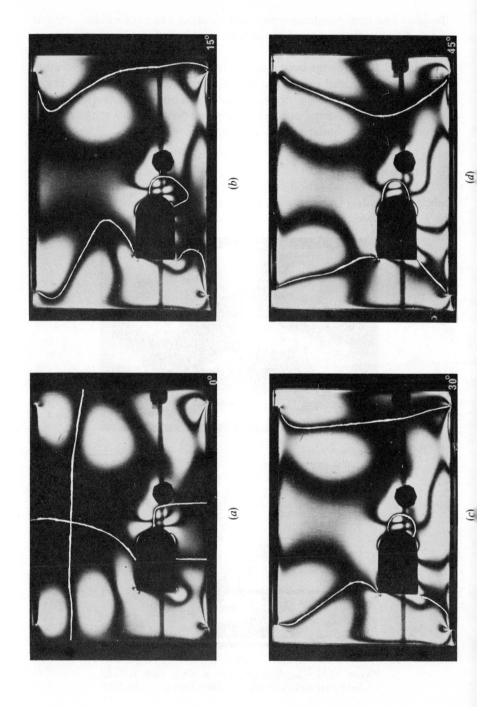

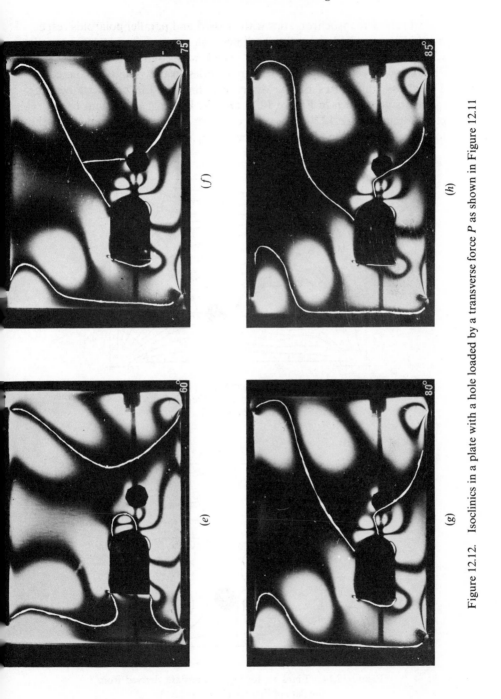

Figure 12.12. Isoclinics in a plate with a hole loaded by a transverse force *P* as shown in Figure 12.11

12.11(*b*) show the isochromatics with crossed and parallel polaroids respectively. Figures 12.12 show the isoclinics which were taken at 15° intervals with additional ones at 80° and 85° which were found to be necessary to obtain the directions of the stress trajectories in the vicinity of the edges. The stress trajectories or the directional lines of the principal bending moments are shown in Figure 12.13 and the lines τ_{xy} = constant or m_{xy} = constant in Figure 12.14. The sign of τ_{xy} or m_{xy} is determined using the following convention: Bending moments m_x, m_y or m_1, m_2 are considered

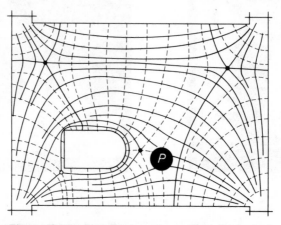

Figure 12.13.   Stress trajectories or lines of principal bending moments derived from the isoclinics shown in Figure 12.12

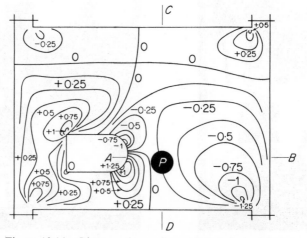

Figure 12.14.   Lines τ_{xy} (or m_{xy}) = constant derived from the data of Figures 12.11 and 12.13

positive when they produce tensile stresses at the lower surface of the plate; twisting moments m_{xy} or shear stresses τ_{xy} are positive when their directions coincide with those of positive shear stresses in plates loaded by coplanar forces. The magnitudes of the shear stresses are determined from the nomogram Figure 9.1, in the same manner as described in Section 9.1.2.

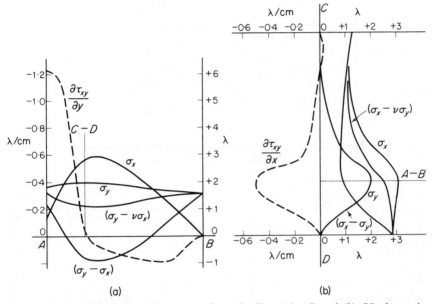

(a)   (b)

Figure 12.15.   Distribution of stresses along the lines (a) *AB* and (b) *CD* shown in Figure 12.14

In Figure 12.15(a) the slope $\partial\tau_{xy}/\partial y$ along the line *AB* (see Figure 12.16) is plotted. For integration, the area below this curve is measured, e.g. by counting small squares if graph paper is used. Then, after multiplying the result by $(1 + v)$ the values of $(\sigma_y - v\sigma_x)$ are found using the formula

$$(\sigma_y - v\sigma_x) = (1 + v)\int \frac{\partial\tau_{xy}}{\partial y}\,dx + c,$$

which is obtained by integrating equation (12.10b). The value of c is given by the boundary value of σ_x or σ_y. The values of $(\sigma_y - v\sigma_x)$ thus obtained together with the corresponding values of $(\sigma_y - \sigma_x)$ allow the individual stresses σ_x and σ_y to be determined. The results are shown in Figure 12.15(a).

The normal stresses along the line *CD* (Figure 12.16) are determined in a similar way from the slopes $\partial\tau_{xy}/\partial x$ using equation (12.10d). These results are shown in Figure 12.15(b). At the point of intersection of the lines *AB* and

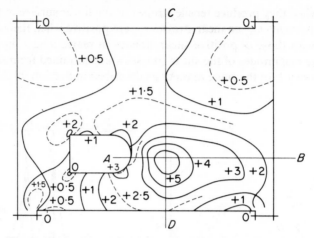

Figure 12.16. Lines of constant sum of bending moments or normal stresses in the plate

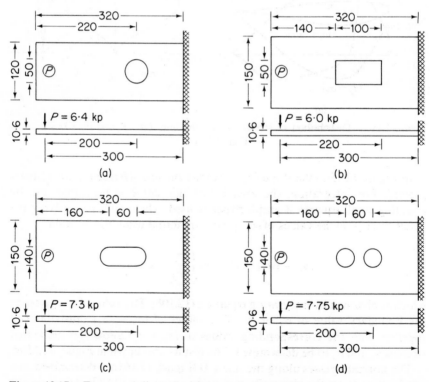

Figure 12.17. Forms and dimensions in mm of models of flat springs investigated by the reflection method

CD the values obtained for σ_x and σ_y should, of course, be identical in both cases. This check is necessary when determining the stresses at a point since, even if the experimental data and the procedure of separation of the principal stresses are exact, the results will be slightly different in certain cases due to the effect demonstrated in Figures 12.8 and 12.9 and to some other reasons resulting from the fact that the theory of transversely bent plates is not an exact one.

Figure 12.16 is a draft of the lines of constant sum of principal stresses.

12.11.2 *Models of the reflection type*

The factors of stress concentration in thin plates used as springs in electric switches were required. The models were made to an enlarged scale of 10:1 and consisted of two identical layers of hot curing epoxy resin containing a central reflecting layer of aluminium powder. These were produced in the manner described in Section 12.6.

Four different forms of spring were investigated. These are shown in Figure 12.17 which gives the model dimensions in mm. The isochromatics for the four models are shown in Figure 12.18. The relations between the different factors of stress concentration can be read from the photographs taking into account the different loads acting. The absolute values of the stresses can be found by determining the fringe value from a calibration test on a specimen cut from the cemented plate or simply by assuming the usual well known fringe value of this material. In the present example, the fringe value as determined was $23\cdot6 \text{ kp/cm}^2$. Separation of principal stresses here was unnecessary since the maximum stresses obviously occur at the load free edges.

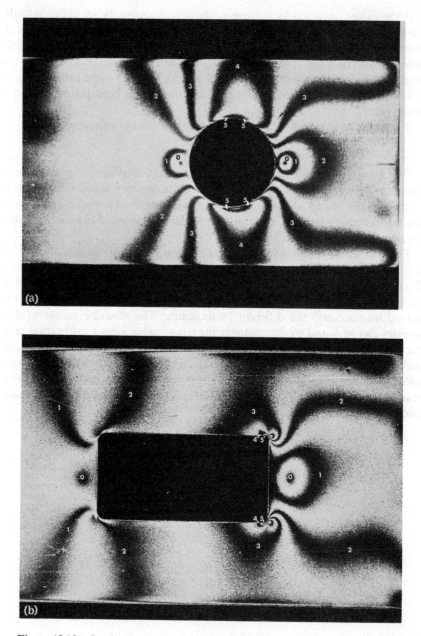

Figure 12.18.   Isochromatic patterns obtained by the reflection method for the
flat springs shown in Figure 12.17

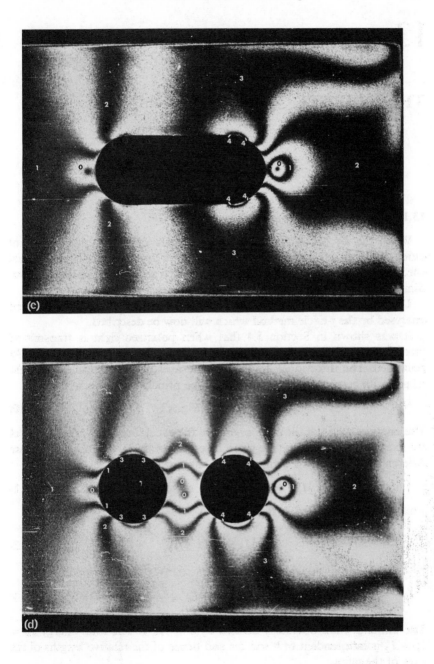

13

THE j-CIRCLE METHOD

13.1 Optical basis of the j-circle

When the directions of the principal stresses vary from point to point along the path of a light beam through a birefringent model, additional effects are introduced which were not taken into account in the simpler two-dimensional analysis of plates loaded in their own planes.

The effects of double refraction on polarized light in general cases can be analysed by the j-circle method which will now be described.

It was shown in Section 5.4 that when polarized light is transmitted through a birefringent body, the emergent light in general is elliptically polarized. The directional angle ψ of the axes of the characteristic light ellipse with respect to the directions of the principal stresses is given by

$$\tan 2\psi = \tan 2\theta \cos \alpha. \tag{3.10}$$

The energies I_1, I_2 of the components of the oscillation in the directions of the axes of the light ellipse are given by equation (5.13); replacing the phase difference α by twice the compensation angle m, this becomes

$$I_1, I_2 = \tfrac{1}{2}I_0(1 \pm \sqrt{[\cos^2 2\theta + \sin^2 2\theta \cos^2 2m]}), \tag{13.1}$$

which can be expressed in the form

$$I_1, I_2 = \frac{I_0}{2}(1 \pm j), \tag{13.2}$$

where

$$j = \sqrt{(\cos^2 2\theta + \sin^2 2\theta \cos^2 2m)} = \frac{I_1 - I_2}{I_0} = \frac{I_1 - I_2}{I_1 + I_2} \tag{13.3}$$

The limits of j are obviously defined by $0 \leqslant j \leqslant 1$. The total energy $I_0 = I_1 + I_2$ is independent of θ and $2m$ and hence of the relative lengths of the axes of the ellipse.

The form of the oscillation as defined by the relative lengths and the directions of the axes of the ellipse is represented in the j-circle method by a vector.

The length of this vector, called the *j*-vector, corresponds to the value of *j* and its direction indicates that of the major axis of the ellipse.

If the *j*-vectors representing all possible forms of oscillation resulting from different path differences are drawn from one pole, the tips of these vectors all lie within a circle of unit radius which is called the *j*-circle.

In the *j*-circle, the *j*-vector is drawn from the centre at an angle equal to twice that made by the major axis of the ellipse with some chosen reference direction, e.g. the *x* axis. All other angles involved in the *j*-circle are similarly doubled. This results in a simplification of the laws governing the relationship between the changes in the form of polarization and the position of the different *j*-vectors when polarized light passes through a birefringent body.

It is obvious that the position of the tip of the vector alone is sufficient to define the nature of the oscillation. Thus a tip on the circumference of the circle represents plane polarized light while one at the centre represents circularly polarized light. A tip at any other point within the circle represents some form of elliptical polarization. An example of an ellipse represented by the *j*-vector in the *j*-circle is shown in Figure 13.1.

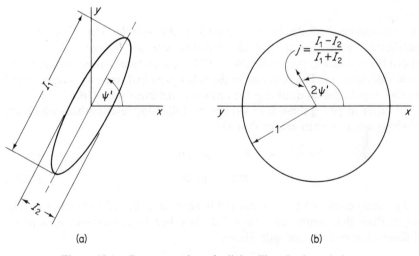

(a)   (b)

Figure 13.1.   Representation of a light ellipse in the *j*-circle

In Figure 13.2, the principal stress σ_1 is inclined at an angle φ to the chosen reference direction Ox while the direction OP of the polarizer and the major axis of the light ellipse include angles θ and ψ respectively with the direction of σ_1; in the *j*-circle, these angles are doubled.

It will be shown that the tips of the *j*-vectors which represent the oscillations at successive points along the path of a light beam through a

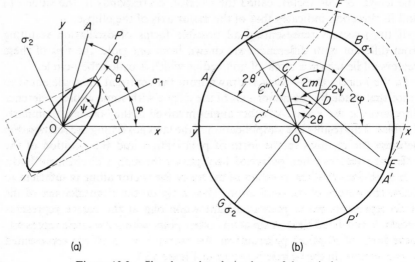

Figure 13.2.   Sketch used to derive laws of the *j*-circle

two-dimensional model lie on a chord in the *j*-circle. This chord is perpendicular to the directions of the principal stresses. In Figure 13.2(*b*), this chord is *PE* which is perpendicular to *OB*, the direction corresponding to the principal stress σ_1. The direction of the other principal stress σ_2 is represented by *OG* which, due to doubling the angles, is collinear with *OB*.

In order to prove these laws we use the following relationships which are obvious when we remember that $OP = 1$:

$$OD = \cos 2\theta, \tag{13.4a}$$

$$PD = \sin 2\theta. \tag{13.4b}$$

The semi-circle *PFE* of radius *DP* is now described about centre *D*. The angle *PDF* then represents $2m$ if 360° is taken to correspond to a phase difference of one wavelength. Hence

$$CD = DF \cos 2m = DP \cos 2m,$$

which upon substituting from equation (13.4*b*), becomes

$$CD = \sin 2\theta \cos 2m. \tag{13.5}$$

We also have $OC = \sqrt{(OD^2 + CD^2)}$ which, together with equations (13.4*a*) and (13.5), gives

$$OC = \sqrt{(\cos^2 2\theta + \sin^2 2\theta \cos^2 2m)} = j,$$

as given by equation (13.3). Further,

$$\frac{CD}{OD} = \frac{\sin 2\theta \cos 2m}{\cos 2\theta} = \tan 2\theta \cos 2m,$$

which is in agreement with equation (3.10) showing that the direction OC coincides with that of the major axis of the ellipse.

The intensity of the light transmitted by the analyser can also be determined from the j-circle. Let the axis of the analyser be inclined at an angle θ' to the direction of the principal stress σ_1. In the j-circle, Figure 13.2, OA represents the direction of the analyser while C' is the projection of the tip of C of the j-vector on OA. It can be read from the diagram that the lines PE and AA' include an angle of $2\theta' - 90°$.

We thus have

$$OC' = CD \cos(2\theta' - 90) - OD \sin(2\theta' - 90)$$

$$= \sin 2\theta \cos 2m \sin 2\theta' + \cos 2\theta \cos 2\theta'$$

and

$$A'C' = 1 + OC' = 1 + \cos 2\theta \cos 2\theta' + \sin 2\theta \sin 2\theta' \cos 2m.$$

Comparing this result with equation (5.11), we see that $A'C'$ represents the intensity transmitted by the analyser when AA' corresponds to I_0.

With crossed polaroids, i.e. with $\theta' = \theta \pm 90°$, the direction of the analyser is represented by OP'. The intensity transmitted is therefore proportional to PC''. Similarly, with parallel polaroids, i.e. with $\theta' = \theta$, OP represents the directions of both polaroids and the intensity transmitted is represented by $P'C''$.

By means of the j-circle method outlined above, all the optical phenomena and techniques of two-dimensional photoelasticity can be represented in a simple and easily understood manner. In addition the geometrical relationships which can be deduced from the j-circle frequently lead to the solution of problems in the most direct way. The greatest value of the method lies, however, in its application to general three-dimensional stress systems in which the directions of the principal stresses vary from point to point along the path of the light beam through the model.

The principle that the path described by the tips of the j-vector, known as the tip curve, is perpendicular to the direction of the principal stresses also applies when the directions of the principal stresses vary. This follows from the fact that the change in shape of the light ellipse depends only upon its instantaneous shape, i.e. upon the position of the tip within the circle at that instant and is independent of the effects which have produced this shape. Thus the direction of the tip curve at any point depends only on the directions of the principal stresses at the corresponding point in the path of the light beam. Hence, while in two-dimensional systems the path traced by a tip is a straight line, in general three-dimensional systems it is curved.

The optical effects occurring in general cases can be readily visualized if the j-circle is regarded as the projection of a sphere which rotates about an axis parallel to the plane of projection. A tip curve then corresponds to the projected path of a point on the surface of the sphere. It has already been shown that the circumference of the circle of radius PD and centre D in Figure 13.2(b) represents a phase difference of one wavelength while sectors of it represent corresponding fractions of a wavelength. When this circle is rotated through 90° about PE it becomes a small circle of the sphere. The sphere must therefore rotate about the axis parallel to BG to produce the tip curve PE by the projection of a point on its surface. This point obviously travels on the front side of the sphere during one half revolution and on the rear side during the other half.

We now consider plane polarized light emerging from the polarizer for different directions of vibration. The tips of the j-vectors lie on the circumference of the j-circle, i.e. on the great circle of the sphere which is parallel to the plane of projection. Since the direction of the axis of rotation of the sphere depends only on the direction of the principal stresses and not on the position of the tip, it is the same for all points on the circumference. The circle is therefore transformed by the rotation into an ellipse, the major axis of which is a diameter of the circle. In two dimensional cases the tip curves starting at all points on the circumference are straight lines parallel to the minor axis of the ellipse, Figure 13.3. The direction of the principal stresses is therefore given by the direction of the major axis of the ellipse.

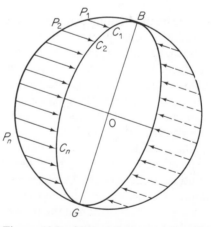

Figure 13.3. Vector tip curves for all possible directions of vibration of plane polarized light

Circularly polarized light, as we have already seen, is represented by a tip at the centre of the *j*-circle. This corresponds to one of the points, depending on the sense of the vibration, in which the axis perpendicular to the plane of projection intersects the surface of the sphere. The position of this point relative to the original great circle is unchanged during rotation so that after rotation it lies at some point along the direction of the minor axis of the ellipse whether the directions of the principal stresses are constant or not.

As can be seen from Figure 13.3 the tip curves in two dimensional cases for the points in which the circle and ellipse touch are of zero length, i.e. the starting and end points coincide. This obviously represents the case when the direction of vibration coincides with the direction of the principal stresses. With crossed polaroids, the intensity transmitted is then zero for any angle of rotation of the sphere, i.e. for any value of the phase difference. This illustrates the phenomenon of extinction due to isoclinics. The other well-known phenomena and procedures of two-dimensional photoelasticity may also be represented in the *j*-circle. Several of these are demonstrated in the following section.

13.2   Representation of two-dimensional photoelasticity

13.2.1   *Formation of isochromatics in the circular polariscope*

Figure 13.4 illustrates the formation of isochromatics in the circular polariscope. Plane polarized light emerging from the polarizer is represented by the point P on the circumference of the *j*-circle. The principal axes of the

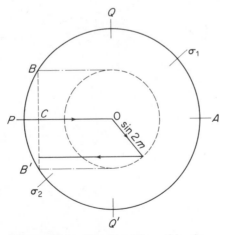

Figure 13.4. The formation of isochromatics in the circular polariscope represented in the *j*-circle

quarter-wave plate, being inclined at 45° to the vibration axis of the polarizer in the polariscope, are represented by OQ, OQ' in the j-circle. Corresponding to the retardation of one quarter wavelength, the sphere rotates through one quarter revolution about the axis QQ' so that the tip of the j-vector moves from P in the direction perpendicular to QQ' to the centre O of the circle. During transmission through the model, the vector tip is displaced from the centre in a direction depending on the directions of the principal stresses. The ends of the tip curves for all points where the stresses produce the same phase difference but have different directions lie on a circle of radius sin $2m$ concentric with the j-circle; rotation of one quarter revolution of the sphere corresponding to the phase difference produced by the second quarter-wave plate transforms this circle into the straight line BB' which is perpendicular to the line OA representing the direction of the analyser. The projections of all points of the line BB' on the line PA coincide at C, showing that the intensity of light transmitted by the analyser (represented by PC) is independent of the directions of the principal stresses and is the same at all points having the same phase difference.

13.2.2   *Compensation by the Tardy method*

The quarter-wavelength retardation introduced by the first quarter-wave plate displaces the vector tip from the point P on the circumference (plane polarized light) to the centre O of the j-circle (circularly polarized light) as shown in Figure 13.5.

The fractional part of the phase difference produced in the model is represented by $OB = \sin 2m$, the direction of which depends on the directions of

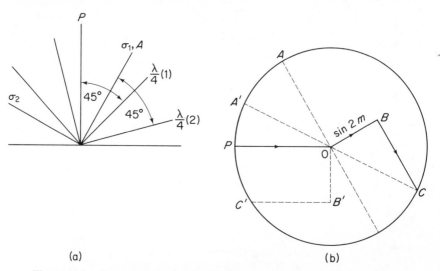

(a)   (b)

Figure 13.5.   Compensation by the Tardy method represented in the j-circle

the principal stresses. The second quarter-wave plate is set with its axes at 45° to the principal axes of stress; in the *j*-circle it causes a displacement of the vector tip from *B* to the point *C* on the circumference, corresponding to a rotation of one quarter revolution of the sphere about the axis parallel to *OB*. The initial direction of the analyser, which is parallel to a principal axis of stress, is represented by *OA*. During compensation the analyser is rotated into the extinction position *OA′*. As can be read from Figure 13.5(*b*), $\widehat{AOA} = \widehat{OCB} = 2m$.

In the application of the Tardy method, the orientation of the principal stresses relative to the directions of the polarizer and first quarter-wave plate is of course immaterial and the tip curve *OBC* in Figure 13.5(*b*) represents a general case. More commonly, the set-up used is that of the standard crossed circular polariscope shown in Figure 5.9 in which the polaroids and the quarter-wave plates are both crossed with the principal axes of the former parallel to, and those of the latter at 45° to, the principal axes of stress. In Figure 13.5(*b*), the corresponding tip curve is represented by *OB′C′*.

13.2.3   Compensation by the Senarmont method

For this procedure, the axes of the polarizer and the single quarter-wave plate are set at 45° to the directions of the principal stresses. The phase difference introduced by the stresses causes the vector tip to move from *P* to *B*, Figure 13.6, where $PB = 1 - \cos 2m$ and $OB = \cos 2m$. The rotation of one quarter revolution corresponding to the retardation produced by the quarter-wave plate displaces the vector tip from *B* to the point *C* on the circumference. The axis of the analyser is initially perpendicular to that of the polarizer and is represented by *OA* in Figure 13.6. During compensation,

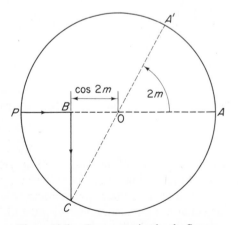

Figure 13.6.   Compensation by the Senarmont method represented in the *j*-circle

the analyser is rotated into the extinction position OA'. Obviously, $A\widehat{O}A' = C\widehat{O}B = 2m$.

13.3   Analysis of cases of changing directions of principal stresses

When the directions of the principal stresses vary along the path of the light beam the tip paths are curved and the starting and end points do not lie on lines parallel to the minor axis. Further, no tip curves exist for which the starting and end points coincide so that complete extinction cannot be obtained for any position of crossed polaroids. In such cases complete extinction corresponding to that produced by isoclinics is obtained when the axis of the polaroids include an angle differing from 90°. The relative rotation of the polaroids from the crossed position necessary to produce extinction will be called the angle of rotation or rotational angle and will be denoted by δ (see Figure 13.7).

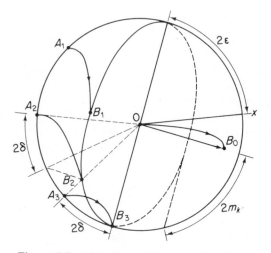

Figure 13.7.   Notation and determination of optical values with varying directions of the principal stresses

The angle between the major axis of the ellipse and the reference direction will be called the angle of inclination and be denoted by ε. It is necessary to differentiate between this angle and the corresponding angle in two dimensional cases which as we have seen is identical with the directional angle φ of the principal stresses, i.e. with the parameter of the isoclinic passing through the point.

The third value which can be determined from the j-circle is called the angle of compensation, denoted by m_k. In two-dimensional cases the corresponding angle m is equal to one half the phase difference and can be measured by any of the usual methods of compensation, e.g. with the aid of a quarter-wave plate as in the Senarmont method described in Section 6.5.2. m_k cannot be defined in terms of phase difference since in general cases the component vibrations of the light do not maintain their separate identities, the light being continuously resolved into different components at every point in its path through the model.

The three independent values ε, δ and m_k can be measured experimentally in the following way. Crossed polaroids are rotated into the position giving minimum intensity. If extinction is not obtained, as will generally be the case, the relative angular position of the polaroids is altered slightly and they are again rotated simultaneously until minimum intensity is obtained. This process is repeated until the position giving complete extinction is obtained. When using monochromatic light, two such positions always exist which are mutually perpendicular. The angle of inclination ε is then given by the direction of the analyser. Which of the two possible directions should be chosen is discussed later. The angle of rotation δ is equal to the difference between the angle included between the axes of the polaroids and 90°, i.e. to their relative rotation from the crossed position.

The setting for extinction having been obtained, the compensation angle m_k can now be measured by the Senarmont method using a procedure similar to that in two-dimensional experiments. The polaroids are first rotated through 45° from the extinction position. The sense of this rotation is discussed later. A quarter wave plate is then inserted between the model and the analyser in such a way that one of its axes is parallel to the axis of the analyser. The intensity of the light transmitted is thereby unaffected except by absorption. The angle through which the analyser must now be rotated until extinction is restored is the angle of compensation m_k. If a photomultiplier or similar photosensitive device is used greater accuracy can be obtained. The setting corresponding to complete extinction can then be obtained even if a certain amount of vagrant light is present while the setting for maximum intensity can also be determined. The latter permits the use of the following simpler procedure for measuring the required optical values. The position of the analyser to produce maximum intensity transmitted with incident circularly polarized light is first determined. As we have seen, the vector tip curve in this case terminates at a point B_0 (Figure 13.7) along the line of the minor axis of the ellipse and maximum intensity is transmitted when the axis of the analyser as represented in the j-circle is parallel to OB_0. Hence the direction of the major axis and the angle of inclination ε are obtained.

The analyser is now rotated through 45°, corresponding to 90° on the j-circle so that its axis coincides with OB_3. After removing the quarter-wave plate thus reverting to plane polarized light, the polarizer is rotated until extinction is achieved. Its axis now coincides with OA_3 so that the actual angle included between the axes of the polaroids is equal to δ. The procedure for determining m_k is then the same as before.

The angle of compensation can obviously have values between $+180°$ and $-180°$. Except when the phase difference is less than one wavelength, it does not give the phase difference in two-dimensional systems but only the amount by which the phase difference differs from an integral number of wavelengths. Thus the compensation angle varies periodically with increasing phase difference or stress. A single value for the stress difference cannot therefore be obtained from the compensation angle. The angle of compensation in three-dimensional systems varies in the same way.

A further difficulty arises from the fact that extinction can be obtained for two positions of the analyser differing by 180°, i.e. by rotating the analyser through a certain angle in one sense or by the difference between this angle and 180° in the opposite sense. Which of these values is the correct one can only be obtained from the condition of continuity of the stresses. This difficulty together with that arising from the two possible values of the angle of inclination (differing by 90°) can be avoided by suitably defining the quantities involved. The angle of compensation m_k is accordingly defined as an angle between the limits 0° and $+90°$. With the positive sign for m_k, the direction indicated by ε corresponds to that of the algebraically greater principal stress in two-dimensional cases when the compensation angle m is less than or equal to 90° or perpendicular to that direction when m is equal to or greater than 90° assuming tension positive and a positive photoelastic effect. A compensation angle of 0° is the same as one of 90° if the angle of inclination is turned through 90°.

The direction of the algebraically greater principal stress, i.e. the direction of vibration of the faster component light beam, can be found in the same way as in two dimensions using a compensator or quarter-wave plate. When using the latter, errors can be avoided if the same initial relative position of the analyser and either the fast or slow axis of the quarter-wave plate is used for all measurements. The correct setting can be verified using a known two-dimensional state of stress.

If the compensation angle is found to exceed 90° for a particular initial setting of the analyser and quarter-wave plate, the polarizer, analyser and quarter-wave plate are all rotated through 90°. The angle of compensation will then lie within the specified limits.

The angle of inclination is defined as an angle between 0° and $+180°$. It is the angle between the x axis and the axis of the analyser at the extinction setting and is measured counterclockwise when looking towards the source.

When the state of stress at every point along the path of a light beam through a model is known, the optical values can be plotted from the laws governing the behaviour of the j-vector within the j-circle. These values can be determined either graphically or analytically. The reverse process of determining the stresses from measured optical values cannot in general be carried out although this can be accomplished in particular cases by the method described for shells in Chapter 17.

13.4 Graphical determination of optical data

This involves the drawing to scale of the necessary tip curves which enable the values of ε, δ and m_k to be obtained.

The first step consists of the determination of the 'path curve'. The length of any element of this curve is proportional to the phase difference produced in the corresponding element of the light path through the model and the tangent to the curve at any point represents the direction of the algebraically greater principal stress at the corresponding point in the model but with all angles relative to the reference direction doubled. If the sphere mentioned previously is rolled along this curve, the projection of a point on its surface forms a tip curve. In so rolling, the axis of rotation of the sphere is parallel to the plane of projection and normal to the path curve at every point. The path curve is drawn to scale such that the circumference of a great circle of the sphere represents one wavelength.

The tip curve can be regarded as a distorted path curve, corresponding elements of both having the same direction but different lengths depending on the position of the generating point on the surface of the sphere.

In order to determine ε, δ and m_k, the ellipse which represents the loci of the end points of all possible tip curves with initially plane polarized light must first be obtained. The values of ε and m_k are determined by drawing the tip curve OB_0 (see Figure 13.7) which starts at the centre of the circle, i.e. the tip curve for incident circularly polarized light. The straight line joining the end point B_0 of this curve with the centre of the circle indicates the direction of the minor axis of the ellipse while the length of this line is obviously equal to $\sin 2m_k$. The angle between the major axis and the reference direction OX is equal to 2ε.

The value of the angle of rotation δ is obtained by drawing a tip curve such as A_1B_1 or A_2B_2 starting at any point on the circumference. The end point of this curve lies on the required ellipse and this point together with the known directions of the axes enable the complete ellipse to be drawn. The angle between the radii through the starting point of this tip curve and the projection perpendicular to the major axis of the end point of the curve on the circumference is equal to 2δ.

The angle of rotation δ can also be found by drawing a tip curve $B_3 A_3$ starting at one of the extremities of the major axis of the ellipse and using the elementary phase differences in the reverse sequence. In this special case the resulting tip curve is identical with that which is produced when starting at A_3 and using the elementary phase differences in their normal order. The end point of this curve should also lie on the circumference thereby providing a check on the accuracy of the drawing. As can be seen from Figure 13.7 the angle between the directions indicated by the starting and end points of this curve is equal to 2δ.

The drawing of tip curves is facilitated by the use of a nomogram of the form shown in Figure 13.8. This contains a number of parallel chords which represent the projections of the corresponding small circles of the sphere.

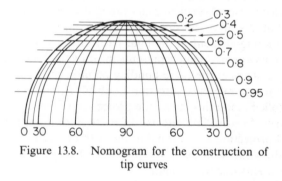

Figure 13.8. Nomogram for the construction of tip curves

The latter are divided into equal segments so that curves of equal angles of compensation or, more accurately, of equal angular positions of the tip of the j-vector form a set of concentric ellipses which intersect the chords.

The tip curve is drawn on tracing paper which is fixed over the nomogram in such a way that it can be rotated about the centre of the circle. The path curve is first drawn on this paper but is turned through 90° since the sphere rolls in the direction perpendicular to that of the principal stresses. The elements of the tip curve to be drawn are then parallel to the corresponding elements of the rotated path curve. A tip curve may be drawn starting at any point, e.g. at the centre of the circle or at any point on its circumference.

The operation of the nomogram is explained by considering the effects of two adjacent elements in the light path through the model. The directions of the principal stresses can be regarded as constant through each element but the stresses in the two elements are mutually inclined at an angle $\Delta\varphi$.

In the first element the directions of the principal stresses are represented by OD_1, Figure 13.9, so that the first element of the tip curve is produced by a movement of the tip along the chord $A_1 B_1$. If C is the end point of this

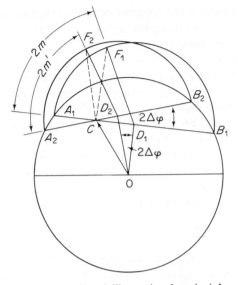

Figure 13.9. Sketch illustrating the principle
of the nomogram of Figure 13.8

element then the state of vibration at exit from the first element is represented
by the j-vector OC while the angle of compensation m is given by $2m = A_1\hat{D}_1F_1$. The vector OC also represents the state of vibration at entry to the
second element in which the directions of the principal stresses are repre-
sented by OD_2. The second element of the tip curve is therefore produced by
a movement of the tip along the chord A_2B_2.

If we now suppose the vector OC also to have been produced by a movement of the tip along A_2B_2 the corresponding angle of compensation would
be represented by the angle $A_2D_2F_2$ which differs from $A_1D_1F_1$. Thus, when
the vibration entering the second element is resolved into components in
the new directions of the principal stresses, these have an apparent angle of
compensation which we shall denote by m' and which is given by $2m' = A_2D_2F_2$.

If we now imagine the chord A_2B_2 to be rotated about the centre through
an angle $2\Delta\varphi$ until it is parallel to A_1B_1 and the vector OC to rotate through
the same angle then obviously the intercept A_2C and hence the apparent
angle of compensation m' will be the same as before. In the nomogram the
chords are drawn parallel and when proceeding from one element to the next
the tip is rotated by rotating the tracing paper on which it is drawn. The
angle $2m'$ can obviously be obtained from the angular scale along the chord
on which the tip then lies.

To draw the first element of the tip curve, the tracing paper is rotated until
the first element of the rotated path curve is parallel to the chords in the

nomogram. The length of the required element is then marked off along the appropriate chord by counting the number of degrees represented by the length of the element of the path curve.

While m_k is independent of the position of the polarizer, etc., the value of m as we have seen depends on the direction to which it is related. To draw the second element of the tip curve the tracing paper is turned until the second element of the path curve is parallel to the chords. After so doing, it will be found that the end point of the first element now lies on a different ellipse from before. Thus the value $2m_1$ has changed into the apparent value $2m_1'$ where the subscript denotes the element to which the value refers. The length and hence the end point of the second element can now be obtained by marking off from the value $2m_1'$ the number of degrees $\Delta2m_2$ represented by the length of the second element of the path curve, i.e.

$$2m_2 = 2m_1' + \Delta2m_2.$$

After rotating the paper into the position corresponding to the third element of the path curve, $2m_2$ is transformed into $2m_2'$. Then

$$2m_3 = 2m_2' + \Delta2m_3,$$

and so on.

The difference between m and m' can be read from the nomogram but the accuracy will be poor especially at points near the circumference of the circle. Greater accuracy is obtained using the relationship

$$\frac{\sin 2m}{\sin 2m'} = \frac{d'}{d}, \tag{13.6}$$

in which d and d' denote respectively the lengths of the chords of m and m'.

Equation (13.6) is derived from the geometrical conditions in Figure 13.9. It can be seen that $CF_1 = CF_2$ since these represent the same half chord of the sphere and coincide when the semicircles $A_1F_1B_1$ and $A_2F_2B_2$ are turned through 90° about their respective diameters. Further, since $\sin 2m = CF_1/A_1D_1$ and $\sin 2m' = CF_2/A_2D_2$ it follows that

$$\sin 2m/\sin 2m' = A_2D_2/A_1D_1 = d'/d.$$

The relative lengths of the chords are indicated in Figure 13.8.

When plotting a tip curve it is convenient to record the results in tabular form as illustrated by the following example:

element no.	$\Delta2m$	$2m$	$2m'$	d	d'	$\sin 2m$	$\sin 2m'$
1	20	20	18·7	0·8	0·85	0·342	0·322
2	17	35·7	35·2	0·85	0·86	0·584	0·577
3	15	50·2	52·8	0·86	0·83	0·768	0·796

13.5 Optical effects of birefringent bodies in series

In order to obtain some general rules, we now consider two birefringent bodies placed between the polarizer and the analyser so that the light emerging from the first passes through the second. The combined effect of the two bodies is characterized by the three optical values m_k, ε and δ which depend on the values m_{k1}, ε_1 and δ_1 produced by the first body and m_{k2}, ε_2 and δ_2 produced by the second. If the bodies are two-dimensional models, $\delta_1 = \delta_2 = 0$ and the angles of inclination correspond to the directions of the principal stresses in the individual models. The resulting angle of compensation then depends only on m_{k1}, m_{k2} and the angle $(\varphi_2 - \varphi_1)$ between the directions of the principal stresses in the two models. For this case, m_k can be derived from the equation

$$\sin^2 2m_k = \sin^2 2m_{k1} \sin^2 2m_{k2} \sin^2 2(\varphi_2 - \varphi_1) + \sin^2 2m_{k1} \cos^2 2m_{k2}$$

$$+ \sin^2 2m_{k2} \cos^2 2m_{k1} \tag{13.7}$$

$$+ 2 \sin 2m_{k1} \sin 2m_{k2} \cos 2m_{k1} \cos 2m_{k2} \cos 2(\varphi_2 - \varphi_1).$$

It will be observed that this equation is unaltered if the indices 1 and 2 are interchanged since $\cos 2(\varphi_2 - \varphi_1) = \cos 2(\varphi_1 - \varphi_2)$. The value of m_k is therefore unaltered if the models are interchanged. It might be supposed that a similar result would be obtained if more than two models were placed in a field of polarized light. It will be shown, however, that a pair of two-dimensional models in which the directions of the principal stresses do not coincide produces an angle of rotation which influences the optical values resulting from succeeding models. This effect is not included in the derivation of equation (13.7).

It will now be shown that the angle of rotation produced by two two-dimensional models in series depends on the order in which they are placed. Figure 13.10(a) shows the rotated path curve for two such models while Figure 13.10(b) shows the corresponding tip curve OAB and also the tip curve $OA'B'$ for the models in the reverse order when placed in circularly polarized light. As will be observed, the end points B and B' do not coincide. This would occur only if $AB = OA'$ and $OA = A'B'$ in which case these four lines would form a parallelogram. These pairs of lines are however of unequal length even although they represent equal angles of rotation of the sphere since they are the projections of points lying on parallel circles of different diameters.

The lengths OB and OB' correspond to the respective values of $\sin 2m_k$. As shown by equation (13.7) however, m_k is the same in each case so that B and B' are at equal distances from the centre.

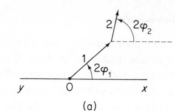

(a)

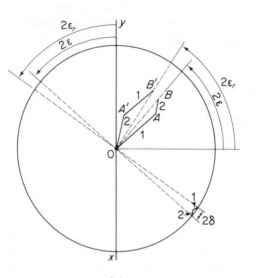

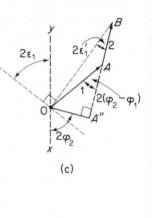

(c)

(b)

Figure 13.10.   Influence of the order of sequence of two two-dimensional models in series

To determine the resulting angle of inclination ε, we use the angle ε' (Figure 13.10(c)) which can be obtained from the equation

$$\sin 2\varepsilon' = \frac{OA''}{OB} = \frac{\sin 2m_{k1} \sin 2(\varphi_2 - \varphi_1)}{\sin 2m_k}, \qquad (13.8)$$

after inserting the value of the resultant angle of compensation m_k calculated from equation (13.7). For the models in the order shown in Figure 13.10(a), the angle of inclination ε is given by

$$2\varepsilon = 2\varphi_2 - 2\varepsilon', \qquad (13.9)$$

as can be seen from Figure 13.10(c). For the models in the reverse order ε' is replaced by ε_r' where

$$\sin 2\varepsilon_r' = \frac{\sin 2m_{k2} \sin 2(\varphi_1 - \varphi_2)}{\sin 2m_k}. \qquad (13.10)$$

Then

$$2\varepsilon_r = 2\varphi_1 - 2\varepsilon_r'. \tag{13.11}$$

The angles of rotation δ and δ_r for the models in the original and reversed orders can be determined from the equations

$$\delta = \varepsilon - \varepsilon_r, \tag{13.12}$$

and

$$\delta_r = \varepsilon_r - \varepsilon, \tag{13.13}$$

as can be read from Figure 13.10(*b*).

We next consider the effect of two general three-dimensional models placed in series between the polarizer and the analyser. To obtain the resultant values m_k, ε and δ due to the combined effect of both models, the sphere must be rotated according to the following sequence of operations for which the notation used is the same as before.

1. Rotation about the axis perpendicular to the plane of projection through the angle $2\delta_1$.
2. Rotation about the axis parallel to the plane of projection and to the direction indicated by $2\varepsilon_1$ through the angle $2m_{k1}$.
3. Rotation about the axis perpendicular to the plane of projection through the angle $2\delta_2$.
4. Rotation about the axis parallel to the plane of projection and to the direction indicated by $2\varepsilon_2$ through the angle $2m_{k2}$.

These rotations can be effected by drawing a tip curve starting for instance from the centre of the *j*-circle (see Figure 13.11). The first rotation does not influence the shape of the curve. The second rotation shifts the tip of the *j*-vector through a distance corresponding to $\sin 2m_{k1}$ from the centre of the circle along the straight line OA which is parallel to the direction of $2\varepsilon_1$. The third rotation causes the end of the tip curve to travel along the circular arc AA' about the centre of the *j*-circle. The fourth rotation produces a movement along the straight line $A'B$ parallel to the direction of $2\varepsilon_2$ and completes the tip curve.

The shape of the triangle $OA'B$ is independent of the angles of rotation but not its position since it is obviously rotated through the angle $2\delta_2$. The resultant angle of compensation can therefore be determined from equation (13.7) if the angle $(\varphi_2 - \varphi_1)$ is replaced by $\varepsilon_2 - (\varepsilon_1 + \delta_2)$.

To obtain the angle of inclination, we take $2\varepsilon'$ to represent the angle OBA' in Figure 13.11 which is given by equation (13.8) after making the above substitution for $(\varphi_2 - \varphi_1)$. The angle δ_1 must be added to the value corresponding to that given by equation (13.9) so that we now have

$$2\varepsilon = 2\varepsilon_2 - 2\varepsilon' + 2\delta_1.$$

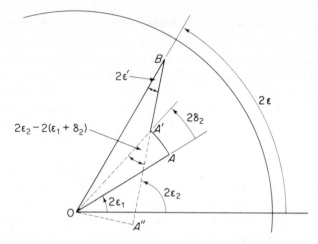

Figure 13.11.. Determination of the optical effects produced by two general three-dimensional models in series

With the models in the reverse order, δ_2 must be added to the value corresponding to that given by equations (13.10) and (13.11) so that

$$2\varepsilon_r = 2\varepsilon_1 - 2\varepsilon_r' + 2\delta_2.$$

The angle of rotation for the models in the original order is obtained by adding the angles of rotation δ_1 and δ_2 of the individual models to the corresponding value given by equation (13.12) so that

$$\delta = \varepsilon - \varepsilon_r + \delta_1 + \delta_2.$$

For the reverse order,

$$\delta_r = -\delta.$$

When three or more two-dimensional models are placed in series in a beam of polarized light, the resultant optical values can be found by first determining those for two adjacent models. The optical effect of this first pair can then be regarded as if it were produced by a single three-dimensional model and combined with the effect produced by the next model and so on. It follows from our previous considerations, however, that the laws are now different from those governing the case of only two two-dimensional models. Thus, the angle of compensation is no longer independent of the order of the models except when they are placed in the reverse sequence. In the latter case, the angle of compensation is the same, the angle of rotation changes in sign but not in magnitude and the angles of inclination differ by the angle of rotation.

13.6 Composite quarter-wave plates

Tuzi and Oosima[46] have given formulae for quarter-wave plates composed of two suitably oriented birefringent plates producing equal phase differences within the limits of one-eighth to three-eighths wavelength. It can easily be seen from the *j*-circle, however, that any two or more component plates for which the end point of the tip curves lies at the centre of the circle will produce circularly polarized light (Figure 13.12.) Thus, the phase differences produced by the component plates may be equal or unequal and may have values outwith the above limits.

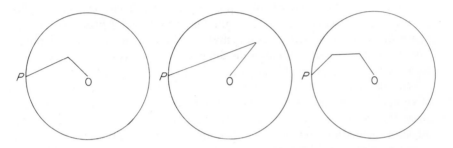

Figure 13.12.   Production of circularly polarized light by birefringent bodies in series

It follows from Section 13.5 that a quarter-wave plate formed of two components will cause a certain angle of rotation. This can be avoided by using three (or more) components which produce a skew symmetrical path curve. As shown in Section 13.7, the angle of rotation is then zero, but since a certain amount of dispersion with respect to m_k is unavoidable, mono-chromatic light only should be used. Quarter-wave plates of this type may be adjusted to suit the wavelength of the light used, the second plate being rotated relative to the first and third. Since this rotation will alter the angle of inclination, the first and third plates must also be rotated in common through a certain angle if the angle of inclination is to remain constant. This can be accomplished by using a suitable drive mechanism.

13.7 Numerical determination of optical data

General cases of birefringence in which the directions of the principal stresses rotate along the path of the light beam can be solved using a system of three differential equations which are derived from the geometrical relationships in the *j*-circle.

These equations are

$$dm_k = \cos 2(\varphi - \varepsilon)dm, \qquad (13.14a)$$

$$d\varepsilon = \sin 2(\varphi - \varepsilon) \cot 2m_k dm, \qquad (13.14b)$$

$$d\delta = \sin 2(\varphi - \varepsilon) \tan m_k dm, \qquad (13.14c)$$

in which φ denotes the directional angle of the principal stresses at an element of the path of the light beam.

13.8 Optical phenomena of special cases

It has been shown that when a tip curve is drawn from the centre of the j-circle, i.e. for circularly polarized light, for a model which produces a given phase difference or angle of compensation, the distance of the end point from the centre is independent of the directions of the principal stresses. The locus of these end points is therefore a circle concentric with the j-circle.

With plane polarized light the end points of tip curves for constant angle of compensation and varying angles of inclination also form a circle. This circle touches the j-circle at the point where the direction indicated by the angle of inclination coincides with the direction of the polarizer.

In two-dimensional cases the phase difference is inversely proportional to the wavelength. Thus for light of a given wavelength, varying phase differences produce a characteristic scale of colours. In general three-dimensional cases with rotating principal stresses the angle of compensation is no longer proportional to the reciprocal of the wavelength as can be shown for instance from equation (13.7). This produces changes in the colours observed as compared with two-dimensional cases and in general they are less distinct. This effect can be called dispersion.

The amount of dispersion depends on the particular case and general rules governing it can only be established when the special conditions regarding the state of stress are known such as in the case of shells (see Chapter 17).

With monochromatic light, no dark lines corresponding to isochromatics normally appear though dark spots can sometimes be observed. With plane polarized light and crossed polaroids there are no isoclinics in the form of continuous dark lines but only dark spots. When the polaroids are rotated out of the crossed position, different spots on the model will appear dark. When white light is used, however, extinction will not occur even at these points since the angles of inclination and rotation depend upon the wavelength of the light.

If a birefringent body is provided with a reflecting layer, light passing normal to the layer will enter and be reflected back along the same path. Obviously the reflected beam passes through all points in the reverse order. The path curve in this case will therefore be skew symmetrical as shown in

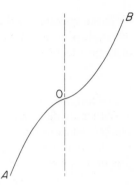

Figure 13.13. Skew symmetrical path curve for light
reflected back along its original path

Figure 13.13. This is true also when light passes normally through a plane
of symmetry in a three-dimensional body. It can be shown that in such cases
there are two points 180° apart on the circumference of the circle for which
the starting and end points of the tip curves coincide. Starting from either
of those points the first half of the tip curve ends at some other point on the
circumference and the second half coincides with the first as illustrated by
curve $AA'A$ in Figure 13.14. One of these halves lies on the front side and the

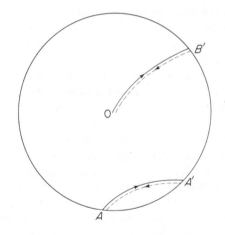

Figure 13.14. Tip curves for which
the starting and end points coincide

other on the rear side of the sphere. As a result of the coincidence of the starting and end points of such tip curves, the angle of rotation vanishes. Thus
isoclinics (i.e. lines of constant angle of inclination) will appear with crossed
polaroids. As can be read from Figure 13.7, the angle of inclination with
double passage of the light due to reflection $(A_3B_3A_3)$ will differ from that

for single passage in the direction of the incident light (tip curve A_3B_3) by the angle of rotation δ for single passage or, with reference to the path curve of Figure 13.13,

$$\varepsilon_{A \to B} = \varepsilon_{A \to O} - \delta_{A \to O}.$$

The angle of compensation is double that with single passage of the light.

With reflection, dark and bright bands corresponding to isochromatics can be observed when using circularly polarized monochromatic light and parallel polarizers. The intensity of light along the bright bands is not constant however and with white light the colours along them vary. With crossed polarizers, no isochromatics of this type will in general be observed when using either monochromatic or white light. This effect can be explained by means of the j-circle in the following way: In a model in which the state of stress varies from point to point there will exist certain points, lying in general on a curve, for which the end of the first half of the tip curve will lie on the circumference of the circle. Here again, the second half of the tip curve will coincide with the first, one half lying on the front side and the other on the rear side of the sphere. Since the light is circularly polarized, the starting and end points coincide with the centre of the circle although lying on opposite sides of the sphere (see curve $OB'O$, Figure 13.14). This obviously represents a compensation angle of 90° which, in a two dimensional system, would correspond to a phase difference equal to an odd number of half wavelengths. With parallel polarizers, lines along which this condition is satisfied will therefore appear dark.

With reflection of the light, the compensation angle m_k may be defined as an angle between 0° and +180° and a value of 90° does not represent an exceptional case. A compensation angle of 180° will only occur in exceptional cases however.

Similar phenomena will obviously be observed in the case of a finite or infinite number of doubly refracting layers in which alternate layers have the same constant direction of principal stresses and phase difference. A path curve for a series of layers of this kind consists of a regularly serrated line which can be subdivided into identical skew symmetrical parts as shown in Figure 13.15. Each of these parts produces the same change in the angle of compensation. The angle of inclination remains constant as the light passes through the layers and the angle of rotation is zero. Dispersion depends on the thickness of the layers or on the absolute magnitude of the phase difference and the difference between the direction of the principal stresses in the alternate layers.

Some particular laws apply to symmetrical path curves. If the line joining the ends of such a curve coincides with the reference direction x in Figure 13.16(a), then as shown in Figure 13.16(b) the tip curve AE which connects two points on the circumference of the j-circle is also symmetrical and the chord

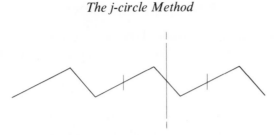

Figure 13.15. Path curve for light transmitted
through an alternation of equal layers

joining the ends A and E is perpendicular to the reference direction x. Since OE represents the direction of the major axis of the ellipse, twice the angle of rotation δ is equal to the angle AOE which is twice the angle BOE. Thus, the angle of inclination ε is equal to $\frac{1}{2}\delta$.

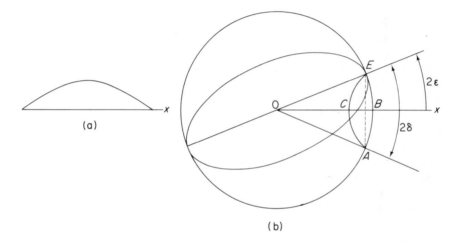

Figure 13.16. (a) Symmetrical path curve, and (b) the corresponding tip curve

In certain particular cases the angle of compensation produced by a birefringent body is zero and the body merely rotates the plane of polarization. This obviously happens when a tip curve for circularly polarized light terminates at the centre of the *j*-circle as shown in Figure 13.17. The corresponding tip curves for plane polarized light vibrating in any direction, i.e. starting from any point on the circumference, also terminate at the circumference. In general, a certain amount of dispersion occurs, however, so that the angle of compensation is zero for only one particular wavelength while any other wavelength produces a finite angle of compensation.

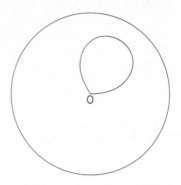

Figure 13.17. Tip curve for circularly polarized light terminating at the centre of the *j*-circle

14

THE FROZEN STRESS METHOD

14.1 Introduction

Three-dimensional states of stress cannot be investigated by the simple two dimensional procedure of observing the isochromatics and isoclinics when polarized light is transmitted through the model even if the model is sufficiently transparent to allow this to be done. This is because the integrated optical effect is generally so complex that it is impossible to analyse it or relate it to the stresses which produce it.

Several methods are available for the investigation of three-dimensional problems. The most widely used of these is the so-called 'frozen stress' method which is restricted in its application, however, to static cases of loading by external forces or constant body forces such as gravitational or centrifugal forces. In this method, advantage is taken of the multiphase (diphase) nature of plastics used as model materials to conserve strain and double refraction indefinitely in the model after the loads have been removed. With care, slices can be cut from the model for analysis without disturbing the frozen-in stress pattern.

14.2 Experimental procedure

The procedure of strain freezing consists of heating the model to a temperature slightly above the softening point or critical temperature followed by slow cooling under load to room temperature. The load may be applied to the model either before or after reaching this temperature. In practice, however, it is usually more convenient to load the model when cold, thus avoiding the difficulty of handling the model and loading device when hot. During the heating process, the rigidity of the model diminishes as the temperature rises and at the softening point is reduced to a level corresponding to that of rubber. As a result, the deformations increase during this period and it is possible that this may cause the load to become displaced. The loading system should therefore be checked after the softening point has been reached to ensure that no undue displacement has taken place. Whenever possible, loading devices should be constructed in a manner which will allow corrections to be made by simple manipulations.

Extreme accuracy is required to ensure that the model is subjected to the correct loading. Because of the low rigidity of the material when hot, spurious bending can often be induced in models which are intended to be subjected to pure tensions or compressions. Undue additional stresses may be induced by gravitational forces if the shape of the model is such as to allow large deformations, e.g. in thin-walled structures.

While the rate of heating is immaterial, adequate time must be allowed before commencing the cooling process to ensure that the softening point has been reached not only at the surfaces of the model but also in the interior. Cooling should be carried out slowly, especially if the model is thick-walled, to avoid introducing thermal stresses. With models having a thickness of 10 cm or more, the rate of cooling of the surrounding air within the oven should not exceed 2 °C per hour.

It is often possible to check the correctness of the distribution of the frozen stress system in the model before it is sliced by observing it in polarized light. In cases of symmetrical stress distribution, for instance, deviations from symmetry can be detected in this way. If any serious deviation is found, the heating, loading and cooling cycle should be repeated.

If the model becomes overheated during the process of cutting slices, the stress pattern may be disturbed or even destroyed. To avoid this, only sharp tools should be used and methods of cutting should be employed which are unlikely to generate an undue amount of heat. Cutting by band saw or circular saw is satisfactory in this respect while milling at low speeds should be avoided. It is advisable to apply a coolant and for this purpose compressed air is preferable to water which may cause edge effect. Rough surfaces of the slices such as are produced by a band saw can be treated carefully with sand paper. If accurate thickness is required, the slices can be milled at high speed after sticking them to the table of the milling machine with double-sided adhesive tape.

To avoid edge effect, models should only be sliced immediately before their optical analysis. When it is necessary to store slices for longer periods, edge effect can be reduced by placing them in a desiccator or in an oven at a temperature of 60 to 80 °C (for hot curing epoxy resin).

The load to be applied to a frozen stress model must be chosen within fairly narrow limits. For hot curing epoxy resin, which is the most suitable of the model materials available, both Young's modulus ($20 \, MN/m^2$) and the maximum allowable tensile stress ($c. \, 1 \, MN/m^2$) are comparatively low. This restricts the magnitude of the load which may be applied if excessive deformation, especially of models of thin-walled structures, and the danger of fracture are to be avoided. On the other hand, the optical sensitivity of the material is rather low ($f = 300$ N/m fr) and this imposes a lower limit on the load, since the optical effect produced must be sufficient to permit accurate analysis. The optical effect depends also upon the thickness of the slices to be cut. All of these factors should be taken into account in choosing the load.

14.3 Determination of optical data

If a slice is cut from a model in a general state of stress where the stresses vary from point to point, the frozen-in strain will not be uniform throughout the thickness. The slice should therefore be thin enough to allow the effect of this variation to be neglected. When it is impossible to satisfy this condition, special methods such as the j-circle method described in Chapter 13 can be applied.

Some of the optical techniques applicable to the frozen stress method are different from those of two-dimensional photoelasticity. Since the optical data provide only five of the six values necessary to characterize the state of stress at a point in a body, some other method must be applied to determine the sixth value except at load free edges where the boundary conditions supply the necessary information. If a slice is cut from a three-dimensional model in such a way that the directions of two of the three principal stresses lie in the plane of the slice, the difference of these two stresses can be determined by the same methods as in two-dimensional photoelasticity. This presupposes that the direction of at least one of the principal stresses is known and this is frequently the case, e.g. on a section of symmetry or a free surface of the body. If such a slice is observed under normal incidence, Figure 14.1,

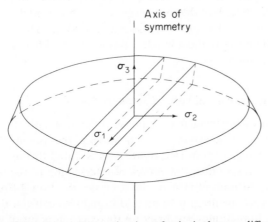

Figure 14.1. Determination of principal stress differences from slices when the direction of one principal stress is known

the birefringence will give the difference $(\sigma_1 - \sigma_2)$ of the principal stresses in the plane of the slice while the isoclinics will indicate their directions. To determine the difference of principal stresses in the other two principal planes, a subslice may be cut from the original slice as indicated, parallel to one of the principal stresses, say σ_1. If this subslice is observed under normal incidence in the direction of σ_2, the birefringence will give the value of $(\sigma_1 - \sigma_3)$. The remaining value, i.e. $(\sigma_2 - \sigma_3)$ may then be found by subtracting the second value from the first. If one of the surfaces of the original slice is part of a free surface of the body, then $\sigma_3 = 0$ and the above measurements will yield the individual magnitudes of σ_1 and σ_2. In regions of high stress gradient normal to the surface, however, the value of σ_2 can be determined with greater accuracy from observations of the subslice under normal and oblique incidence in the plane of σ_1 and σ_2.

In general, however, the directions of the principal stresses are oblique with respect to the plane of the slice and it is necessary to apply the general law relating the birefringence and the direction of observation. This relation can be represented in terms of either the ellipsoid of strain (see Section 1.1) or the index ellipsoid (Section 4.2). The primary light vibration is split into two component vibrations, the directions of which coincide in each case with the axes of the central elliptical section of the ellipsoid perpendicular to the direction of propagation. These directions are those of the secondary principal stresses in the plane of the wave front and the birefringence is proportional to the difference in length of the two axes of the section. The absolute maximum birefringence is obviously observed in the direction perpendicular to the principal section of the ellipsoid containing the greatest and least axes, i.e. perpendicular to the greatest and least principal stresses.

Both ellipsoids can be characterized by six values, for example the directions and lengths of each of the three axes. These three lengths are obviously independent of each other and can assume any value. There are only two independent differences of these lengths, however, one difference being equal to the sum of the other two.

14.4   Observation in conical light: The conoscope

The variation of birefringence or phase difference with the direction of transmission can be determined by observing the isochromatics when convergent or divergent polarized light is transmitted through a slice. The isochromatics correspond to the curves of intersection of the isochromatic surfaces and the second surface of the slice (see Section 4.10). In crystallography, a microscope fitted with an additional lens called a Bertrand lens is used. This additional lens produces an image, each point of which corresponds to a certain direction within the (homogenous) crystal. A microscope is not suitable, however, for the investigation of slices of more than 2 mm thickness. As most slices used in photoelasticity are thicker than this, a simple optical setup based on the same principle but having larger lenses can be used. This setup, known as a conoscope, is shown in Figure 14.2 and

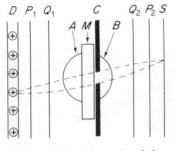

Figure 14.2. Schematic of the conoscope

consists of two stress-free plano-convex lenses A and B and a diaphragm C having a small aperture. The model slice M is placed between the first lens and the diaphragm. Light from the diffuse source D passes through this system which is placed between polarizers P_1, P_2 and quarter-wave plates Q_1, Q_2. Light rays which are parallel within the slice intersect in the rear focal plane of the lens B where they form an image on the screen S. The phase difference observed at a point on this image thus relates to a particular direction of transmission through the slice and is the same as that which would be observed with collimated light falling on the slice at an angle of incidence producing the same direction of transmission. Such an interference pattern therefore gives simultaneously the birefringence for all angles of incidence, the value at the centre corresponding to that for normal incidence.

The images formed on the screen differ in every individual case but conform to certain patterns depending on the state of stress and the directions of the principal stresses relative to the plane of the slice. If the slice is cut perpendicular to the direction of the single principal stress in a uniaxial stress system, the isochromatic surfaces correspond to those of a uniaxial crystal cut perpendicular to its optic axis (see Figure 4.10). In this case the isochromatic lines form a system of concentric circles. In general, when the three principal stresses are all different, the isochromatic surfaces correspond to those for a biaxial crystal and have the form shown in Figure 4.11(a). If the slice is cut in such a way that two of the principal stresses lie in its plane, the isochromatic lines will assume one or other of the forms indicated in Figures 4.11(a) and 4.11(b).

For a point source O in the interior of a medium, the isochromatic surfaces are symmetrical about the xy plane (Figure 4.11(a)). The x, y and z axes are then axes of symmetry of the isochromatic surfaces and will be referred to as such in the following discussion. The axes and planes of symmetry of the isochromatic surfaces clearly coincide with those of the ellipsoids and of the principal stresses.

Figure 14.3 shows the isochromatic pattern of a slice cut perpendicular to an axis of symmetry in circularly polarized conical light. Both the birefringence and the directions of the axes of symmetry parallel to the plane of the

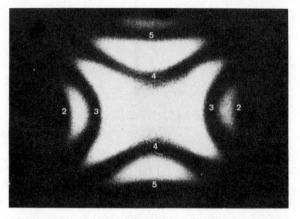

Figure 14.3.   Isochromatic pattern of a slice cut perpendicular to an axis of symmetry in conical light

slice can be read from this photograph. Figure 14.4 shows the corresponding isochromatic pattern of a slice cut from the same model perpendicular to one of the optic axes.

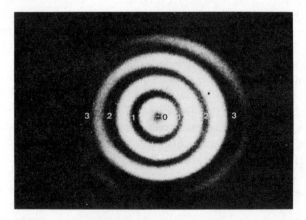

Figure 14.4.   Isochromatic pattern of a slice cut perpendicular to an optic axis in conical light

If plane polarized conical light is used, a second system of fringes known in crystallography as isogyres is observed superimposed on the isochromatics. The isogyres are the loci of points in the interference field where the directions of vibration of the light waves are parallel to the axes of the crossed polarizers and consequently vary as the latter are rotated. The directions of the principal or secondary principal stresses at the point of the slice to which the interference pattern relates are indicated by the parameter of the isogyre passing through the centre of the field. Figures 14.5 and 14.6 show the

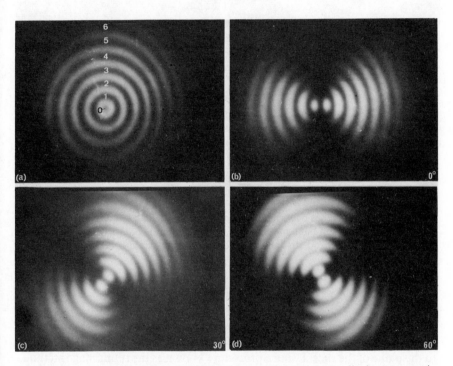

Figure 14.5. Interference patterns obtained from a slice cut perpendicular to an optic axis in convergent light. (*a*) isochromatics in circularly polarized light, (*b*), (*c*), (*d*) isochromatics and isogyres in plane polarized light

isochromatics in circularly polarized light and the isochromatics and isogyres of various parameters in plane polarized light obtained from slices cut perpendicular to an optic axis and perpendicular to an axis of symmetry respectively.

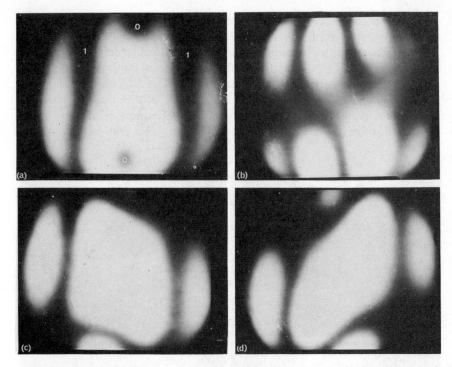

Figure 14.6.   Interference patterns obtained from a slice cut perpendicular to an axis of symmetry in convergent light. (*a*) isochromatics in circularly polarized light, (*b*), (*c*), (*d*) isochromatics and isogyres in plane polarized light

14.5   The tilting stage method

If a slice is cut containing points at which the directions of the principal stresses are unknown, these directions in general will be oblique with respect to the plane of the slice. Using a procedure known as the tilting stage method, it is possible to orient the slice in such a way that one of the principal axes of the ellipsoid at the point under observation coincides with a known axis in the plane of the wavefront. In crystallography, a polarizing microscope fitted with a universal tilting stage is used and this equipment is suitable for frozen stress slices also provided they are not too thick. Figure 14.7 shows a typical three-axis universal stage. This contains an inner turntable on which the slice is mounted between two glass hemispheres of matching refractive index with immersion fluid between the contact surfaces to eliminate the effects of refraction. This turntable can be rotated in its own plane about its axis A_1 (vertical in the zero position) and is mounted on gimbals within a ring allowing rotation of its plane about the north-south axis A_2.

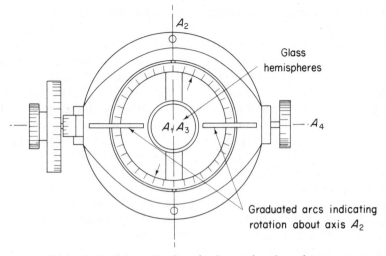

A_2

Glass
hemispheres

A_4

$A_1 A_3$

Graduated arcs indicating
rotation about axis A_2

Figure 14.7. Schematic plan of a three-axis universal stage

The ring can be rotated about the main horizontal east-west axis A_4 of the stage. The various angles of rotation can be read from graduated circles or arcs.

The procedure consists of manipulating the stage systematically until one of the principal axes of the ellipsoid at the observed point in the slice coincides with the horizontal axis A_4. Since a principal axis of the ellipsoid is also a principal axis of all sections containing it, the necessary condition to be fulfilled is that, with crossed polaroids parallel to the axes A_2 and A_4 respectively, extinction will occur at the point observed and will be maintained when the stage is rotated about the axis A_4. The stage and slice are first adjusted so that the desired point on the slice coincides with the intersection of the crosswires of the ocular and remains stationary during rotation about the various axes. The following is one of several procedures which may then be followed to achieve the desired orientation:

1. Commencing with the stage set in the zero position, rotate the inner turntable about the vertical axis A_1 until the light is extinguished at the centre point of the field. This occurs when one of the two principal axes of the central section of the ellipsoid parallel to the plane of the slice at the point coincides with the axis A_4.

2. Rotate the stage about the axis A_4. If extinction is maintained, the axis parallel to A_4 is a principal axis of the ellipsoid. In general, rotation about A_4 will relieve extinction. In this case, proceed as follows:

3. Rotate the stage about the axis A_4 through a small angle in the direction in which the light is most rapidly restored.

4. Rotate about the axis A_2 until extinction is restored.

5. Rotate about axis A_4 through a small angle of the opposite sense to the rotation in step 3.

6. Rotate the inner turntable about the axis A_1 until extinction is restored.

The sequence of operations 2 to 6 is repeated until rotation about the axis A_4 fails to relieve extinction. When this condition is satisfied, one of the principal sections of the ellipsoid lies in the vertical N-S plane and one of the principal axes coincides with the axis A_4. If this axis is the intermediate axis OB (Figure 4.3(b)), it is possible to bring one and perhaps both of the optic axes into coincidence with the axis of the microscope tube. To investigate this, the microscope stage is now rotated about its own vertical axis A_3 into the 45° position. The universal stage is then rotated about the axis A_4. If a position producing extinction is found, one of the optic axes will then coincide with the axis of the microscope tube and the axis OB of the ellipsoid coincides with the axis A_4. If extinction cannot be obtained, then either axis OA or OC of the ellipsoid coincides with A_4. Which of these two is the one concerned can be determined by inserting a compensator in the slot in the microscope tube.

In step 1, rotation about the axis A_1 brings one of the two principal axes of the elliptical section into coincidence with A_4. If this initial rotation is given in the opposite direction, extinction will occur when the other secondary principal axis coincides with A_4. The sequence of operations 1 to 6 will then lead to the location of one of the other principal axes of the ellipsoid.

In order to determine the phase difference and hence the difference of two principal stresses, the slice is first oriented so that the corresponding principal section of the ellipsoid lies in the plane of the wave front. This process can be simplified by plotting the positions of the principal sections and principal axes on a stereographic net as described by Jessop and Wells[47] from the rotations recorded. The microscope stage is then rotated to bring the principal axes into the 45° position for measurement of the phase difference with a compensator. To avoid errors due to excessive variation of the stresses along the path of light, the permissible tilt of the slice must be limited to about 45°. In many cases, however, the loading conditions will give some indication of the direction in which the slice should be cut so that this requirement can be satisfied.

The operation described above can be performed in the polariscope using a tilting stage specially constructed for the purpose. This allows the investigation of larger slices than is possible using a microscope. In such stages, the glass hemispheres are usually dispensed with, the effects of refraction being avoided by immersing the stage and cell in a liquid having the same refractive index as the material of the slice contained in a strain-free glass cell. Extinction at a point in the slice will then be indicated by the passage of an isoclinic through the point.

14.6 Interpretation of optical data

In the following discussion the principal stresses are denoted by $\sigma_1, \sigma_2, \sigma_3$ where σ_1 is the algebraically greatest and σ_3 the algebraically smallest stress (i.e. $+\sigma_1 \geqslant +\sigma_2 \geqslant +\sigma_3$). If n_y is the intermediate refractive index, the optic axes will lie in the xz plane. The z axis will be taken as that which includes angles of less than $45°$ with the optic axes and will coincide with either the greatest or the least principal axis of the index ellipsoid depending on the value of n_y in relation to those of n_x and n_z or the value of σ_2 in relation to those of σ_1 and σ_3. To a close approximation, the z axis coincides with the greatest principal axis if $n_y < (n_x + n_z)/2$ as can readily be derived from equation (14.1) or if $\sigma_2 < (\sigma_1 + \sigma_3)/2$. In this case, n_z is the greatest refractive index and n_x the least; also $\sigma_1 = \sigma_z$ and $\sigma_3 = \sigma_x$ (Figure 14.8(a)). These conditions are reversed if $\sigma_2 > (\sigma_1 + \sigma_3)/2$ (Figure 14.8(b)). Hence, any

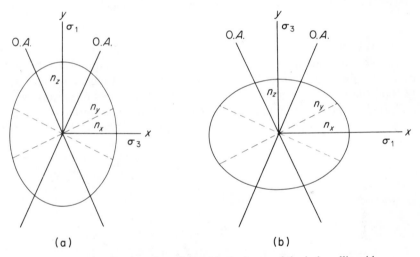

(a)   (b)

Figure 14.8. Designation of the principal axes of the index ellipsoid
(a) $n_y < (n_x + n_z)/2$, (b) $n_y > (n_x + n_z)/2$

particular triplet of phase differences observed or determined indirectly in the directions of the three axes of symmetry can correspond to one of two different systems of Mohr's circles, that of either Figure 14.9(a) or Figure 14.9(b) depending on the sign of the corresponding phase difference or difference of principal stresses. This difference can be determined by means of a compensator in case of doubt.

The interrelation between the signs of the birefringence in the planes of symmetry can be deduced from the index ellipsoid. If the two refractive indices corresponding to every direction of observation are plotted about a point O, a surface of refractive indices of two sheets is obtained similar to

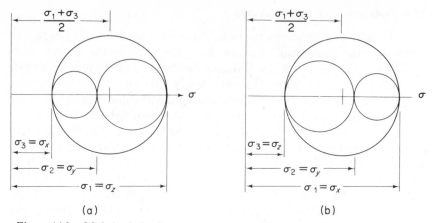

Figure 14.9.   Mohr's circles for stresses corresponding to the ellipsoids of Figures
14.8(*a*) and 14.8(*b*) respectively

the ray velocity or wave surfaces in Figure 4.5. The distance between the two
sheets measured along any straight line through O corresponds to the birefringence observed in the direction of that line. In Figures 14.10(*a*) and
14.10(*b*) the variations in sign with direction in the planes of symmetry are
indicated by the traces in these planes of the birefringence plotted over a unit
sphere. The systems of Figure 14.10(*a*) and Figure 14.10(*b*) correspond to the

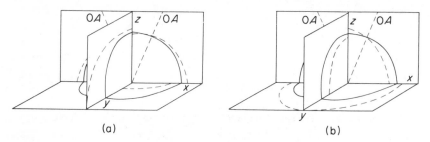

Figure 14.10.   Surfaces of refractive indices corresponding to the ellipsoids of
Figures 14.8(*a*) and 14.8(*b*) respectively

circle systems of Figure 14.9(*a*) and Figure 14.9(*b*) respectively. As can be seen,
the sign of the birefringence is constant in the xy and zy planes but changes
in the xz plane at the optic axes. Further, since the order of magnitude of the
refractive indices is reversed, the birefringence is of opposite sign in the two
cases.

For any direction of transmission in the planes of symmetry, the directions
of the secondary principal stresses lie within those planes. For any other
direction of transmission, the direction of the secondary principal stresses

can be derived from the ellipsoid of strain or the index ellipsoid. In photo-elastic materials the difference of the refractive indices is between 10^{-4} and 10^{-3} which is very small compared with their absolute values. Under these conditions the birefringence D for any arbitrary direction of propagation can be expressed by the approximate formula

$$D = D_y \sin \theta_1 \sin \theta_2, \qquad (14.1)$$

in which D_y is the maximum birefringence $(n_z - n_x)$ which is observed in the direction of the y axis and θ_1, θ_2 are the angles between the direction considered and those of the optic axes. As can be derived from equation (14.1) or equation (5.6), the birefringence D_i in any direction within a plane of symmetry is given by

$$D_i = D_1 + (D_2 - D_1) \sin^2 i, \qquad (14.2)$$

where D_1, D_2 are the birefringences observed in the directions of the two axes of symmetry in that plane and i is the angle between the direction of D_i and that of the axis corresponding to D_1. Equation (14.2) shows that the curves of Figure 14.10 are $\sin^2$ curves plotted over circles. This relation can be used to determine by extrapolation the birefringence D_2 in the direction of an axis which cannot be observed: Equation (14.2) is written in the form

$$D_2 = \frac{D_i - D_1}{\sin^2 i} + D_1. \qquad (14.3)$$

This equation can be solved in a simple graphical way using the nomogram given in Figure 14.11 in which the birefringence D is plotted against the angle i. The scale of the abscissa is subdivided in proportion to $\sin^2 i$ so that the $\sin^2$ curves become straight lines. The lines $i = 0$ and $i = 90°$ represent the axes of D_1 and D_2 respectively. To apply the nomogram, a straight line is drawn through the points representing D_1 on the line $i = 0$ and D on the line $i = i$ and projected to intersect the line $i = 90°$. This point of intersection represents D_2.

It can be seen from equation (14.3) that the determination of D_2 is simplified if $i = 30°$ or $45°$. Then

$$D_2 = 4D_i - 3D_1 \qquad \text{for } i = 30°,$$

and

$$D_2 = 2D_i - D_1 \qquad \text{for } i = 45°.$$

If the directions of both optic axes or of one optic axis and one axis of symmetry (x or z) can be determined, the unknown birefringence for the direction of the other axis (z or x) can be calculated from the birefringence observed using the following relations derived from equation (14.1):

$$\sin^2 \beta = D_z/D_y, \qquad \cos^2 \beta = D_x/D_y, \qquad \tan^2 \beta = D_z/D_x,$$

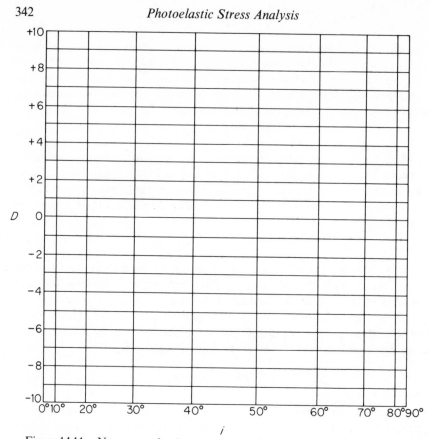

Figure 14.11. Nomogram for determination of the birefringence in the direction of an axis which cannot be observed

where β is the angle between the z axis and the optic axis. The directions of both optic axes and an axis of symmetry can be determined by conoscopic observation of a slice cut perpendicular to the z axis or of a slice cut perpendicular to the x axis if the angle between this axis and the optic axes is not too great.

14.7 Oblique incidence

Though the method of conoscopic observation of slices is the most rapid for the complete determination of the optical data at a point, other methods which yield the optical data over a whole field are often preferred. Such methods, in which parallel light is employed, give the double refraction for one direction only with each observation. Several observations are therefore usually required in order to obtain all the information needed. It will be shown in the following section, however, that in many cases observation with normal incidence alone will suffice to determine all of the stresses at a point, on a line, or within a restricted area.

Since each observation yields only two optical values at a point, namely, a phase difference given by the isochromatic and an angle given by the isoclinic, a minimum of three observations in three different directions are necessary in three-dimensional problems to determine the five values which it is possible to obtain optically. It is most convenient to choose these three directions in one plane with adjacent directions including an angle of 45°. For example, if a slice cut parallel to the xy plane is observed in the direction normal to its plane, i.e. in the z-direction, the difference of the secondary principal stresses $(\sigma'_1 - \sigma'_2)_{xy}$ in the xy plane and the directional angle of these stresses φ_{xy} are found from the isochromatics and the isoclinics. From these data,

$$(\sigma_x - \sigma_y) = (\sigma'_1 - \sigma'_2)_{xy} \cos 2\varphi_{xy},$$

and

$$\tau_{xy} = \tfrac{1}{2}(\sigma'_1 - \sigma'_2)_{xy} \sin 2\varphi_{xy},$$

are evaluated. A second observation in the direction bisecting the angle between the x and z axes is now made (Figure 14.12). This direction will be designated as x' while the direction perpendicular to x' in the xz plane will be designated as z'. This observation provides $(\sigma'_1 - \sigma'_2)_{yz'}$ and $\varphi_{yz'}$, from which

$$(\sigma_y - \sigma_{z'}) = (\sigma'_1 - \sigma'_2)_{yz'} \cos 2\varphi_{yz'},$$

and

$$\tau_{yz'} = \tfrac{1}{2}(\sigma'_1 - \sigma'_2)_{yz'} \sin 2\varphi_{yz'}.$$

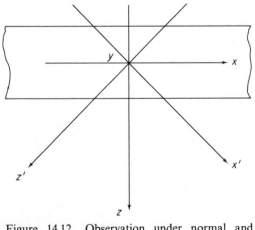

Figure 14.12. Observation under normal and oblique incidence

A third observation in the direction of z' produces $(\sigma'_1 - \sigma'_2)_{x'y}$ and $\varphi_{x'y}$ from which

$$(\sigma_{x'} - \sigma_y) = (\sigma'_1 - \sigma'_2)_{x'y} \cos 2\varphi_{x'y},$$

and

$$\tau_{x'y} = \tfrac{1}{2}(\sigma'_1 - \sigma'_2)_{x'y} \sin 2\varphi_{x'y}.$$

Two of the five values obtainable, namely $(\sigma_x - \sigma_y)$ and τ_{xy}, are obtained directly from the first observation. The remaining values, i.e. $(\sigma_x - \sigma_z)$, τ_{xz} and τ_{yz} can be extracted from the data provided by the three observations. To obtain $(\sigma_x - \sigma_z)$ we have

$$(\sigma_x - \sigma_z) = 2\sigma_x - (\sigma_x + \sigma_z).$$

As can be read from Mohr's circle for stresses in the xz plane given in Figure 14.13, $(\sigma_x + \sigma_z) = \sigma_{x'} + \sigma_{z'}$ from which it follows that

$$(\sigma_x - \sigma_z) = 2\sigma_x - (\sigma_{x'} + \sigma_{z'})$$
$$= 2(\sigma_x - \sigma_y) - (\sigma_{x'} - \sigma_y) + (\sigma_y - \sigma_{z'}).$$

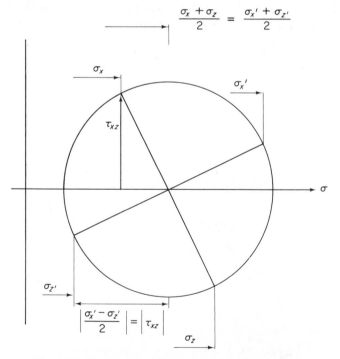

Figure 14.13.   Mohr's circle for stresses in the xz plane

The value of τ_{xz} can also be deduced from Mohr's circle, Figure 14.13; it is given by

$$\tau_{xz} = \tfrac{1}{2}(\sigma_{x'} - \sigma_{z'}) = \tfrac{1}{2}[(\sigma_{x'} - \sigma_y) + (\sigma_y - \sigma_{z'})].$$

The shear stresses acting on the plane parallel to the xz plane are τ_{xz} and τ_{yz}. These have the same resultant as $\tau_{x'y}$ and $\tau_{yz'}$ so that, as can be read from Figure 14.14,

$$\tau_{xy} = (\tau_{x'y} - \tau_{yz'})/\sqrt{2}, \tag{14.4}$$

$$\tau_{yz} = (\tau_{x'y} + \tau_{yz'})/\sqrt{2}, \tag{14.5}$$

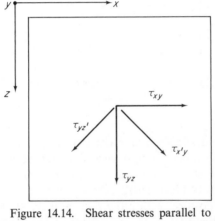

Figure 14.14. Shear stresses parallel to the xz plane

where the signs depend on the directions in which the shear stresses act. Equation (14.5) produces the shear stress τ_{yz} required while the value of τ_{xy} calculated from equation (14.4) can be compared with that obtained for this stress in the first observation as a check on the results.

To avoid the effects of refraction when optical values are measured with oblique incidence, the slice can be immersed in a liquid having the same

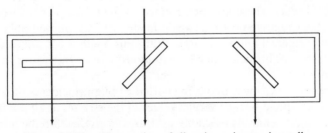

Figure 14.15. Observation of slices in an immersion cell

refractive index as the model material contained in a transparent strain free vessel, Figure 14.15. Alternatively, the slice can be placed between two prisms of the appropriate angle having the same refractive index as the material, Figure 14.16. Here again the correct immersion liquid should be introduced between the surfaces of contact.

When a slice is photographed or viewed under oblique incidence, the image corresponds to the projection of the slice on a plane normal to the direction of observation so that the dimensions in the photograph are distorted. In

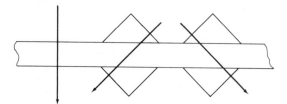

Figure 14.16.   Oblique incidence observations through prisms

cases when this is undesirable, a different procedure can be adopted. In this, a slice cut from the model is first viewed under normal incidence only, Figure 14.17(*a*). The slice is then cut transversely into subslices of square cross-section, Figure 14.17(*b*). These subslices are then rotated through 90° from their original position within the slice, and are again observed under normal incidence, Figure 14.17(*c*). For the purpose of photographing, the subslices can be fixed to a strain-free transparent plate in their correct positions relative to the slice. The photograph obtained is then identical in shape with the original slice except, of course, for the gaps resulting from the saw cuts.

The four sets of optical values obtained from the above two observations will often suffice to allow for the complete solution by integration of the equation of equilibrium. When it is necessary to determine the fifth value optically, the subslices can be cut at an angle of 45° in the manner indicated in Figure 14.17(*d*). These are then rotated in their own planes through 45° and with their relative positions again unchanged are photographed under normal incidence, Figure 14.17(*e*). With this method of subslicing, the optical data can be determined continuously over the whole plane by interpolation.

The optical values measured using both the oblique incidence and subslice methods correspond to the average values of the stresses at all points in the path of the light beam. Appreciable inaccuracy can therefore be introduced if the slope of the stresses changes considerably along the path of the beam.

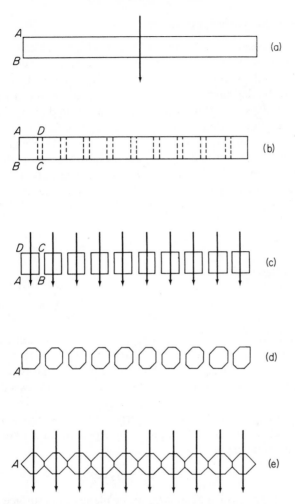

Figure 14.17. Method of subslicing for the production
of photographs in which the original shape of the slice
is retained

This can be overcome, while avoiding the difficulties involved with very thin
slices, using the technique of destructive removal of successive fine layers of
material from the model or from a thick slice cut from it by milling or otherwise. The optical values are measured initially and after the removal of each
layer. By this means the double refraction of each layer is obtained indirectly.
In such cases, analysis of the optical data is accomplished by procedures
based on the j-circle method as described in Section 15.7.

14.8. Separation of principal stresses

In three-dimensional problems the principal stresses can be separated using either the equations of equilibrium or the equations of compatibility.

If the slopes of the shear stresses on two mutually perpendicular planes are known or can be determined, the equation of equilibrium in the direction of the line of intersection of the planes can be integrated along that line. The required slopes can be determined from the appropriate lines of constant shear stress. If the intersection planes are the xy and xz planes, the lines $\tau_{xy} = $ constant in the first and $\tau_{xz} = $ constant in the second are plotted. The corresponding slopes $\partial\tau_{xy}/\partial y$ and $\partial\tau_{xz}/\partial z$ can then be obtained in the same way as in two-dimensional problems. If we integrate the first of equations (1.11) with respect to x, substituting $X = 0$ we obtain

$$\sigma_{x1} = \sigma_{x0} - \int_0^1 \frac{\partial\tau_{xy}}{\partial y}\,dx - \int_0^1 \frac{\partial\tau_{xz}}{\partial z}\,dx, \tag{14.6}$$

or, in finite difference form:

$$\sigma_{x1} = \sigma_{x0} - \sum_0^1 \frac{\Delta\tau_{xy}}{\Delta y}\,\Delta x - \sum_0^1 \frac{\Delta\tau_{xz}}{\Delta z}\,\Delta x. \tag{14.7}$$

The normal stress σ_{x1} at a point 1 on the line of integration can therefore be found from equation (14.7), starting at a point O where σ_{x0} is known, e.g. at a point on a load free surface. Except for the fact that two integrations are involved, this procedure is identical with that in two-dimensional problems.

If O is a point on a load free surface perpendicular to the x axis, the value of σ_{x0} is zero. If the surface is oblique with respect to the x axis, σ_{x0} can be determined from the equation

$$\sigma_{x0} = \sigma_1 \cos^2\varphi_1 + \sigma_2 \cos^2\varphi_2,$$

where σ_1, σ_2 are the principal stresses in the plane of the surface and φ_1, φ_2 are the angles between σ_1, σ_2 and the x direction.

The signs of τ_{xy} and $\partial\tau_{xy}/\partial y$ or τ_{xz} and $\partial\tau_{xz}/\partial z$ respectively can be determined from the conditions of equilibrium of the shear forces and normal forces acting in the x-direction on the faces of a rectangular element with edges parallel to the axes in the line of integration, Figure 1.2.

Instead of using the lines of constant shear stress, the necessary values can be calculated from the isochromatics and the isoclinics as in two-dimensional problems. The slope of the shear stress in the plane of the slice, say the xy plane, can be derived with good accuracy. That for the direction normal to the slice, i.e. $\partial\tau_{xz}/\partial z$, can only be obtained however from the difference of shear stresses at corresponding points on adjacent parallel slices cut from the model. Since the closeness of such slices is restricted to the thickness of the slice plus that of the saw, the accuracy of this method in general is poor. When it is possible to reproduce accurately all the experimental conditions, this difficulty can be overcome by cutting a slice in the xz plane from a second model.

The procedure of separating the principal stresses is simplified in cases of multiple symmetry, particularly rotational symmetry. With double symmetry, the shear stresses τ_{xz} and τ_{yz} are equal at corresponding points if an axis of symmetry is parallel to the z axis. Hence, at points on this axis (or on planes bisecting the angle between the planes of symmetry) $\partial\tau_{xz}/\partial x = \partial\tau_{yz}/\partial y$ so that only one integration has to be accomplished:

$$\sigma_{z1} = \sigma_{z0} - 2\int_0^1 \frac{\partial\tau_{xz}}{\partial z}\,dz = \sigma_{z0} - 2\int_0^1 \frac{\partial\tau_{yz}}{\partial z}\,dz. \tag{14.8}$$

Thus, one slice cut from the model containing the axis of symmetry is sufficient to allow this integration.

With rotational symmetry, it is convenient to use the equations of equilibrium in cylindrical co-ordinates (1.12) and we shall take the axis of symmetry to be the z axis. Since the stresses are independent of θ, all derivatives with respect to θ vanish. The shear stresses $\tau_{r\theta}$ and $\tau_{\theta z}$ also vanish so that, with no body-forces, these equations reduce to

$$\frac{\partial\sigma_r}{\partial r} + \frac{\partial\tau_{rz}}{\partial z} + \frac{\sigma_r - \sigma_\theta}{r} = 0, \tag{14.9a}$$

$$\frac{\partial\tau_{rz}}{\partial r} + \frac{\partial\sigma_z}{\partial z} + \frac{\tau_{rz}}{r} = 0. \tag{14.9b}$$

In such cases, all of the stresses can be determined from one slice containing the axis of symmetry provided the value of σ_z at a point on the line of integration is known. The shear stress τ_{rz} in the plane of the slice can be determined in the usual way from the isochromatics and the isoclinics with the help of the nomogram given in Figure 9.1. The lines $\tau_{rz} = $ constant can then be plotted and from this system of lines the slope $\partial\tau_{rz}/\partial r$ is obtained. This allows the normal stress σ_z to be found by integrating equation (14.9b)

$$\sigma_{z1} = \sigma_{z0} - \int_0^1 \frac{\partial\tau_{rz}}{\partial r}\,dz - \int_0^1 \frac{\tau_{rz}}{r}\,dz. \tag{14.10}$$

For determining σ_z at points on the axis of symmetry, equation (14.8) is used. The values of σ_r can then be determined by adding those of σ_z and $(\sigma_r - \sigma_z)$. The latter can be read from the isochromatics and isoclinics by means of the nomogram in Figure 9.1 or calculated from the relation

$$(\sigma_r - \sigma_z) = (\sigma_1 - \sigma_2)\cos 2\varphi.$$

From the values of σ_r, the slope $\partial\sigma_r/\partial r$ can be found. All of the values necessary to calculate σ_θ have now been obtained. This is accomplished by inserting these values in the equation

$$\sigma_\theta = r\left[\frac{\partial\sigma_r}{\partial r} + \frac{\partial\tau_{rz}}{\partial z} + \frac{\sigma_r}{r}\right], \tag{14.11}$$

which is equation (14.9a) rewritten to express σ_θ.

If there is no point on the line of integration of known σ_z to provide the required value of σ_{z0}, the radial stress σ_r must first be determined by integrating equation (14.9a). The values of $(\sigma_r - \sigma_\theta)$ to be inserted in this equation can be read from the isochromatics in slices cut perpendicular to the axis; because of the axial symmetry of the stress distribution, σ_r and σ_θ are principal stresses so that $(\sigma_r - \sigma_\theta) = (\sigma_1 - \sigma_2)$. Alternatively, $(\sigma_r - \sigma_\theta)$ can be determined by observing the axial slice under oblique incidence. From these values together with those of $\partial\tau_{rz}/\partial z$, σ_r can be plotted if the value of σ_{r0} at a point on the line of integration is known:

$$\sigma_{r1} = \sigma_{r0} - \int_0^1 \frac{\partial\tau_{rz}}{\partial z}\,dr - \int_0^1 \frac{(\sigma_r - \sigma_\theta)}{r}\,dr. \tag{14.12}$$

The tangential stress σ_θ can then be obtained from

$$\sigma_\theta = \sigma_r - (\sigma_r - \sigma_\theta). \tag{14.13}$$

It is possible to apply the conditions of equilibrium of a finite section of a body instead of those of an infinitesimal element to separate the principal stresses in a manner similar to that described in Section 9.1.3 for two-dimensional problems. The determination of the shear forces acting on the faces of such a section is more laborious, however, since it requires knowledge of the distribution of shear stresses over finite areas instead of merely along certain lines. These shear stresses are obtained from slices cut perpendicular to the corresponding faces of the element. It is obvious that this procedure in general is fairly complicated.

The equation of equilibrium of an element cut in such a way that its edges are parallel to the stress trajectories reads

$$\frac{\partial\sigma_1}{\partial s_1} + \frac{\sigma_1 - \sigma_2}{\rho_2'} + \frac{\sigma_1 - \sigma_3}{\rho_3'} = 0, \tag{14.14}$$

for the direction of σ_1. Similar equations relating to the directions of σ_2 and σ_3 can be written by cyclic permutation of the indices. In equation (14.14) ρ_2', ρ_3' are not the true radii of curvature of the stress trajectories s_2, s_3 but those of the projections of s_2 and s_3 on the tangent planes of s_1, s_2 and s_1, s_3 respectively; thus they are identical with the radii of curvature of the trajectories of the corresponding secondary principal stresses. It can be shown that in certain cases, equation (9.14) is identical in practice with equation (14.6). Let us suppose, for example, that the direction of the principal stress σ_1 coincides with that of σ_x, i.e. with the x direction, and that the data necessary for evaluating equation (14.6) are obtained from two slices intersecting on the line of integration and parallel to the xy and xz planes respectively. With $1/\rho'_2 = \partial\varphi_{xy}/\partial y$, etc., equation (14.14) can be transformed into

$$\frac{\partial\sigma_1}{\partial s_1} = \frac{\partial\sigma_x}{\partial x} = (\sigma_1 - \sigma_2)'_{xy}\frac{\partial\varphi_{xy}}{\partial y} + (\sigma_1 - \sigma_2)'_{xz}\frac{\partial\varphi_{xz}}{\partial z}, \qquad (14.15)$$

where $(\sigma_1 - \sigma_2)'_{xy} = $ difference of secondary principal stresses in the xy plane, $\varphi_{xy} = $ parameter of isoclinic in the xy plane, etc.

Corresponding to equation (1.7) we have

$$\tau_{xy} = \frac{(\sigma_1 - \sigma_2)'_{xy}}{2}\sin 2\varphi_{xy}.$$

Differentiating with respect to y and substituting $\varphi_{xy} = 0$ in accordance with the stated conditions we obtain

$$\frac{\partial\tau_{xy}}{\partial y} = (\sigma_1 - \sigma_2)'_{xy}\frac{\partial\varphi_{xy}}{\partial y}.$$

Similarly,

$$\frac{\partial\tau_{xz}}{\partial z} = (\sigma_1 - \sigma_2)'_{xz}\frac{\partial\varphi_{xz}}{\partial z}.$$

With $\partial\sigma_1/\partial s_1 = \partial\sigma_x/\partial x$, equation (14.15) is thus seen to be identical with equation (14.6).

As an example of the application of equation (14.14), let us consider an axially symmetrical body containing a notch and suppose that the slope of the meridional surface stress in the notch with respect to the direction of the normal is required. Let the meridional surface stress be σ_1 and let s_3 be the stress trajectory in the direction of the normal to the surface. The required slope $\partial\sigma_1/\partial s_3$ can be read from a slice containing the axis of symmetry since s_1 and ρ_1 lie in its plane while s_2, which is the trajectory of the tangential stress σ_2, is a circle with its centre on the axis. Obviously,

$$\frac{\partial\sigma_1}{\partial s_3} = \frac{\partial(\sigma_1 - \sigma_3)}{\partial s_3} + \frac{\partial\sigma_3}{\partial s_3},$$

from which, expressing $\partial\sigma_3/\partial s_3$ in the form of equation (14.14),

$$\frac{\partial\sigma_1}{\partial s_3} = \frac{\partial(\sigma_1 - \sigma_3)}{\partial s_3} - \frac{(\sigma_3 - \sigma_1)}{\rho_1} - \frac{(\sigma_3 - \sigma_2)}{\rho_2}.$$

At the surface, s_1 coincides with the profile of the notch so that ρ_1 is equal to the radius of curvature of the profile at the point considered; ρ_2 is the distance of this point from the axis. The stress σ_3 normal to the surface is zero. The tangential stress $\sigma_2 = \sigma_\theta$ can be determined from equation (14.11), observing that here σ_r is the radial component of the surface stress σ_1; $\partial \tau_{rz}/\partial z$ can be obtained from the lines of constant shear stress and $\partial \sigma_r/\partial r$ from the relation

$$\frac{\partial \sigma_r}{\partial r} = \frac{\partial}{\partial r}(\sigma_r - \sigma_z) + \frac{\partial \sigma_z}{\partial r},$$

the required slopes being determined from the respective values of $(\sigma_r - \sigma_z)$ obtained in the usual way from the isochromatics and isoclinics and of σ_z obtained by integration as described previously.

The equations of compatibility represented by equations (2.23) and (2.24) can be applied to the separation of principal stresses in three-dimensional problems in a manner similar to that employed in two-dimensional photoelasticity. If the body forces are zero or constant, these equations when expressed in terms of stresses produce

$$\left(\frac{\partial^2}{\partial x^2} + \frac{\partial^2}{\partial y^2} + \frac{\partial^2}{\partial z^2}\right)(\sigma_x + \sigma_y + \sigma_z) = 0,$$

or

$$\nabla^2 \Sigma = 0,$$

so that Liebmann's formula and relaxation procedure can again be applied. In three dimensions, the following approximate formula suggested by Schumann can be used:

$$\Sigma_0 = \frac{\Sigma_1 + \Sigma_2 + \Sigma_3 + \Sigma_4 + \Sigma_5 + \Sigma_6}{6},$$

where Σ_0 is the sum of the three normal stresses at the centre of a cube while $\Sigma_1 - \Sigma_6$ are the sums of these stresses at the centres of the six faces of the cube.

The sum of the stresses at points on the surfaces of a body which are load free or on which the load distribution is known can be determined from the optical data alone. At other points, the required values must be determined by some other method, e.g. by integrating the equations of equilibrium.

The following method enables the stresses in the plane of a section of a three-dimensional body to be determined from the optical data provided by a slice containing the section alone. No other slice parallel or perpendicular to the first is required.

Using Cartesian co-ordinates and considering a section parallel to the xy plane, the following two of the three equations of equilibrium are used:

$$\frac{\partial \sigma_x}{\partial x} + \frac{\partial \tau_{xy}}{\partial y} + \frac{\partial \tau_{xz}}{\partial z} = 0,$$

$$\frac{\partial \sigma_y}{\partial y} + \frac{\partial \tau_{xy}}{\partial x} + \frac{\partial \tau_{yz}}{\partial z} = 0.$$

For simplicity, these equations may be written in the form

$$\frac{\partial \sigma_x}{\partial x} + \frac{\partial \tau_{xy}}{\partial y} + T_x = 0, \tag{14.16}$$

$$\frac{\partial \sigma_y}{\partial y} + \frac{\partial \tau_{xy}}{\partial x} + T_y = 0, \tag{14.17}$$

where T_x, $T_y = (\partial \tau_{xz}/\partial z), (\partial \tau_{yz}/\partial z)$ respectively.

Differentiating equation (14.16) with respect to y, equation (14.17) with respect to x and subtracting produces

$$\frac{\partial T_x}{\partial y} = \frac{\partial T_y}{\partial x} + \left(\frac{\partial^2}{\partial x^2} - \frac{\partial^2}{\partial y^2} \right) \tau_{xy} - \frac{\partial^2}{\partial x\, \partial y} (\sigma_x - \sigma_y). \tag{14.18}$$

Values of the second derivatives of τ_{xy} and $(\sigma_x - \sigma_y)$ in this equation can be derived from the lines $\tau_{xy} = $ constant and $(\sigma_x - \sigma_y) = $ constant. These lines may be drawn using the information provided by the isochromatics and the isoclinics in the manner described in Section 9.1.2.

The following identity is also used:

$$\sigma_y = \sigma_x - (\sigma_x - \sigma_y). \tag{14.19}$$

The field within which the stresses are to be determined is divided by lines drawn parallel to the x and y axes. For the application of the method, certain initial values are necessary at the edges or in the interior of the field. Sufficient values for example are those of σ_y, T_x and T_y on the line $y = 0$ and σ_x on the line $x = 0$. With these values, separation of the stresses is accomplished progressing in the direction parallel to the y axis. Using values read from the systems of lines $(\sigma_x - \sigma_y) = $ constant and $\tau_{xy} = $ constant, curves of $\partial(\sigma_x - \sigma_y)/\partial y$ and $\partial \tau_{xy}/\partial x$ are plotted on the lines $y = $ constant, and $\partial \tau_{xy}/\partial y$ on the lines $x = $ constant, Figure 14.18. The slopes of the curves provide the values of the second derivatives required for the solution of equation (14.18). The following steps then produce σ_x, σ_y, T_x and T_y over the whole field, the line $y = $ constant to which each quantity refers being indicated by the index:

1. Values of $(\partial T_y/\partial x)_0$ are read from the curve of T_{y0}.
2. These values together with those of the second derivatives of $(\sigma_x - \sigma_y)_0$ and $(\tau_{xy})_0$ are inserted in equation (14.18) to obtain values of $(\partial T_x/\partial y)_0$.

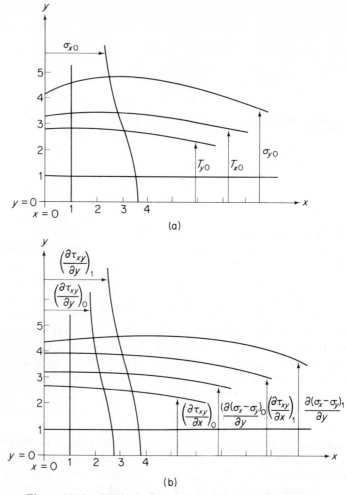

Figure 14.18.   Method of separation of principal stresses in
a section of a three-dimensional model from observations of
a single slice

3. Values of T_{x1} can now be determined. The difference ΔT_x between the
lines $y = 0$ and $y = 1$ is obviously

$$\Delta T_x = \int_0^1 \frac{\partial T_x}{\partial y} \, dy \backsimeq \left(\frac{\partial T_x}{\partial y}\right)_0 \Delta y,$$

where Δ_y is the distance between the lines $y = 0$ and $y = 1$.
   Then, $T_{x1} = T_{x0} + \Delta T_x$.

4. With the values of T_{x1} known, those of σ_{x1} can now be obtained by integrating equation (14.16), proceeding from the line $x = 0$ where, as postulated, the initial value of σ_{x1}, i.e. σ_{x10}, is known:

$$\sigma_{x1} = \int_0^x \left(\frac{\partial \tau_{xy}}{\partial y} + T_x \right)_1 dx + \sigma_{x10}$$

5. Values of σ_{y1} are then calculated using equation (14.19).

6. Approximate values of $(\partial \sigma_y / \partial y)_1$ can now be determined using the values of σ_{y0} and σ_{y1}:

$$\left(\frac{\partial \sigma_y}{\partial y} \right)_1 \simeq (\sigma_{y1} - \sigma_{y0}) / \Delta y.$$

7. The values of $(\partial \sigma_y / \partial y)_1$ and $(\partial \tau_{xy} / \partial x)_1$ are then inserted in equation (14.17) to obtain T_{y1}. The complete cycle is then repeated between the lines $y = 1$ and $y = 2$, and so on.

If the values of T_x, T_y are also known along the line $x = 0$, the whole operation can be repeated interchanging x and y and progressing in a direction parallel to the x axis. The two sets of results then provide a mutual check of the accuracy. If both operations are performed simultaneously, the results may be improved during the process. The accuracy may also be checked by comparing the results obtained at the last line with the edge values there if these are known.

If T_x, T_y are known only in a small region, say about the point $(0, 0)$, the procedure can be applied to determine first the values of T_x, T_y along narrow strips bordered by the lines $x = 0$ and $y = 0$ and then all values over the entire field.

The above method can be applied to determine the stresses in other types of problem when additional terms in the equations of equilibrium prevent simple integration in the manner possible with plates loaded by forces within their planes. For example, the method may be applied to problems of transversely bent plates, shells, dynamic stresses and thermal stresses. It should be noted, however, that the meaning of the terms T_x, T_y will differ according to the nature of the problem. For example, in ordinary three-dimensional problems they are gradients of shear stress; in two-dimensional dynamic problems they are body forces due to acceleration while in three-dimensional dynamic problems they are sums of acceleration forces and shear stress gradients.

When dealing with curved surfaces, it is more convenient to employ a variation of the above method. In this, the equations of equilibrium in rectangular co-ordinates are replaced by those for an element bounded by stress trajectories, and the stress trajectories are used instead of the lines $\tau_{xy} = $ constant and $(\sigma_x - \sigma_y) = $ constant.

The equations of equilibrium

$$\frac{\partial \sigma_1}{\partial s_1} + \frac{(\sigma_1 - \sigma_2)}{\rho_2'} + \frac{(\sigma_1 - \sigma_3)}{\rho_3'} = 0,$$

$$\frac{\partial \sigma_2}{\partial s_2} - \frac{(\sigma_1 - \sigma_2)}{\rho_1'} + \frac{(\sigma_2 - \sigma_3)}{\rho_3''} = 0,$$

which are the three-dimensional equivalents of equations (1.16), with the substitutions

$$T_1 = \frac{(\sigma_1 - \sigma_3)}{\rho_3'}, \; T_2 = \frac{(\sigma_2 - \sigma_3)}{\rho_3''},$$

become

$$\frac{\partial \sigma_1}{\partial s_1} + \frac{(\sigma_1 - \sigma_2)}{\rho_2'} + T_1 = 0, \tag{14.20}$$

$$\frac{\partial \sigma_2}{\partial s_2} - \frac{(\sigma_1 - \sigma_2)}{\rho_1'} + T_2 = 0, \tag{14.21}$$

where ρ_1', ρ_2' are the radii of curvature of the projections of the stress trajectories s_1, s_2 respectively on the tangent plane to s_1 and s_2. The radii ρ_3', ρ_3'' are those of s_3 (perpendicular to the surface) projected into the tangent plane of s_1 and s_3, and s_2 and s_3, respectively. These values can be read from the isoclinics and the curvature of the model surface.

Differentiating equation (14.20) with respect to s_2, equation (14.21) with respect to s_1 and subtracting produces

$$\frac{\partial T_1}{\partial s_2} = \frac{\partial T_2}{\partial s_1} - \frac{\partial^2}{\partial s_1 \, \partial s_2}(\sigma_1 - \sigma_2) - \frac{\partial}{\partial s_1}\left[\frac{\sigma_1 - \sigma_2}{\rho_1'}\right] - \frac{\partial}{\partial s_2}\left[\frac{\sigma_1 - \sigma_2}{\rho_2'}\right].$$

To obtain the values of the terms involving $(\sigma_1 - \sigma_2)$ in this equation, curves of

$$\frac{\sigma_1 - \sigma_2}{\rho_1'} \quad \text{and} \quad \frac{\sigma_1 - \sigma_2}{\rho_2'},$$

are plotted against s_1 and s_2 respectively while curves of

$$\frac{\partial}{\partial s_2}(\sigma_1 - \sigma_2) \quad \text{or} \quad \frac{\partial}{\partial s_1}(\sigma_1 - \sigma_2),$$

are plotted against s_1 or s_2 respectively.

Finally, equation (14.19) is replaced by the identity

$$\sigma_2 = \sigma_1 - (\sigma_1 - \sigma_2).$$

The subsequent procedure is completely analogous to that using rectangular co-ordinates.

14.9 Experimental methods of separation of principal stresses

When slices cut from a model with frozen-in deformations are heated to the softening temperature of the model material, the frozen-in strains are released. By measuring this recovery, an additional independent experimental value is obtained which can be applied in the solution of the problem. The five values determined optically together with this sixth value fully characterize the state of stress at a point in the interior of a body.

Several mechanical and optical methods (e.g. comparators and interferometers respectively) have been suggested for determining the recovery of a frozen slice on reheating. All of these methods suffer, however, from the fact that at the softening temperature, the value of Poisson's ratio for all model materials is almost 0·5. For this limiting value, the stresses produce constant volume deformation. Thus, the action of an additional hydrostatic pressure in this case will change neither the volume nor shape of an element. This pressure therefore cannot be determined by such methods. In the case of a cube in a general state of stress, the sum of the elongations of its edges will be zero though it may suffer angular distortion. The latter values are those which are determined optically, however, so that measurement of recovery does not produce an additional value.

Some investigators have attempted to measure the small change of volume associated with the small amount by which the actual value of Poisson's ratio v differs from 0·5. It has been found, however, that this difference is in fact much smaller than has previously been assumed so that such methods are scarcely practicable.

The most accurate method known for determining the change of volume (and simultaneously that of v) is by measurement of the change of specific weight which is obviously associated with it. Small particles of the frozen model are immersed in a liquid having approximately the same specific weight as the model material. The particles sink or float depending on the specific weight of the material in relation to that of the liquid. By employing mixtures of different liquids, e.g. trichlorethylene and methylalcohol, a wide range of specific weights can be obtained. The rate of change of specific weight of these liquids with temperature is different from that of the model material. Hence, by varying the temperature a specific weight can be found at which the particle just begins to float or sink as the case may be. It would be difficult, of course, to determine the absolute values of the specific weights by this means. By comparing particles of unknown stress sum with those of known stress sum, however, the problem under consideration can be solved.

Using the above method, which is obviously extremely sensitive as to specific weight, it has been found that the change of specific weight or the difference between v and 0·5 is much smaller than has been suggested by authors of the methods mentioned earlier.

When a model is unloaded after cooling in the frozen stress procedure, a certain though small relaxation of strain occurs. Thus, part of the small change of volume or specific weight vanishes at the instant of unloading. For this recovery or relaxation, Poisson's ratio of the material when cold (approximately 0·38) is valid.

14.10   Examples of the frozen stress method

14.10.1   *Surface stresses in a body having multiple symmetry: Stresses in pipe flanges*

In order to determine the distribution of surface stresses in pipe flanges subjected to bolt loading and internal pressure, a series of tests was performed on flanges integral with a short length of pipe. The dimensions of the models are given in Figure 14.19. Dead weight loading simulating that of the bolts was applied at the bolt holes through the agency of steel balls and buttons.

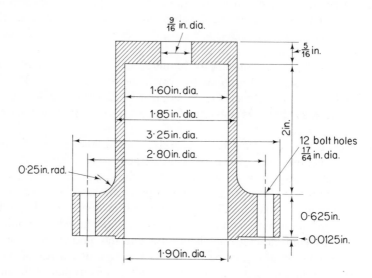

Figure 14.19.   Dimensions of models of pipe flanges

The buttons were provided with heads of a diameter intermediate between that of the width across the flats and the width across the corners of the corresponding bolt head and projected a short distance into the holes with a diameter corresponding to that of the bolts. Internal pressure was controlled by means of a mercury column and was transmitted to the interior of the models by water.

To eliminate radial friction at the joint ring, tests were carried out using two identical models as in the normal pipe joint.

At the end of the stress freezing cycle, the models were cut into tangential, radial and axial slices as indicated in Figure 14.20. To obtain the surface stresses in the fillet an additional number of thin slices was cut normal to the contour of the fillet.

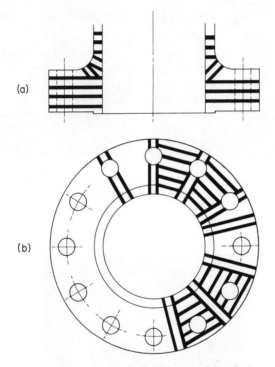

Figure 14.20. Method of slicing models of pipe
flanges

Due to the multiple symmetry of the models, a number of identical slices of each type could be cut. This provided a method of improving the stress pattern without impairing the accuracy as would be the case if this were achieved by increasing either the load and hence the deformation or the thickness of the slices. By carefully aligning a pile of such slices in the field of the polariscope so that the light passed through each in succession, the fringe order at every point was multiplied by the number of slices in the pile. Figure 14.21 shows the isochromatic patterns of radial sections passing midway between and through the axes of the bolt holes obtained using four

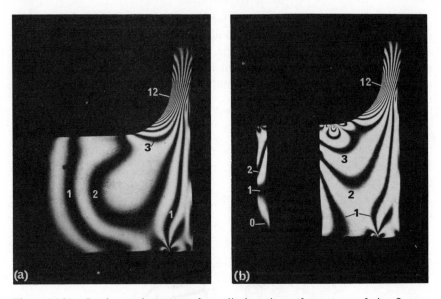

Figure 14.21.   Isochromatic patterns for radical sections of symmetry of pipe flanges obtained using four identical slices in series

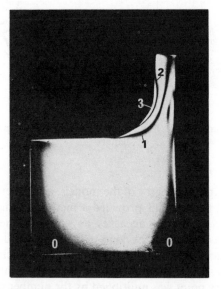

Figure 14.22.   Isochromatic pattern corresponding to that of Figure 14.21(a) obtained using a single slice

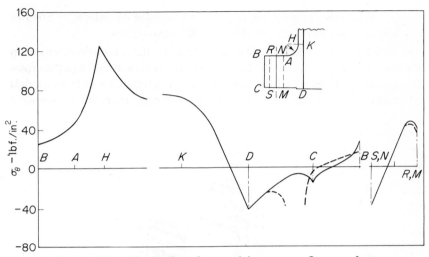

Figure 14.23. Distribution of tangential stress over flange surfaces

identical slices in the manner described. Figure 14.22 shows the isochromatic pattern corresponding to Figure 14.21(*a*) obtained using a single slice.

The stress patterns shown were obtained with a total bolt load of 30·7 lbf. and internal pressure of 10·2 lbf./in.^{2}, representing a value of the ratio bolt load/axial hydrostatic load of 1·5. The corresponding distributions of the tangential and radial contour stresses are shown in Figures 14.23 and 14.24 respectively.

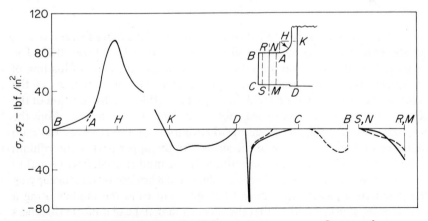

Figure 14.24. Distribution of radial contour stresses over flange surfaces

14.10.2 *Separation of principal stresses in Cartesian co-ordinates: Stresses in columns loaded over areas smaller than the cross-section*

In order to find a general law applicable to the stress distribution in columns loaded by axial forces distributed over areas smaller than the cross-sectional areas of the columns, three models were tested. In all three models the height h (Figure 14.25) and the width b of the square cross-section were

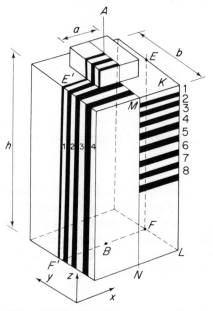

Figure 14.25. Notation and location of slices in models of columns

the same. The loads were applied through the smaller square sections $a \times a$, the dimension a being varied to give ratios a/b of 0·25, 0·5 and 0·75. The magnitude of the load was varied to give a constant intensity of loading of 0·8 kp/cm^2 within the area $a \times a$ in each case. The bottom faces of the columns were supported on rigid metal plates so that the distribution of the reactive forces on these faces was not controlled. As will be seen from the results, however, differences in the distribution of these forces have no influence on the distribution of the stresses in the upper part of the column if the height h is greater than the width b. In the models used, h was equal to $2b$. After freezing the stresses, 4 vertical slices and 8 horizontal ones occupying the positions indicated in Figure 14.25 were cut from the models using a circular saw. It was intended that the slices should all be of uniform thickness but after slicing minor differences were found to exist.

The isochromatics of the slices of one of the models ($a/b = 0.5$) are shown in Figure 14.26. The isoclinics are not shown but the stress trajectories plotted from them are given in Figure 14.27. These are lines of true principal stress in the vertical slice designated by the number 1 which contains one

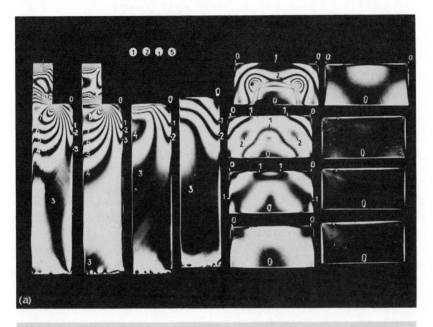

Figure 14.26.   Isochromatic patterns for slices of a column. $a/b = 0.5$

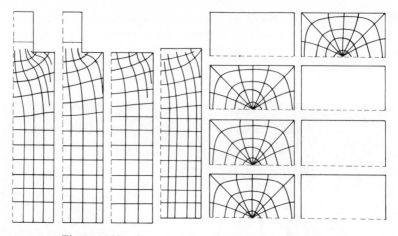

Figure 14.27. Stress trajectories for slices of the column

of the planes of symmetry and in slice 4 which can be taken to represent conditions at the load-free surface; in all other slices they are directional lines of secondary principal stress in the plane of the slice only.

Figure 14.28 shows the lines of constant shear stress, i.e. lines along which the shear stress within the plane of the slice is constant. In the vertical slices, this stress is τ_{xz} while in the horizontal ones it is τ_{xy} corresponding to the directions of the axes shown in Figure 14.28.

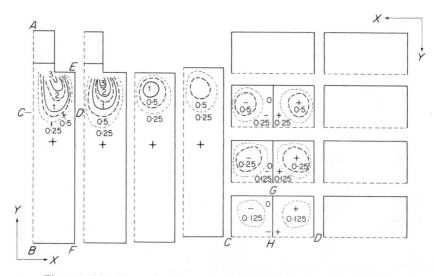

Figure 14.28. Lines of constant shear stresses in the planes of the slices

The measured thicknesses of the slices at different points are registered in mm in Figure 14.29. These values were used to determine the absolute values of the stresses instead of merely the fringe orders and to reduce the fringe orders to correspond to slices of uniform thickness. The numbers encircled identify the slices in accordance with the numbering in Figure 14.25. In Figures 14.26 to 14.29, the slices are depicted in the same relative positions.

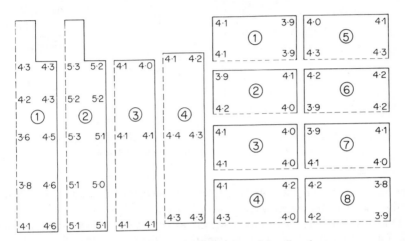

Figure 14.29.   Measured thicknesses of the slices in mm

The slope $\partial \tau_{xz}/\partial x$ along the axis of symmetry AB is represented in Figure 14.30(a) by the dotted line. The second slope required to determine σ_z by integration is $\partial \tau_{yz}/\partial y$. From the condition of double symmetry, however, this slope is identical with the former so that equation (14.8) could be used:

$$\sigma_{z1} = \sigma_{z0} - 2 \int_0^1 \frac{\partial \tau_{xz}}{\partial x} \, dz.$$

Since neither of the ends of the line AB were load-free, it was necessary to determine the value of σ_z at some point on this line from which the integration could be started. The necessary value of σ_{z0} was determined here by performing a similar integration along a line CD perpendicular to AB and intersecting it in the point H. The location of this line is indicated in Figure 14.28. The slopes $\partial \tau_{xz}/\partial z$ and $\partial \tau_{xy}/\partial y$ were required for this integration, equation (14.6), which could be started at the point D on the load-free surface where $\sigma_x = 0$. For clarity, the partial values of σ_x found by integrating each of these two slopes are plotted separately in Figure 14.30(b) where they are represented by σ_x' and σ_x''. The sum of these values gave σ_x, which enabled σ_z at the point H to be determined from the relation $\sigma_z = \sigma_x + (\sigma_z - \sigma_x)$.

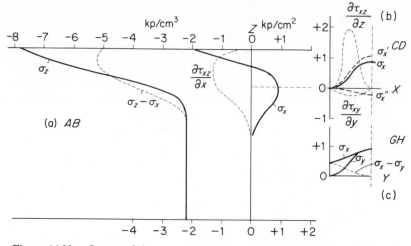

Figure 14.30. Curves of the normal stresses on the axis of symmetry *AB* and auxiliary curves necessary for their determination

The value of $(\sigma_z - \sigma_x)$ was derived from the isochromatics and the isoclinics of the slice 1 in the usual way.

Due to symmetry, the results obtained for the line *HD* obviously hold true also for the line *HG* although the designations of the stresses are different. Thus, the curve of σ_y along *HG* is identical with that of σ_x along *HD*. The curve of σ_x along *HG* is found from the values of σ_y and $(\sigma_x - \sigma_y)$, Figure 14.30(*c*).

The stresses σ_x and σ_z along the line *EF* (Figure 14.25) are plotted in Figure 14.31. Since this line lies on a load-free surface, these stresses can be

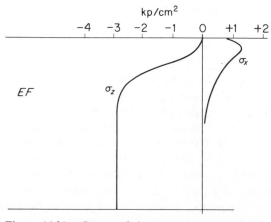

Figure 14.31. Curves of the normal stresses on the line *EF* of Figure 14.25

determined directly from the isochromatics, σ_z being read from those of the vertical slice 1 and σ_x from those of the horizontal slices.

The final generally applicable result is shown in Figure 14.32. Here the ratios of the maximum tensile stress $\sigma_{x\,max}$ occurring on each of the lines AB and EF to the mean axial stress σ_{zm} in the column are plotted against the ratio a/b. As can be seen, the maximum tensile stress on the axis of symmetry

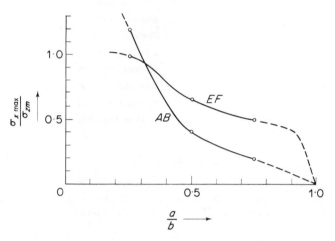

Figure 14.32. Curves of the ratio, maximum tensile stress/mean axial stress versus the ratio a/b

is roughly inversely proportional to a/b; the maximum tensile stress on the outer surface of the column is smaller than that on the axis of symmetry with small area of loading but greater with large areas.

14.10.3 *Separation of principal stresses in cylindrical co-ordinates: Shrinkage stresses in cylinders*

The stresses in a compound cylinder formed by shrinking one cylinder on to another were investigated. The dimensions in mm are given in Figure 14.33. The initial inner diameter of the outer cylinder was 59 mm, giving a shrinkage allowance of 1.5%.

To enable the cylinders to be assembled, the outer one was heated to 150 °C while the inner one was cooled to -10 °C. The assembly was then brought to a uniform temperature of 135 °C to obtain the correct distribution of stresses for freezing. Before the cylinders had achieved equality in temperature, the length of the outer one exceeded that of the inner due to their different thermal expansions. Friction between the cylinders prevented full recovery of the longitudinal dimensions with the result that a difference in

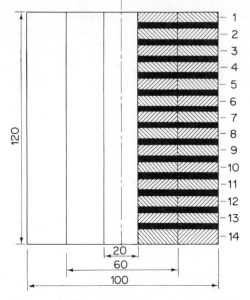

Figure 14.33. Dimensions of compound
cylinder and location of transverse slices

length of 0·4 mm = 0·33 % existed between them at the end of the freezing cycle.

The possibility of avoiding the above effect by compressing the outer cylinder axially was investigated. This was done using a second pair of cylinders having the same dimensions as before and following the same procedure with the exception that the outer cylinder was subjected to axial compression which was maintained for some time after the cylinders had been assembled. Although this resulted in the cylinders being of identical length when the axial pressure was removed, it was found that the same difference in length as before existed at the end of the freezing cycle. The isochromatics and isoclinics of corresponding slices cut from models were also identical in both cases.

One slice was cut from the model in a plane containing the axis while 14 transverse slices were cut from one half of the model as indicated in Figure 14.33. The thickness of the vertical (axial) slice was 6 mm. That of the horizontal slices numbered 1 to 13 was 5·6 mm while no. 14 was wedge shaped and of irregular thickness.

Figure 14.34 shows the isochromatics of the vertical slice, Figures 14.35 the isoclinics (parameters 0°, 15°, 30°, 45°, 60° and 75°), Figure 14.36 the stress trajectories and lines of constant shear stress derived from the isochromatics

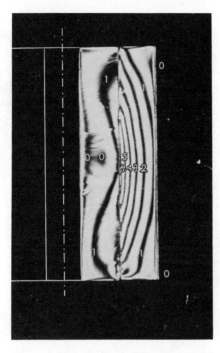

Figure 14.34. Isochromatics in an axial
slice of the compound cylinder shown in
Figure 14.33

and isoclinics in the usual way. Figure 14.37 shows the isochromatics of the
horizontal slices numbered in accordance with Figure 14.33.

Figure 14.38 shows the variations of the stresses along the line of contact
AB (Figure 14.36(b)) of the two parts of the vertical slice, i.e. along the common
surface of the two cylinders. The results given in Figure 14.38(a) relate to
the outer surface of the inner cylinder and those in Figure 14.38(b) to the
inner surface of the outer cylinder. Figure 14.39 gives the distribution of the
stresses in the horizontal plane of symmetry, i.e. over the central transverse
section CD (Figure 14.36(b)) of the assembly.

All of the values necessary for determining the stresses at points on the
vertical line AB by the procedure described in Section 14.8 can be read from
the vertical slice alone. The shear stress τ_{rz} and the stress difference $(\sigma_r - \sigma_z)$
can be obtained from the isochromatics and isoclinics in Figures 14.34 and
14.35 respectively with the assistance of the nomogram given in Figure 9.1
or by calculation. The axial stress σ_z is given by equation (14.10) and the
radial stress σ_r by the sum of σ_z and $(\sigma_r - \sigma_z)$.

For determining the stresses at points on the line CD both the vertical
slice and the central horizontal one (no. 8) were used. Since the latter contains

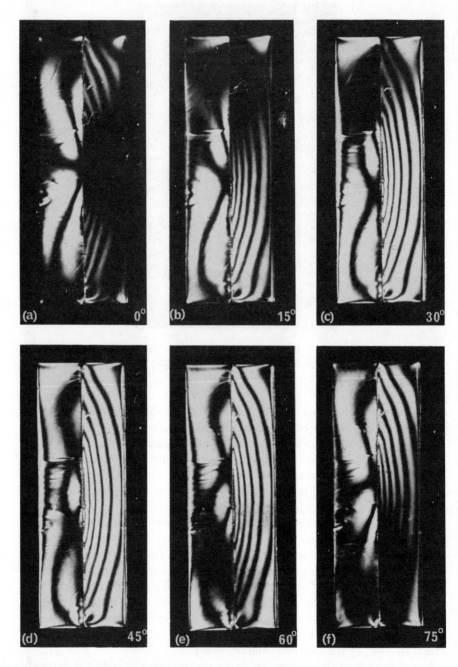

Figure 14.35. Isoclinics of various parameters in an axial slice.

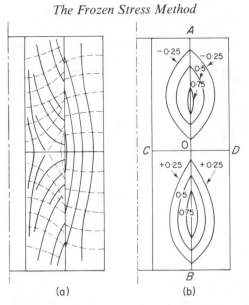

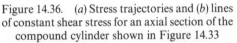

Figure 14.36. (a) Stress trajectories and (b) lines of constant shear stress for an axial section of the compound cylinder shown in Figure 14.33

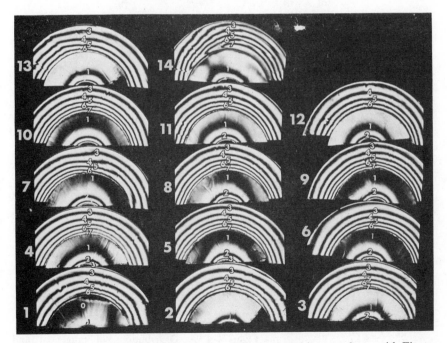

Figure 14.37. Isochromatics in transverse slices numbered in accordance with Figure 14.33

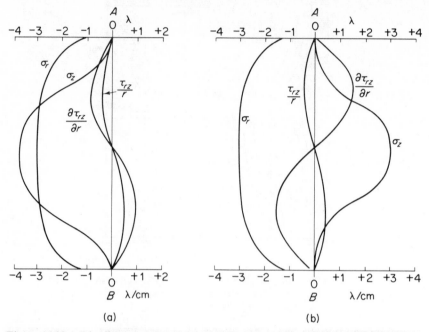

Figure 14.38. Distributions of radial and axial stresses along the contacting surfaces and auxiliary curves used for their determination. (*a*) Outer surface of inner cylinder. (*b*) Inner surface of outer cylinder

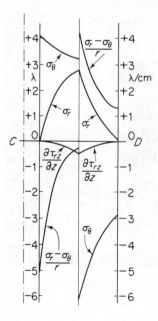

Figure 14.39. Distributions of tangential and radial stresses over the central transverse section *CD* and curves used for their determination

a plane of symmetry, the stresses σ_r and σ_θ are here both true principal stresses and their difference is given directly by the order of the isochromatics in Figure 14.37, no. 8. The values of $\partial \tau_{rz}/\partial z$ were derived from the lines of constant shear stress, Figure 14.36(b). The stresses σ_r and σ_θ could then be determined from equations (14.12) and (14.13) respectively. For analysing the results, the fringe orders of Figure 14.37 were reduced to correspond to the same thickness as that of the vertical slice.

The results obtained on each of the lines AB and CD should, of course, be identical at their point of intersection.

As can be read from the results of Figure 14.38, friction prevents the inner cylinder from expanding to its free length. This introduces shear stresses which are equal at corresponding points on the surface of contact of the two cylinders. The radial stress σ_r which is also equal at points of contact in the two cylinders diminishes towards the ends of the cylinders.

15

SPECIAL APPLICATIONS OF
THE FROZEN STRESS METHOD

15.1 Application of the frozen stress method to torsion problems

If a prismatic bar of any cross-section is twisted by a pure couple acting about its axis, St Venant's theory of torsion predicts that the state of stress at any point of the bar will be biaxial. The two principal stresses at any point are equal in magnitude but opposite in sign, Figure 15.1, and are inclined at 45° to the axis of the bar and the normal cross-section. Thus, one of the two

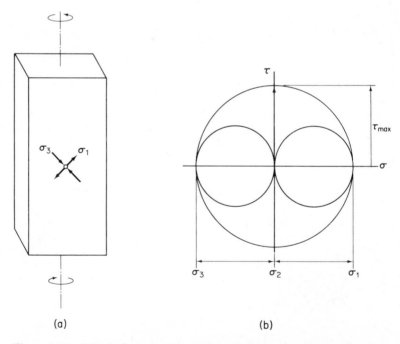

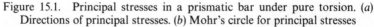

(a)   (b)

Figure 15.1.   Principal stresses in a prismatic bar under pure torsion. (a) Directions of principal stresses. (b) Mohr's circle for principal stresses

directions of maximum shear stress is parallel to the normal cross section while the other is parallel to the axis of the bar.

The directions of maximum shear stress within the plane of the normal cross-section form lines which obey laws mathematically identical with those encountered in some other problems. This allows the application of certain analogies to the solution of problems by experimental means.

One analogy is the circulation of an ideal fluid with uniform vorticity in a prismatic vessel, the shape of which corresponds to the cross-section of the bar. In such a vortex, the velocity of the fluid is inversely proportional to the spacing of the streamlines and so also, by analogy, is the shear stress in the bar.

In the membrane analogy, a soap film or rubber membrane is extended over a frame, the shape of which corresponds to the cross section of the bar. If air pressure is applied to one side of the membrane, it forms a hill of a characteristic shape. The directions of the lines of constant altitude or contour lines of the hill coincide with those of the maximum shear stress. It can be shown that the slope of the hill in any direction is proportional to the shear stress component in the perpendicular direction while the volume of the hill is proportional to the twisting moment. It follows that the maximum shear stress at any point is proportional to the slope measured in the direction perpendicular to the contour line passing through that point.

A visible pattern of lines of constant phase difference corresponding to the streamlines or membrane contour lines mentioned above can be produced photoelastically by the scattered light method (Section 16.8.2) or by the

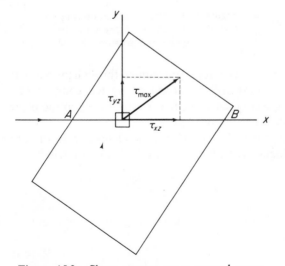

Figure 15.2. Shear stresses on a normal cross-section of a prismatic bar under pure torsion

following simple procedure, reported by Nisida,[48] based on the frozen stress method.

15.2 Wedge method

Let Figure 15.2 represent a normal cross-section of a prismatic bar subjected to pure torsion. The x, y axes are parallel to the plane of the section while the z axis is parallel to the axis of the bar. Let τ_{xz}, τ_{yz} be the components parallel to the x, y axes of the maximum shear stress τ_{max} at any point. The shear stress components acting on an element having faces normal to the x, y, z axes are then as shown in Figure 15.3.

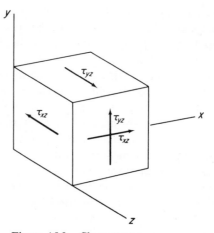

Figure 15.3. Shear stresses on a rectangular element of a prismatic bar under pure torsion

If a beam of polarized light is transmitted through the bar in the x-direction, a photoelastic effect will be produced by the component τ_{yz} only. The corresponding secondary principal stresses σ'_1, σ'_3 in the plane of the wavefront are as shown in Figure 15.4.

From equations (5.20) and (5.21), the change dn in the order of the isochromatic fringes over a length dx of the light path will be

$$dn = \frac{\tau_{yz}\,dx}{f_\tau} = \frac{(\sigma'_1 - \sigma'_2)\,dx}{f}. \tag{15.1}$$

The total fringe order accumulated between the point of entry A (Figure 15.2) and the point considered is

$$n_x = \frac{1}{f_\tau}\int_{x_A}^{x} \tau_{yz}\,dx.$$

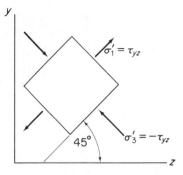

Figure 15.4. Secondary principal stresses in the plane of the wavefront for light transmitted normal to the axis

At exit we have

$$n_B = \frac{1}{f_\tau} \int_{x_A}^{x_B} \tau_{yz} \, dx.$$

Since the axial stress σ_z is everywhere zero, however, the condition of equilibrium in the axial direction of that part of the cross-section above or below the line AB requires that

$$\int_{x_A}^{x_B} \tau_{yz} \, dx = 0.$$

The fringe order at exit, i.e. the net photoelastic effect, is therefore zero. This results because τ_{yz} has a certain sign over part of the length of the light path but diminishes to zero and has the opposite sign over the remainder.

In the wedge method, the bar in which torsional stresses have been frozen is cut through at an angle of 45° to its axis. One of the parts of the bar is then placed in an immersion cell to avoid the effects of refraction and plane polarized light is passed through it in the direction normal to the axis within the plane perpendicular to the cut surface as shown in Figure 15.5(a). In this manner, a finite phase difference is obtained between the component light vibrations at exit except on the boundary. The accumulated phase difference at the point of exit B', Figure 15.5(a), is

$$n_{B'} = \frac{1}{f_\tau} \int_{x_A}^{x_{B'}} \tau_{yz} \, dx.$$

When viewed in the direction of transmission looking towards the source, an isochromatic pattern will be visible over the projection of the cut surface of the wedge which, in this direction, has the same shape as the normal cross-

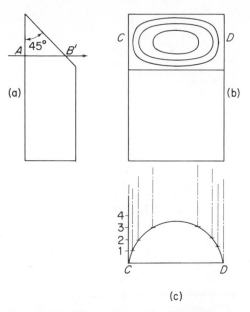

(a)

(b)

(c)

Figure 15.5. The wedge method for prismatic bars under torsion (a) Direction of observation. (b) Isochromatic pattern. (c) Equivalent membrane profile

section, Figure 15.5(b). The isochromatics represent lines along which $\int \tau_{yz}\,dx$ is constant. From equation (15.1) we have

$$dn/dx = \tau_{yz}/f_\tau,$$

Figure 15.6. Isochromatic pattern of a rectangular bar obtained by the wedge method

i.e. the slope of the fringes in the x direction is proportional to the shear stress component τ_{yz} in the perpendicular direction. This holds true for any azimuth of the 45° cutting plane so that the fringes are equivalent to the membrane contour lines, Figure 15.5(c). Moreover, since

$$\mathrm{d}n/\mathrm{d}x = \mathrm{d}n/\mathrm{d}z,$$

τ_{yz} is proportional to the gradient in the z direction of the fringes in the pattern projected in the yz plane. Likewise, the stresses τ_{xz} and τ can be

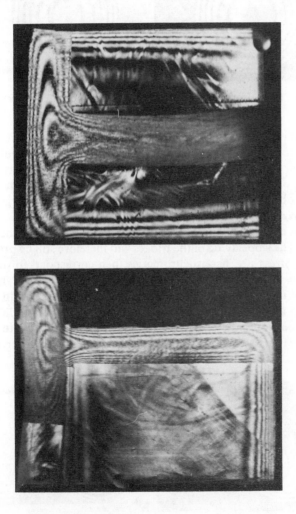

Figure 15.7. Isochromatic pattern of a bar of T-section obtained with the section cut at 45° to (a) the flange, (b) the web

determined from the slope of the fringes in the directions parallel to the y axis and normal to the fringes respectively.

Isochromatic patterns obtained by the wedge method are identical with those observed in scattered light. Figures 15.6 to 15.8 illustrate isochromatic

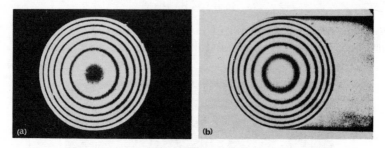

Figure 15.8.   Isochromatic patterns of a bar of circular cross-section

patterns obtained by the wedge method for bars of various cross-sections subjected to torsion. In Figure 15.7, the isochromatics for a bar of T-section are shown for two different locations of the cut section, namely, at 45° to (*a*) the plane of the flange and (*b*) the plane of the web. As can be observed, the patterns obtained differ slightly; this is mainly due to the large deformations produced.

If the bar is cut at some angle other than 45°, x remains everywhere linearly related to z. The slope dn/dz of the fringes will therefore still be proportional to the shear stress τ_{yz}. The isochromatics then correspond to the contour lines for a twisted bar having a normal cross-section of the same shape as the projection of the cut surface of the wedge in the yz plane. It can be shown that this also applies if the model is rotated through some angle about the z axis. It is therefore possible to use a single model to investigate the distribution of shear stresses in bars of related cross-sections of different shapes,

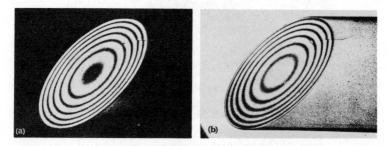

Figure 15.9.   Isochromatic patterns for a bar of elliptical section obtained by rotating the circular bar of Figure 15.8 about its axis

e.g. a circular bar can be employed to determine the distributions relating to bars of elliptical section with different eccentricities. This application is illustrated in Figures 15.8 and 15.9. Figure 15.8 shows the isochromatic patterns of the normal cross-section of a circular bar; Figure 15.9 shows the corresponding patterns obtained after rotating the bar about its axis until the projection of the cut surface in the plane of the wavefront assumed the form of an ellipse having principal axes in the ratio of $2:1$.

15.3 Slicing method

In this method by Frocht,[49] a slice of uniform thickness is cut from the frozen model. This slice is then observed in polarized light with oblique incidence.

Let us assume that the plane of the slice is normal to the axis of the bar. The shear stresses acting parallel to the plane of the slice are indicated in Figure 15.2 while those acting on an element with its faces normal to the x, y, z axes are shown in Figure 15.3. Let us now assume that the direction of transmission is parallel to the xz plane and includes an angle i with the z axis. This direction will be denoted by z' and the perpendicular direction in the xz plane by x'.

The photoelastic effect produced will be determined by the stress components acting in the plane of the wavefront, i.e. the $x'y$ plane. As can be read from Figure 15.10a, τ_{yz} produces a component of shear stress $\tau_{x'y} = \tau_{yz} \sin i$ acting parallel to the $x'y$ plane, Figure 15.10(b). Considering now the effect of τ_{xz}, Figure 15.10(c) shows that only the normal stress $\sigma_{x'}$ acts in the $x'y$ plane. From Mohr's circle for the xz plane, Figure 15.11, this component is

$$\sigma_{x'} = \tau_{xz} \sin 2i.$$

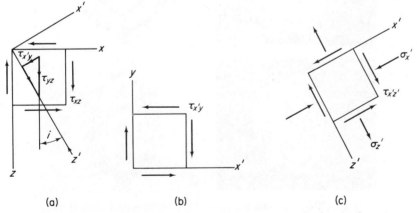

(a)   (b)   (c)

Figure 15.10. Determination of the stress components in the $x'y$ plane

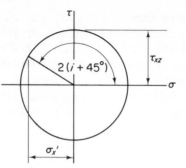

Figure 15.11. Mohr's circle for
stresses in the xz plane

From equation (1.4) the secondary principal stresses in the plane of the wavefront are then

$$\sigma'_1, \sigma'_2 = \tfrac{1}{2}\{\tau_{xz} \sin 2i \pm \sqrt{(\tau_{xz}^2 \sin^2 2i + 4\tau_{yz}^2 \sin^2 i)}\}.$$

From the stress optic law we then have

$$\sqrt{(\tau_{xz}^2 \sin^2 2i + 4\tau_{yz}^2 \sin^2 i)} = nf/d', \qquad (15.2)$$

where d' is the length of the path of light within the slice. Equation (15.2) shows that the components τ_{xz}, τ_{yz} can be determined from the order of the

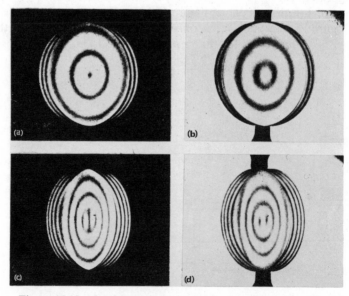

Figure 15.12. Isochromatic patterns of a transverse slice of a circular bar. (*a*), (*b*) Angle of incidence $i = 30°$; (*c*), (*d*) $i = 45°$

isochromatics measured with two different values of the angle *i*. The maximum shear stress can then be evaluated from

$$\tau^2 = \tau_{xz}^2 + \tau_{yz}^2.$$

Since no photoelastic effect is produced when the direction of transmission is parallel to the axis of the bar, both observations must be made with oblique incidence for a slice cut perpendicular to the axis of the bar. If the plane of the slice is inclined to that of the normal cross-section, however, one of the observations may be made with normal incidence. In the latter case, the value of *i* to be inserted in equation (15.2) is the angle between the direction of propagation and the axis of the bar for each observation.

Isochromatic patterns relating to a bar of circular cross-section obtained by the slicing method are given in Figures 15.12 and 15.13. Figure 15.12

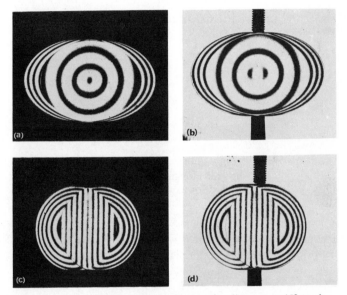

Figure 15.13. Isochromatic patterns of a slice cut at 45° to the axis of a circular bar. (*a*), (*b*) Normal incidence; (*c*), (*d*) Observation at 90° to the axis of the bar

shows the isochromatics of a slice cut normal to the axis of the bar for angles of incidence of 30° and 45°. Figure 15.13 shows the isochromatics of a slice cut at 45° to the axis of the bar viewed under normal incidence and at 45° incidence in the direction normal to the axis of the bar.

15.4 Longitudinal halving method

This method suggested by Nisida[48] can be applied to problems of circular shafts of variable diameter. When a shaft of this type is subjected to pure torsion about its axis, the theory of elasticity predicts that, in polar co-ordinates, only the stress components $\tau_{r\theta}$ and $\tau_{\theta z}$ differ from zero.

The frozen model is cut longitudinally in such a way that the cut surface contains the axis of the shaft. The longitudinal half of the shaft thus obtained is then placed in an immersion cell and polarized light is passed through it in the direction normal to the cut surface. The component $\tau_{r\theta}$ acting in the plane of the wave normal does not influence the light during its passage through the model. Considering the effect of $\tau_{\theta z}$, Figure 15.14 shows that

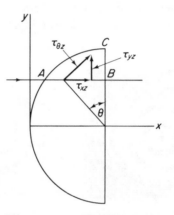

Figure 15.14. Sketch illustrating the longitudinal halving method for axially symmetrical shafts

only the component $\tau_{yz} = \tau_{\theta z} \sin \theta$ will produce a photoelastic effect. The secondary principal stresses in the plane of the wavefront are

$$\sigma'_1, \sigma'_2 = \pm \, \tau_{yz}.$$

The change $\mathrm{d}n$ in the isochromatic fringe order produced in a length $\mathrm{d}x$ of the light path is

$$\mathrm{d}n = \frac{(\sigma'_1 - \sigma'_2)\,\mathrm{d}x}{f} = \frac{2\tau_{yz}\,\mathrm{d}x}{f}.$$

The isochromatic order n at the point of exit B is therefore

$$n = \frac{2}{f} \int_{x_A}^{x_B} \tau_{yz}\,\mathrm{d}x.$$

From the condition that the net axial force over the section *ABC* must be zero, we have

$$\int_{x_A}^{x_B} \tau_{yz} \, dx = \int_{y_B}^{y_C} \tau_{xz} \, dy = \int_{r_B}^{r_C} \tau_{\theta z} \, dr,$$

so that

$$n = \frac{2}{f} \int_{r_B}^{r_C} \tau_{\theta z} \, dr,$$

or

$$\tau_{\theta z} = \frac{f}{2} \frac{dn}{dr} = f_\tau \frac{dn}{dr}.$$

The shear stress $\tau_{\theta z}$ is thus given by the slope of the isochromatics in the radial direction multiplied by the material fringe value in shear of the material.

The other component, $\tau_{r\theta}$, can also be determined from the isochromatic pattern. From equations (2.3), we have for the shear strains

$$\gamma_{r\theta} = \frac{\tau_{r\theta}}{G} = \frac{\partial v}{\partial r} - \frac{v}{r} = r \frac{\partial}{\partial r}\left(\frac{v}{r}\right),$$

$$\gamma_{\theta z} = \frac{\tau_{\theta z}}{G} = \frac{\partial v}{\partial z} = r \frac{\partial}{\partial z}\left(\frac{v}{r}\right).$$

Dividing these equations by r then differentiating the first with respect to z and the second with respect to v and combining yields

$$\frac{\partial}{\partial z}\left(\frac{\gamma_{r\theta}}{r}\right) = \frac{\partial^2}{\partial r \, \partial z}\left(\frac{v}{r}\right) = \frac{\partial}{\partial r}\left(\frac{\gamma_{\theta z}}{r}\right),$$

or, since $\gamma = \tau/G$,

$$\frac{\partial}{\partial z}\left(\frac{\tau_{r\theta}}{r}\right) = \frac{\partial}{\partial r}\left(\frac{\tau_{\theta z}}{r}\right).$$

Integrating, we obtain

$$\tau_{r\theta} = r \int \frac{\partial}{\partial r} \frac{\tau_{\theta z}}{r} \, dz = f_\tau r \int \frac{\partial}{\partial r}\left(\frac{1}{r} \frac{dn}{dr}\right) dz$$

$$= f_\tau \int \left(\frac{d^2 n}{dr^2} - \frac{1}{r} \frac{dn}{dr}\right) dz + (\tau_{r\theta})_{z_0}. \qquad (15.3)$$

Equation (15.3) can be solved provided the value of $\tau_{r\theta}$ is known at some point z_0 on the line of integration. Such a point will often be available either on the free surface of the shaft where the tangent to the profile is parallel to the axis or anywhere within a parallel portion of the shaft remote from any change of section. Since the solution involves second order graphical differentiation, the accuracy may be rather low. The magnitude of $\tau_{r\theta}$ is usually small however compared with that of $\tau_{\theta z}$.

When using the above method, a very small difference of the refractive indices of the immersion fluid and the model material causes considerable error in the location of the fringes near the edges. The refractive indices must therefore be accurately matched.

15.5 Warping stresses

Another effect involved in torsion problems can be demonstrated and investigated by the frozen stress method. If the cross section of a twisted bar is not rotationally symmetrical, it becomes distorted longitudinally, i.e. sections originally plane and normal to the axis do not remain so after twisting. This effect is called warping.

If free warping of the cross-sections is prevented, additional stresses are introduced. One example is that of a bar clamped at its centre and twisted at both ends in the same sense. Since the longitudinal displacements at the centre associated with the warping of each half of the bar are exactly equal but opposite in direction, warping of this section will be completely prevented. The resulting additional stresses are called warping stresses.

In order to separate warping stresses from those due to pure torsion, two identical models of the bar are twisted in opposite senses and the stresses frozen in. Both are then cut at their centres exactly perpendicular to their axes. If these parts of the models were now annealed, their plane cut faces would become distorted showing the warping effect. Instead, before annealing, one half of one of the models is cemented to one half of the other using cold setting adhesive. The composite bar is then annealed by heating to the softening temperature. By this annealing, St Venant's torsion vanishes while recovery of the warping effect is prevented as the displacements of the two halves at their junction are in opposite directions.

The effect of applying this procedure to a bar of rectangular section (identical with that of Figure 15.6) is illustrated in Figure 15.15. This shows the isochromatics for longitudinal slices cut parallel to the longer and shorter edges of the cross-section respectively.

The warping effect is much greater in the torsion of thin-walled sections. Figure 15.16 shows the isochromatics produced by warping in an I-section, the direction of observation being normal to the plane of the flanges.

15.6 Problems of centrifugal and gravitational stresses

The stress freezing procedure may be applied to the investigation of problems of the stresses due to centrifugal forces in bodies rotating at constant

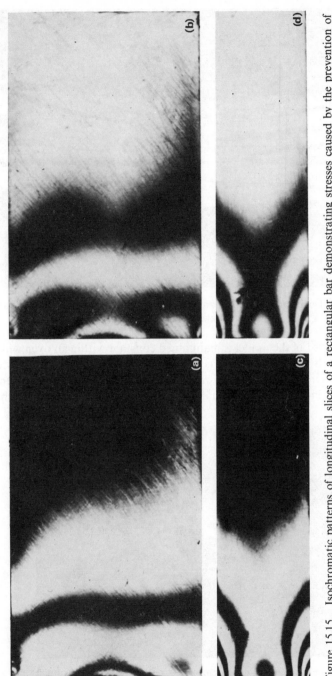

Figure 15.15. Isochromatic patterns of longitudinal slices of a rectangular bar demonstrating stresses caused by the prevention of warping. (*a*), (*b*) Slice parallel to the longer edge. (*c*), (*d*) Slice parallel to the shorter edge

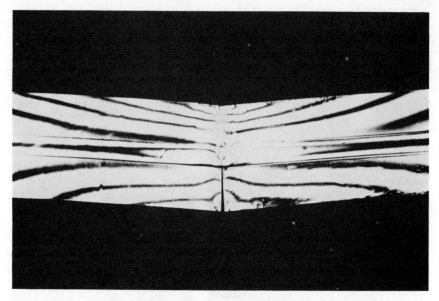

Figure 15.16.   Isochromatics in the flanges of a bar of I-section due to stresses caused
by the prevention of warping

speed. The speed of the model is maintained constant within an oven while it
is subjected to the normal stress freezing cycle. Figure 15.17, kindly provided
by Professor Dr Gotterbarm, shows the isochromatic pattern due to centrifugal stresses in a gear wheel obtained in this way. This method has been

Figure 15.17.   Frozen isochromatic pattern due to centrifugal
forces in a gear wheel. Courtesy Gotterbarm

used to study centrifugal stresses in various forms of rotating bodies such as discs with holes, turbine discs, etc., by Hetenyi,[50] Newton,[51] Reichner,[52] Guernsey,[53] Fessler and Thorpe,[54] and others.

Centrifugal forces may also be employed to simulate gravitational forces for the investigation of problems of self-weight, etc. With this procedure, the model is located within an oven mounted at the end of the arm of a centrifuge. If the dimensions of the model are small compared with its distance from the axis of rotation, variation of the body forces will be small and the stress distribution will approximate to that produced in a uniform gravitational field. This method has been applied by Hendry[55,56] and Rydzewski[57] to the study of the distribution of stresses in dams and by Saad and Hendry[58] to determine self-weight stresses in beams of deep section.

15.7   Application of the *j*-circle to the frozen stress method

The conventional methods of analysing the birefringence in slices cut from frozen models are based on the assumptions that the stresses are either approximately constant or else vary linearly along lines normal to the plane of the slice and maintain constant directions. In certain cases, the *j*-circle method allows a more precise analysis of the optical effects produced.

One procedure is to apply the same methods as used for the analysis of the stresses in shells, etc., using models of uniform material (Section 17.4). By this means, the changes of directions and the magnitudes of the stresses can be determined on the assumption that the stresses through the thickness of the slice vary in a similar manner to those in shells.

By means of another procedure free from any assumption of the above kind, it is possible to determine the directions and magnitudes of the birefringence at a point not only in one direction but, similar to the methods described in Section 14.6, for any direction of observation. Thick slices or, if possible, the whole model are observed in circularly polarized light for three different directions of transmission along lines intersecting at the point of interest, Figure 15.18 and the values of m_k and ε obtained for each as

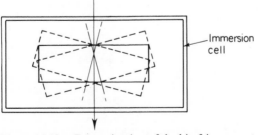

Figure 15.18.   Determination of the birefringence at a point in a frozen stress model or slice by the *j*-circle method

described in Section 13.3 (the values of δ are not required). The position within the j-circle of the tip of the j-vector corresponding to each of the three pairs of values is plotted. The angles $2m_k$ are equal to the angular distances of the tips from the centre of the j-circle and the required points can be plotted with the aid of the nomogram, Figure 13.8. In this way the end points of the tip curves for incident circularly polarized light passing through the whole model or slice in the three directions are obtained.

The thickness of the slice is then reduced in small steps, e.g. by milling. After each step, the values of m_k and ε are determined and the corresponding points marked in the corresponding j-circle. In this way, the three tip curves are traced back to their starting points.

From the tip curves, the magnitude and direction of the birefringence of each element of the light path can be determined. The value of $2m_k$ for any element is equal to the angular distance between the points representing the conditions before and after removal of the element and can be evaluated from the nomogram, Figure 13.8. The corresponding phase difference is that produced within the length of the light path within the layer removed. The direction of the birefringence within the element coincides with that of the corresponding element of the tip curve. Finally, the stresses are determined from the values of the birefringence in the three directions as described in Section 14.7.

16

THE SCATTERED LIGHT METHOD

16.1 Scattered light

It is well known that when a beam of light passes through a transparent isotropic medium, light is scattered from each point along its path. This scattering may be produced by small particles in suspension or by the molecules of the medium itself. If the primary beam is plane polarized, the intensity of the scattered or secondary light varies according to the direction of observation. If the scattering is due to extremely fine particles or to the molecules of the medium, the scattered light is more or less completely plane polarized. In general, however, some depolarization occurs and the secondary light includes an appreciable amount of ordinary light. The intensity of the ordinary light scattered from any point is the same in all directions within the plane normal to the axis of propagation of the primary beam while that of the polarized light in any direction is proportional to the square of the component of amplitude of the primary vibration in the perpendicular direction. The actual intensity of secondary light observed or measured from points outside the medium will be affected by absorption; this is proportional to the length of the light path within the medium which, in general, will vary according to the direction of observation.

The possibility of determining stresses from observations of the light scattered from a photoelastic model was first suggested by Weller[59] in 1939. Later developments and refinements of the method are due to Menges,[60] Drucker and Mindlin,[61] Frocht and Srinath,[62-64] Jessop,[65] Robert and Guillemet,[66] Cheng[67-69] and others.

When a beam of polarized light is transmitted through a stressed photo-elastic model, each of the component primary waves into which it is resolved produces its own secondary waves. Since the phase difference between these components in general varies from point to point along the path of the primary beam and the same phase difference will exist between the secondary polarized waves, the latter will interfere to produce a pattern of fringes. As shown in Section 3.20, complete interference occurs only when the amplitudes of the interfering waves are equal. If the principal stresses maintain constant directions along the path of the primary beam, this condition will exist when the direction of polarization and the direction of observation are both at 45° to the directions of the principal stresses. In general, the ratio of the amplitudes and hence the degree of contrast exhibited by the interference pattern will vary according to the direction of observation.

The degree of contrast will also be diminished by the presence of some vagrant ordinary light. This superimposes a more or less uniform background intensity on the interference pattern. As a result, the dark fringes observed under even the most favourable conditions for interference will not appear black.

16.2   Formation of the fringe pattern

The component vibrations in the directions OP, OQ parallel to the principal stresses of the primary beam at any point of its path through a stressed photoelastic model may be represented by

$$u = a \cos \theta \sin(\omega t + \alpha); \qquad v = a \sin \theta \sin \omega t,$$

where a is the amplitude of the incident plane polarized wave, θ is the inclination of the axis of the polarizer to the direction OP (Figure 16.1), and α is

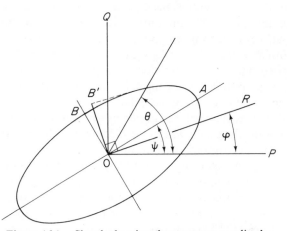

Figure 16.1.   Sketch showing the apparent amplitude
OB' of polarized light scattered in the direction OR

the phase difference accumulated between the point of entry and the point considered. As shown in Section 3.6, these components combine to produce an elliptical vibration; the axes of this ellipse are inclined at an angle ψ to the direction OP where

$$\tan 2\psi = \tan 2\theta \cos \alpha, \tag{3.10}$$

while the lengths of the semi-axes are given by

$$OA, OB = \sqrt{\left\{\frac{a^2}{2}(1 \pm \sqrt{[\cos^2 2\theta + \sin^2 2\theta \cos^2 \alpha]})\right\}}. \tag{3.11}$$

The intensity of the polarized light scattered in the direction OR inclined at an angle φ to OP is proportional to the square of the apparent amplitude OB'. This amplitude is the same as that of a wave transmitted by an analyser with its axis inclined at an angle $90° + \varphi$ to OP. The intensity I_φ of the scattered light can therefore be obtained from equation (5.9) if θ' is replaced by $90° + \varphi$. Hence,

$$I_\varphi = ka^2(1 - \cos 2\theta \cos 2\varphi - \sin 2\theta \sin 2\varphi \cos \alpha). \tag{16.1}$$

If either the direction of observation or the direction of the polarizer is rotated through 90° from this position the intensity becomes

$$I_{90} + \varphi = ka^2(1 + \cos 2\theta \cos 2\varphi + \sin 2\theta \sin 2\varphi \cos \alpha).$$

We thus have

$$I_\varphi + I_{90+\varphi} = 2ka^2 = I_0$$

where I_0 represents the total intensity of polarized scattered light.

As α varies from point to point along the path of the primary beam, the intensity of light scattered in the direction φ will vary from

$$(I_\varphi)_{min} = ka^2(1 - \cos 2\theta \cos 2\varphi - \sin 2\theta \sin 2\varphi)$$

$$= ka^2[1 - \cos 2(\theta - \varphi)],$$

at points where $\alpha = 2n\pi$, ($n = 0, 1, 2$, etc.), to

$$(I_\varphi)_{max} = ka^2(1 - \cos 2\theta \cos 2\varphi + \sin 2\theta \sin 2\varphi)$$

$$= ka^2[1 - \cos 2(\theta + \varphi)]$$

at points where $\alpha = (2n - 1)\pi$.

We thus have

$$(I_\varphi)_{max} - (I_\varphi)_{min} = 2ka^2 \sin 2\theta \sin 2\varphi.$$

This equation shows that for any given position of the polarizer, the greatest variation of intensity occurs when $\sin 2\varphi = 1$, i.e. when the direction of observation is inclined at 45° to the directions of the principal stresses. The greatest possible variation of intensity occurs when $\sin 2\varphi = \sin 2\theta = 1$, i.e. when both the direction of observation and the direction of polarization are inclined at 45° to the directions of the principal stresses as deduced previously. The minimum intensity of the polarized light is then zero while the maximum is $2ka^2$. If the direction of observation coincides with the direction of the axis of the polarizer so that $\varphi = \theta = \pm 45°$, zero intensity occurs at points where $\alpha = 2n\pi$ and maximum intensity where $\alpha = (2n + 1)\pi$; these conditions are reversed if the direction of observation is perpendicular to the direction of polarization, i.e. if $\pm\varphi = \mp\theta = 45°$.

Considering any general point on the path of the primary beam where the phase difference is α, the directions in which the maximum and minimum intensities of scattered light will be observed for any given position of the polarizer are obtained from

$$dI_\varphi/d\varphi = 0,$$

which yields

$$\tan 2\varphi = \tan 2\theta \cos \alpha.$$

The directions of observation so defined coincide with those of the axes of the light ellipse at the observed point. The corresponding maximum and minimum intensities are

$$I_{max}, I_{min} = ka^2(1 \pm \sqrt{[\cos^2 2\theta + \sin^2 2\theta \cos^2 \alpha]}).$$

Similarly, the directions of the axis of the polarizer which will produce maximum and minimum intensities in any given direction of observation are defined by

$$dI_\varphi/d\theta = 0,$$

from which

$$\tan 2\theta = \tan 2\varphi \cos \alpha.$$

The corresponding values of the intensities are

$$I_{max}, I_{min} = ka^2(1 \pm \sqrt{[\cos^2 2\varphi + \sin^2 2\varphi \cos^2 \alpha]}).$$

16.3 Incident circularly polarized light

If the incident light is circularly polarized in the positive (anticlockwise) sense, the component vibrations of the primary beam parallel to the principal stresses σ_1, σ_2 at any point along its path can be represented by

$$u = \frac{a}{\sqrt{2}} \sin\left(\omega t + \frac{\pi}{2} + \alpha\right) = \frac{a}{\sqrt{2}} \cos(\omega t + \alpha),$$

$$v = \frac{a}{\sqrt{2}} \sin(\omega t + \alpha),$$

respectively. The resultant vibration in the direction OB' normal to the direction of observation OR is

$$u' = u \sin \varphi - v \cos \varphi$$

$$= \frac{a}{\sqrt{2}} \sin \varphi \cos(\omega t + \alpha) - \frac{a}{\sqrt{2}} \cos \varphi \sin \omega t$$

$$= \frac{a}{\sqrt{2}} \sin \varphi \cos \alpha \cos \omega t - \frac{a}{\sqrt{2}} (\sin \varphi \sin \alpha + \cos \varphi) \sin \omega t.$$

The intensity of the light scattered in the direction OR is proportional to the square of the amplitude of this vibration.

Hence,

$$I = ka^2[(\sin \varphi \cos \alpha)^2 + (\sin \varphi \sin \alpha + \cos \varphi)^2]$$

$$= ka^2(1 + \sin 2\varphi \sin \alpha),$$

which varies according to the value of α between the limits $ka^2(1 - \sin 2\varphi)$ and $ka^2(1 + \sin 2\varphi)$. If $\varphi = +45°$, the minimum intensity is zero and occurs at points where $\alpha = (4n + 3)\pi/2$ while the maximum intensity is $2ka^2$ and occurs at points where $\alpha = (4n + 1)\pi/2$. These conditions are reversed if $\varphi = -45°$ or if the incident light is circularly polarized in the opposite sense.

16.4 Interpretation of the fringe pattern

We first consider the case where the principal stresses vary in magnitude but not in direction along the path of the primary beam. Let dR be the change of relative retardation of the two component vibrations produced in an elementary length dx of the light path within which the secondary principal stresses σ_1', σ_2' may be regarded as constant. From the stress optic law we have

$$dR = C(\sigma_1' - \sigma_2') \, dx,$$

or

$$(\sigma_1' - \sigma_2') = \frac{1}{C} \frac{dR}{dx}.$$

Expressed in terms of the fringe order and material fringe value, this equation becomes

$$(\sigma_1' - \sigma_2') = f(dn/dx). \tag{16.2}$$

The difference of the secondary principal stresses at any point is therefore proportional to the gradient or inversely proportional to the spacing of the fringes observed in scattering light.

Equation (16.2) may also be applied without introducing appreciable error when the directions of the principal axes vary slowly along the path of the primary beam. If these directions vary rapidly, the methods of the *j*-circle can be applied.

16.5   Determination of fringe orders

If the direction of observation is parallel to that of the axis of the polarizer, the zero fringe will coincide with the point of entry of the light beam. The phase difference between the component primary vibrations will vary continuously along the path of the beam so that the orders of fringes at internal points can be determined by counting from the zero fringe. The position of half-wave fringes can be obtained after rotating the polarizer through 90°.

The possibility of errors in fringe counting due to a change of sign of the difference of the secondary principal stresses at any point may be avoided by introducing an additional phase difference by means of a compensator: the fringes on opposite sides of such a point will move in opposite directions. Alternatively, a change of sign may be indicated by a reversal of the colour sequence of the fringes when using white light or with the colour filter removed.

In cases where the number of whole- and half-wave fringes is insufficient to allow a satisfactory curve to be drawn, it becomes necessary to measure fractional fringe orders at intermediate points. This may be accomplished by means of a compensator, e.g. of the Babinet type. The compensator is placed in the path of the incident plane polarized beam with its axes parallel to the principal axes in the model. It is then adjusted until an adjacent fringe moves to the point observed. If Δn is the measured phase difference introduced by the compensator and n' the order of the fringe moved to the point, then the fringe order n at the point is

$$n = n' \pm \Delta n,$$

the sign depending on whether the fringe moved to the point is that of next lower or higher order.

Fractional phase differences can be measured in scattered light by means of a variation of the Senarmont method of compensation commonly used with transmitted light. In this procedure, reported by Cheng, a quarter-wave plate is inserted between the polarizer and the model. Let us consider the orientation shown in Figure 16.2 in which the direction of observation OR and the fast axis of the quarter wave plate include an angle β with the axis of the polarizer and angles of 45° with the secondary principal axes in the model. The plane polarized wave $u = 0$, $v = a \sin \omega t$ emerging from the polarizer is converted by the quarter-wave plate into an elliptical vibration

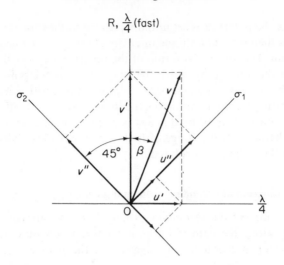

Figure 16.2. Compensation in scattered light by
the modified Senarmont method

represented by the components

$$u' = a \sin \beta \cos \omega t,$$

$$v' = a \cos \beta \sin \left(\omega t + \frac{\pi}{2} \right) = a \cos \beta \cos \omega t.$$

The components of this vibration parallel to the directions of the principal stresses are

$$u'' = \frac{(u' + v')}{\sqrt{2}} = \frac{a}{\sqrt{2}} \cos(\omega t - \beta),$$

$$v'' = \frac{(v' - u')}{\sqrt{2}} = \frac{a}{\sqrt{2}} \cos(\omega t + \beta).$$

If the phase difference accumulated between these components from the point of entry is α, then at the observed point we have

$$u'' = \frac{a}{\sqrt{2}} \cos(\omega t - \beta + \alpha),$$

$$v'' = \frac{a}{\sqrt{2}} \cos(\omega t + \beta).$$

These equations show that when $\beta = \alpha/2$, the resultant is a plane polarized wave vibrating at 45° to the secondary principal stresses, in which direction the intensity will be then a minimum.

The axis of the polarizer is set initially parallel to the direction of observation *OR*. It is then rotated until the intensity at the observed point is reduced to a minimum. The angle of rotation of the polarizer expressed as a fraction of 180° gives the fraction by which the fringe order at the point differs from the next higher or lower integral order. Minimum intensity can obviously be achieved rotating the polarizer through either of two angles differing by 180°. The appropriate angle to be used in any particular case can be ascertained by observing whether it is the fringe of higher or lower order which moves to the point as the polarizer is rotated.

16.6 Determination of secondary principal axes

If the directions of the secondary principal stresses are constant or only rotate slowly along the path of the primary beam they may be determined by the following procedures. In other cases, the *j*-circle method may be applied.

It can be read from equation (16.1) that the intensity of the scattered light observed in the direction of either of the principal stresses ($\varphi = 0$ or $\varphi = 90°$) is independent of the phase difference. No interference effects will then be observed along the path of the primary beam. Moreover, if $\theta = \varphi$, i.e. if the direction of observation coincides with the direction of the axis of the polarizer, the uniform intensity along the light path becomes a minimum. The directions of the principal stresses can therefore be determined by observing in a direction parallel to the axis of the polarizer and rotating the model about an axis coinciding with the axis of propagation until a uniform minimum intensity is observed. A more precise procedure is to vary the phase difference at the observed point by means of a compensator while the intensity of the secondary light is measured by means of a photomultiplier. The direction of observation coincides with one of the principal stresses when the intensity remains unchanged while the phase difference at the point is altered.

The directions of the principal stresses may be determined by means of another procedure based on the fact that when $\alpha = 2n\pi$ and $(2n - 1)\pi$, the light ellipse degenerates into straight lines including angles of $\pm\theta$ respectively with the direction of σ_1 as can be deduced from equations (3.10) and (3.11). The intensity will be a minimum when the direction of observation coincides with the directions of these linear vibrations.

The phase difference at the observed point is varied by means of a compensator to produce in turn minimum and maximum intensity in an arbitrary direction of observation. The corresponding phase differences are $2n\pi$ and $(2n - 1)\pi$. In each case the direction of the resulting linear vibration is determined by rotating the direction of observation until minimum intensity is obtained. The bisector of the angle between these two directions coincides with the direction of one of the principal stresses.

16.7 The scattered light polariscope

The optical conditions necessary for the production of satisfactory inter-ference patterns with scattered light differ considerably from those suitable for conventional observations in transmitted light. For this reason, investiga-tions involving scattered light observations are usually performed using polariscopes specially designed for the purpose.

The basic optical requirement is a fine pencil or thin ribbon of highly collimated, monochromatic, polarized light. Since the efficiency of the scattering process is low, a source of high intensity is required; in order to combine reasonably high intensity with high collimation, it should also be as compact as possible. These requirements are met by certain types of mercury arc lamp while a laser is particularly suitable.

Figure 16.3 shows diagrammatically the optical system of a scattered light polariscope used by Cheng. Light from the mercury arc source A passes

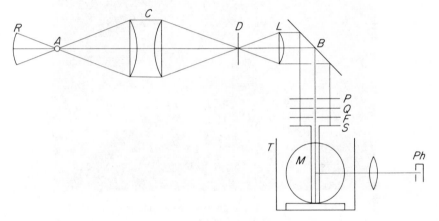

Figure 16.3. Schematic of a scattered light polariscope. After Cheng

through a condenser C and is brought to a focus on the aperture of an iris diaphragm D. The light then passes through a collimator L after which it is projected vertically downwards by means of an inclined mirror B. It then passes through a polarizer P, quarter-wave plate Q, monochromatic filter F and adjustable slit S and finally through the model M which is contained in an immersion tank T. Provision is made to allow rotational and transla-tional movement of the model in the tank. Figure 16.4 shows a scattered light polariscope designed by the same investigator in which a laser is em-ployed as the light source; full field interference patterns are obtained by moving the laser beam at a uniform speed through the model in the direction parallel to the plane of the film.

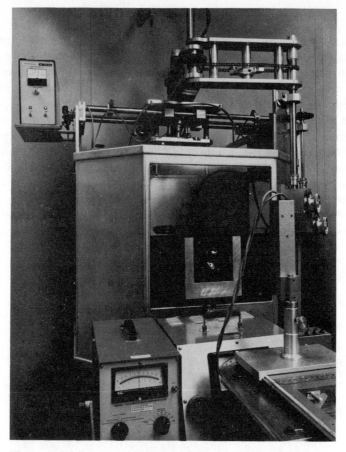

Figure 16.4.   A scattered light polariscope with laser source used by
Cheng

16.8   Applications of the scattered light method

16.8.1   *Two-dimensional problems*

In the investigation of problems of plates loaded in their planes, the
conventional observations in transmitted light may be supplemented by
observations of the scattered light. By this means, additional data are obtained which allow the separation of the principal stresses. It is also possible
to determine the individual principal stresses and their directions from
scattered light observations alone.

Let σ_1, σ_2 be the principal stresses at a point O of a plate in a two-dimensional state of stress and let the direction of transmission Ox of the primary
beam within the plane of the plate include an angle φ with σ_1, Figure 16.5.

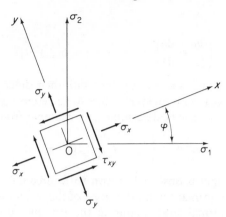

Figure 16.5. Stress components acting on an element with edges parallel and perpendicular to the direction of transmission

The stress components acting on an element of the plate having edges parallel and perpendicular to the direction of transmission are as shown. Of these components, only σ_y will produce a photoelastic effect. The secondary principal stresses in the plane of the wavefront are

$$\sigma_1' = \sigma_y; \; \sigma_3' = 0.$$

The slope of the fringes observed in scattered light normal to the plane of the plate therefore indicates the value of σ_y, i.e.

$$f(\mathrm{d}n/\mathrm{d}x) = \sigma_y = \sigma_1 \sin2\,\varphi + \sigma_2 \cos^2\varphi$$

$$= \sigma_1 - (\sigma_1 - \sigma_2)\cos^2\varphi. \tag{16.3}$$

By adding to this value that of $(\sigma_1 - \sigma_2)\cos^2\varphi$ derived from the isochromatics and isoclinics observed in transmitted light, the value of σ_1 is obtained.

With observation of the scattered light for two mutually perpendicular directions of transmission of the primary light, only the isochromatic pattern in transmitted light is necessary to enable the state of stress to be completely determined. Let the primary light be transmitted in turn in the x and y directions, Figure 16.5. We have

$$f(\mathrm{d}n/\mathrm{d}x) = \sigma_y,$$

$$f(\mathrm{d}n/\mathrm{d}y) = \sigma_x,$$

from which

$$f\left(\frac{dn}{dx} + \frac{dn}{dy}\right) = (\sigma_x + \sigma_y) = (\sigma_1 + \sigma_2).$$

The sum of the principal stresses together with their difference given by the isochromatics allows their individual values to be determined. The directions of the principal stresses can be determined, for example, from equation (16.3):

$$\cos^2 \varphi = \left(\sigma_1 - f\frac{dn}{dx}\right) \Big/ (\sigma_1 - \sigma_2).$$

If the scattered light is observed in turn with three different directions of transmission of the primary light, the slope of the fringes in each case gives the value of the normal stress acting in the perpendicular direction. The principal stresses and their directions can then be determined by drawing Mohr's circle in the manner described in Section 1.4.

16.8.2   *Pure torsion of prismatic bars*

As explained in Section 15.2, when a beam of polarized light is transmitted through a prismatic bar subjected to pure torsion in any direction within the plane normal to the axis of the bar, the fringe order accumulated between the point of entry A, Figure 16.6, and an internal point B is

$$n = \frac{1}{f_\tau} \int_{x_A}^{x_B} \tau_{yz} \, dx,$$

where τ_{yz} is the component of shear stress normal to the direction of transmission, i.e. the x axis. From the condition that the longitudinal thrust on the section ABC is zero, it follows that

$$\int_{x_A}^{x_B} \tau_{yz} \, dx + \int_{x_C}^{x_B} \tau_{xz} \, dy = 0,$$

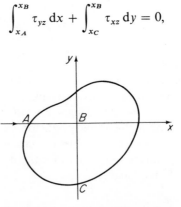

Figure 16.6.   Analysis of the optical
effects produced in a prismatic bar
subjected to torsion

so that we also have

$$n = -\frac{1}{f_\tau}\int \tau_{xz}\,dy.$$

From these equations we obtain

$$\partial n/\partial x = \tau_{yz}/f_\tau, \qquad \partial n/\partial y = -\tau_{xz}/f_\tau.$$

The slope of an isochromatic $n = $ constant is given by

$$\tan\beta = -\frac{\partial n/\partial x}{\partial n/\partial y} = \frac{\tau_{yz}}{\tau_{xz}}. \tag{16.4}$$

In St Venant's theory of torsion, the shear stress components are related to the stress function ϕ by

$$\tau_{xz} = \partial\phi/\partial y, \qquad \tau_{yz} = -\partial\phi/\partial x.$$

The slope of a line $\phi = $ constant is therefore

$$\tan\beta'' = -\frac{\partial\phi/\partial x}{\partial\phi/\partial y} = \frac{\tau_{yz}}{\tau_{xz}}. \tag{16.5}$$

From equations (16.4) and (16.5) we see that the isochromatics correspond to lines $\phi = $ constant or to the equivalent membrane contours. The direction of the maximum shear stress at any point is indicated by the tangent to the isochromatic passing through that point while its magnitude is given by the fringe gradient in the direction of the normal multiplied by the material fringe value in shear.

In practice, the torsional stresses are usually frozen into the material to avoid difficulties of observation. The whole cross-section of the bar is then illuminated by a ribbon of plane polarized light, the direction of vibration and the direction of observation being parallel to the axis. Since the stresses do not vary in the axial direction, the interference pattern observed in this case is independent of the thickness of the primary beam. Alternatively, the cross-section may be traversed by a laser beam moving with uniform velocity.

16.9 Application of the *j*-circle to the scattered light method

The *j*-vector pertaining to any point of the path of a primary beam of light passing through a birefringent body can be determined from measurements of the intensity of the light scattered at the point. It follows that the optical effects can be analysed by tracing the tip curve along the path of the primary beam in the *j*-circle by means of photometric procedures using a photometer or photomultiplier.

As shown in Section 16.1, the intensity of polarized light scattered in a given direction is proportional to the intensity that would be transmitted by an analyser with its principal axis in the perpendicular direction. Let OC, Figure 16.7, be the j-vector characterizing the state of vibration of the primary light at any point within the model. It was shown in Section 13.1 that the intensity transmitted by an analyser set with its axis in the direction represented by OA is proportional to $A'C'$ where C' is the projection of C on AA'.

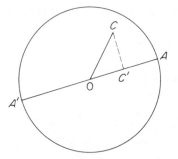

Figure 16.7. The intensity of scattered light represented in the j-circle

It follows that the intensity of polarized light scattered in the same direction is proportional to AC'.

There are several procedures for measuring and analysing the secondary light, the one to be followed depending on the nature of the problem and on the shape and material of the model used. In some cases, the secondary light originating at a point of the path of the primary light beam can be observed and measured in any direction within the plane perpendicular to the direction of the primary beam. In other cases the possibilities of observation may be restricted to some extent by loading devices, etc., or by thick parts of the model. In the most simple procedure, applicable only when observation of the scattered light is unrestricted, the first step consists of determining the maximum intensities I'_{max} and I'_{min} respectively and their directions with incident circularly polarized light. These mutually perpendicular directions indicate the direction of the j-vector within the circle. The measured intensity of the scattered light includes a considerable amount of vagrant light I' which must be taken into account when analysing the results. For this reason, the net values of I_{max} and I_{min} cannot be determined directly from the measured values which, assuming the intensity of vagrant light to be the same in all directions (Figure 16.8(a)) are

$$I'_{max} = I_{max} + I',$$

$$I'_{min} = I_{min} + I'.$$

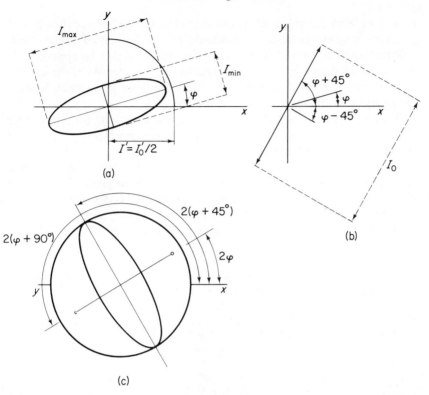

Figure 16.8.   Determination of the *j*-vector with unrestricted observation of scattered light

The total intensity I_0 of scattered polarized light is

$$I_0 = I_{max} + I_{min},$$

where $I_0/2$ corresponds to unit length of the *j*-vector.

From these equations we obtain

$$I'_{max} + I'_{min} = I_{max} + I_{min} + 2I' = I_0 + I'_0$$

and

$$I'_{max} - I'_{min} = I_{max} - I_{min} = 2I_{max} - I_0,$$

where $I'_0 = 2I'$ represents the total intensity of vagrant light. The intensity of light is then observed in a direction at 45° to those of I'_{max} and I'_{min} with incident plane polarized light and the polarizer is rotated until the measured intensity of scattered light is a minimum, Figure 16.8(*b*). In this position the *j*-vector is of unit length, i.e. $I = 0$ and the measured value gives I'. Finally, the polarizer is rotated through 90° in which position the measured intensity

is $I_0 + I'$. With the values of I_0 and I' known, the position of the tip of the
j-vector for incident circularly polarized light can be plotted, Figure 16.8(c).

With problems in which the above procedure cannot be applied, the j-
vector can be determined from observations of the intensity of the scattered
light in one or a few directions only. With the latter procedures, however,
the analysis of the observations is more complex.

If the secondary light can be observed in two directions, including preferably an angle of 45° or at least an angle differing considerably from 0°, 90°
or 180°, the following method can be applied.

Let the directions of observation be denoted by 1 and 2 and include an
angle of 45°. In the j-circle, these directions include an angle of 90°, and the
unknown angle between the direction 1 and the minor axis of the ellipse will
be denoted by 2φ (Figure 16.9). The intensities of the vagrant light in the
directions 1 and 2 may differ and may be represented by I' and $I' + \Delta I'$.

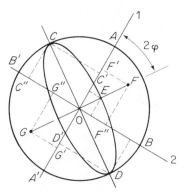

Figure 16.9. Determination of the
j-vector from observations of the
light scattered in two directions
mutually inclined at 45°

With circularly polarized primary light, the maximum and minimum
intensities of the secondary light appear when measuring with one position
of the polarizer and its quarter-wave plate and after rotating one of them by
90°. The corresponding j-vectors are represented by OF, OG in Figure 16.9.

Observing in the directions 1 and 2, the components of these intensities
are measured. These are:

$$(I'_1)_{\max} = (I_1)_{\max} + I' = AG' + I',$$

$$(I'_1)_{\min} = (I_1)_{\min} + I' = AF' + I',$$

$$(I'_2)_{\max} = (I_2)_{\max} + I' + \Delta I' = BG'' + I' + \Delta I',$$

$$(I'_2)_{\min} = (I_2)_{\min} + I' + \Delta I' = BF'' + I' + \Delta I'.$$

From these equations we obtain:

$$(I'_1)_{max} + (I'_1)_{min} = I_0 + 2I',$$

$$(I'_2)_{max} + (I'_2)_{min} = I_0 + 2(I' + \Delta I'),$$

$$\Delta I' = \tfrac{1}{2}[(I'_2)_{max} + (I'_2)_{min} - (I'_1)_{max} - (I'_1)_{min}],$$

and

$$(I'_1)_{max} - (I'_1)_{min} = (I_1)_{max} - (I_1)_{min} = I_0 \sin 2m \cos 2\varphi,$$

$$(I'_2)_{max} - (I'_2)_{min} = (I_2)_{max} - (I_2)_{min} = I_0 \sin 2m \sin 2\varphi,$$

as can be read from Figure 16.9.

Hence,

$$\tan 2\varphi = \frac{(I'_2)_{max} - (I'_2)_{min}}{(I'_1)_{max} - (I'_1)_{min}}.$$

The intensities I_0 and I' can now be determined by means of a trial and error process: Using plane polarized primary light, the intensities I'_1 and I'_2 of the secondary light in the directions 1 and 2 respectively are measured for any random direction of the polarizer. The polarizer is then rotated through 90°, thus producing intensities I''_1 and I''_2. These values are determined in turn for varying directions of the polarizer. In that position where $(I'_2 - I''_2)/(I'_1 - I''_1) = \tan 2\varphi$, the direction of the j-vector and its length coincide with those of the minor axis OE of the ellipse of Figure 16.9. The polarizer is now rotated 45° from this position. The j-vector will now be of unit length (OC or OD) and the intensities measured in the directions 1 and 2 will be $(I'_0)_1$ and $(I'_0)_2$ or $(I'_{90})_1$ and $(I'_{90})_2$ depending on the direction of rotation of the polarizer. Each of these values is the sum of the intensity of true scattered light and the intensity of vagrant light in the direction of observation. Hence, as can be read from Figure 16.9:

$$(I'_0)_1 = I' + AC' = I' + (I - \sin 2\varphi)I_0/2,$$

$$(I'_{90})_1 = I' + AD' = I' + (I + \sin 2\varphi)I_0/2,$$

$$(I'_0)_2 = I' + \Delta I' + BC'' = I' + \Delta I' + (1 + \cos 2\varphi)I_0/2,$$

$$(I'_{90})_2 = I' + \Delta I' + BD'' = I' + \Delta I' + (1 - \cos 2\varphi)I_0/2.$$

From these equations we have

$$(I'_0)_1 + (I'_{90})_1 = I_0 + 2I',$$

$$(I'_{90})_2 + (I'_0)_2 = I_0 + 2I' + 2\Delta I',$$

$$(I'_{90})_1 - (I'_0)_1 = I_0 \sin 2\varphi,$$

$$(I'_0)_2 - (I'_{90})_2 = I_0 \cos 2\varphi,$$

which yield

$$\Delta I' = \tfrac{1}{2}[(I'_{90})_2 + (I'_0)_2 - (I'_0)_1 - (I'_{90})_1],$$

$$I_0^2 = [(I'_{90})_1 - (I'_0)_1]^2 + [(I'_0)_2 - (I'_{90})_2]^2,$$

$$\tan 2\varphi = \frac{(I'_{90})_1 - (I'_0)_1}{(I'_0)_2 - (I'_{90})_2}.$$

With I_0 known, the length of the j-vector with circularly polarized light can be plotted using the relation

$$\sin 2m = \frac{I_{max} - I_{min}}{I_{max} + I_{min}} = \frac{2I_{max} - I_0}{I_0} = \frac{2[(I'_1)_{max} - I'] - I_0}{I_0}.$$

In cases where the secondary light can be observed merely in one direction, the analysis is more complicated. In Figure 16.10, the ellipse to be determined can be defined in terms of co-ordinates u and v by

$$(v^2/a^2) + u^2 = 1,$$

where a is the length of the minor axis while that of the major axis is unity.

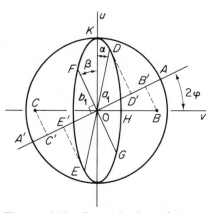

Figure 16.10. Determination of the j-vector from observations of the light scattered in a single direction

The direction of observation OA includes the angle φ with the x axis or 2φ on doubling the angle in the j-circle.

The three unknowns a, φ and I' can be determined from three measured values, namely, the maximum and minimum intensities of the secondary light with circularly polarized primary light and the maximum or the minimum intensity with plane polarized primary light.

The *j*-vectors with the polarizer and quarter-wave plate oriented to produce positive and negative circularly polarized light in turn are represented by *OB*, *OC* in Figure 16.10. The corresponding maximum and minimum intensities of true scattered light measured in the direction *OA* are represented by *AC′*, *AB′* respectively. On these are superimposed the intensity *I′* of the vagrant light. The intensities of secondary light measured are therefore:

$$I'_1 = I' + AC' = I' + \frac{I_0}{2} + OC \cos 2\varphi$$

$$= I' + \frac{I_0}{2}(1 + \sin 2m \cos 2\varphi),$$

$$I'_2 = I' + AB' = I' + \frac{I_0}{2} - OB \cos 2\varphi$$

$$= I' + \frac{I_0}{2}(1 - \sin 2m \cos 2\varphi).$$

With plane polarized primary light, the *j*-vector for that position of the polarizer producing minimum intensity in the direction *OA* is represented by *OD*; that producing maximum intensity after rotating the polarizer through 90° is *OE*. The corresponding intensities are proportional to *AD′*, *AE′* respectively. The maximum and minimum intensities of secondary light measured are therefore:

$$I'_3 = I' + AE' = I' + \frac{I_0}{2} + OE \cos(90 - \alpha - 2\varphi)$$

$$= I' + \frac{I_0}{2} + a_1 \sin(\alpha + 2\varphi),$$

$$I'_4 = I' + AD' = I' + \frac{I_0}{2} - OD \cos(90 - \alpha - 2\varphi)$$

$$= I' + \frac{I_0}{2} - a_1 \sin(\alpha + 2\varphi).$$

It can be read from Figure 16.10 that the diameters *DE*, *FG* are conjugate. The following well known geometrical relations can therefore be used to determine the additional unknowns:

$$a^2 + b^2 = a_1^2 + b_1^2$$

$$ab = a_1 b_1 \sin(\alpha + \beta)$$

$$b^2/a^2 = \tan \alpha \tan \beta$$

where $a = OH = \frac{1}{2}I_0 \sin 2m$, $b = OK = I_0/2$ while a_1, b_1, α, β are as indicated in Figure 16.10. Moreover, it can be seen from the diagram that $\beta = 2\varphi$. There is thus a sufficient number of equations to allow the determination of all the unknowns. As the system includes transcendental equations, however, its solution can only be obtained by trial or by means of an electronic computer.

16.10   The scattering property used as a polarizer

In the procedure just described, the scattering property of the material acts as an analyser. It is also possible, however, to utilize this property as a polarizer. In this case, a beam of ordinary monochromatic light is directed into the model, Figure 16.11. A proportion of the light scattered in any direction ox

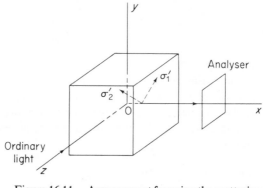

Figure 16.11.   Arrangement for using the scattering
property as a polarizer

perpendicular to the direction oz of the primary beam from any point O of its path is initially plane polarized, the direction of vibration being perpendicular to the axes of transmission of the primary and secondary beams. Each point thus acts as a secondary source of plane polarized light and, if the model is stressed, this light will be influenced by the stresses between the scattering point and the point of exit from the model. If the scattered light emerging from the model is viewed through an analyser, an interference pattern will be visible which has exactly the same meaning as that which would be produced in a plane transmission polariscope.

From the stress optic law, the relative retardation dR produced in an elementary length dx of the path of the secondary beam within which the secondary principal stress difference $(\sigma'_1 - \sigma'_2)_{yz}$ may be assumed constant is

$$dR = C(\sigma'_1 - \sigma'_2)_{yz}\, dx,$$

from which

$$(\sigma_1' - \sigma_2')_{yz} = \frac{1}{C}\frac{dR}{dx},$$

or in terms of the isochromatic fringe order,

$$(\sigma_1' - \sigma_2')_{yz} = f(dn/dx).$$

This equation can be used to determine the difference of the secondary principal stresses provided the latter remain fixed in direction or do not vary too rapidly along the path of the secondary beam.

To apply the method, the primary beam is moved parallel to itself in small steps along the line of observation from the surface towards the interior of the model, e.g. from A to B, C, D, etc., Figure 16.12. For each position, the

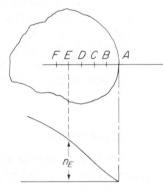

Section 16.12. Method of determining the phase difference when scattering acts as a polarizer

phase difference of the emergent light components is measured. If the results are plotted against the position of the scattering point in the manner indicated, the height at any point of the curve obtained corresponds to the phase difference accumulated in reaching that point by a beam of polarized light entering the model at A; the slope of the curve multiplied by the material fringe value gives the difference of the secondary principal stresses at the point.

When appreciable rotation of the secondary principal axes occurs along the path of the secondary light, the j-circle method may be applied. Let the points A, B, C, D, Figure 16.13, represent the states of vibration of the secondary light in the j-circle observed with progressively increasing distances of the primary beam from the surface of the model. The vector tip

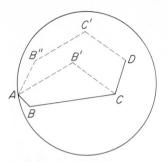

Figure 16.13. Determination of the vector tip curve with varying directions of the principal stresses

curve relating to an internal point can be determined from the line joining the points *A*, *B*, *C*, *D* by tracing the elements of the curve in the reverse order. For example, to determine the tip curve relating to the point *C*, a line *CB'* is drawn from *C* parallel to *AB* and of such a length that it represents the same angle of rotation of the sphere, i.e. the same phase difference, as *AB*. This length can be determined from the nomogram, Figure 13.8. The line *CB'A* then represents the tip curve for the point *C*. To determine the tip curve relating to the point *D*, a line *DC'* is drawn from *D* parallel to *AB* and representing the same phase difference. From *C'*, a line *C'B''* is drawn parallel to and representing the same angle of rotation as *B'A*. The line *DC'B''A* is the required tip curve.

16.11 Examples of the scattered light method

The authors are indebted to Dr Y. F. Cheng for the following examples illustrating practical applications of the scattered light method.

16.11.1 *Residual stresses in tempered glass plates*

Experiments were carried out on a sample of aircraft windshield having a glass-vinyl-glass sandwich structure to determine the distribution of residual stresses resulting from thermal tempering.

A horizontal laser beam having its direction of vibration in a vertical plane entered the immersion cell normally through a window and was focussed by means of a lens at a point *O* on the glass-vinyl interface of the plate, Figure 16.14. The plate was oriented such that the normal to the plane of incidence was inclined at 45° to the vertical. The direction of observation was horizontal and normal to the light path *AO*. The angle of incidence was such that the light entering the plate at *A* was totally reflected at *O* and emerged at the point *B*.

Figure 16.15 shows the fringe pattern observed in scattered light with an angle of incidence of 84°. Sharp fringes are visible except near the point of entrance. In this region, the results could easily be obtained by extrapolation since the curve is antisymmetrical about the middle plane.

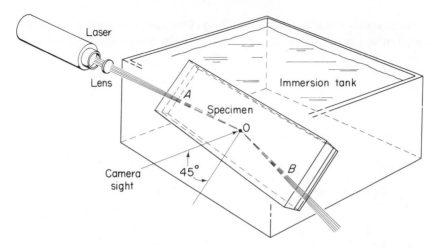

Figure 16.14.   Arrangement used by Cheng for the investigation of residual stresses in aircraft windshields

Figure 16.15.   Scattered light fringe pattern produced by residual stresses in an aircraft windshield. Courtesy Cheng

In plates of the type investigated the state of stress in planes parallel to that of the plate is isotropic while experiment shows that the stress in the direction of the normal is negligible. The principal stresses at a point are therefore $\sigma_1 = \sigma_2 = \sigma$, $\sigma_3 = 0$ while the secondary principal stresses in the plane of the wavefront are $\sigma_1' = \sigma$, $\sigma_2' = \sigma \cos^2 i$ where i is the angle of incidence. Hence, from equation 16.2

$$(\sigma_1' - \sigma_2') = \sigma(1 - \cos^2 i) = f(dn/dx),$$

or

$$\sigma = \frac{f}{\sin^2 i} \frac{dn}{dx}.$$

The variations of fringe order and fringe gradient along the light path OA are given in Figures 16.16 and 16.17 respectively. Since the thickness of the

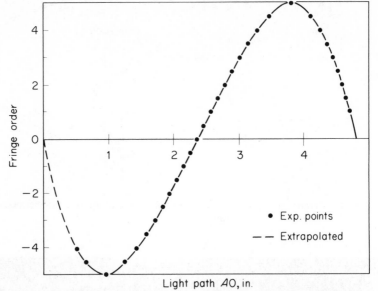

Figure 16.16.   Variation of fringe order along the light path OA. Courtesy Cheng

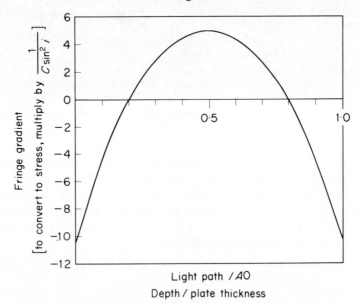

Figure 16.17.   Curve of fringe gradient and stress distribution through the plate thickness. Courtesy Cheng

glass layer is $AO/\sin i$, the latter curve also represents the stress distribution across the thickness of the layer.

16.11.2   *Stresses in a thick circular plate*

A thick circular Castolite plate of diameter 3·5 in. and thickness 1 in. was simply supported around a 3 in. diameter circle on a Castolite block mounted on an aluminium frame as shown in Figure 16.18. A concentrated

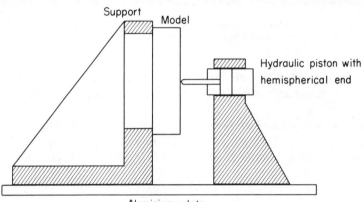

Figure 16.18.   Method of loading thick circular plates. Courtesy Cheng

load was applied at the centre of the plate through the hemispherical end of a hydraulic piston. The assembly was attached to the platform of the polariscope so that the axis of the plate was horizontal and was contained in an immersion tank to eliminate refraction. The primary beam from the laser source was vertical and the laser tube and model could be rotated about the axis of the beam to allow the determination of the directions of the secondary principal stresses. Minimum intensity was detected with the aid of a photomultiplier and electronic photometer.

Scattered light observations in the rz plane yielded values of the difference of the secondary principal stresses and their directions within this plane. These values allowed the shear stress τ_{rz} and the difference of normal stresses $(\sigma_r - \sigma_z)$ to be determined. The remaining stress components were then determined following a procedure similar to that described in Section 14.8 for bodies having axial symmetry.

Stress patterns relating to planes parallel to the rz plane at distances of 0·15, 0·50 and 0·85 in. from the loaded surface under normal observation are given in Figure 16.19. These patterns were obtained by traversing the model

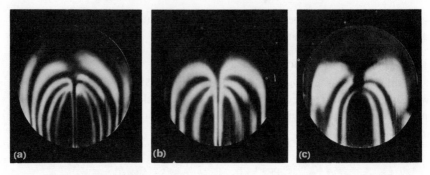

Figure 16.19.   Scattered light fringe patterns in thick circular plates at various distances from the loaded surface. (*a*) 0·15 in. (*b*) 0·05 in. (*c*) 0·85 in. Courtesy Cheng

with the laser beam linearly polarized in the direction of observation and travelling horizontally at uniform speed parallel to the plane of the plate and the camera screen.

The complete state of stress was determined for two concentric cylindrical sections *A* and *B* having diameters of 0·6 and 1·68 in. respectively. The results are given in Figures 16.20 and 16.21. The circled test points of Figure

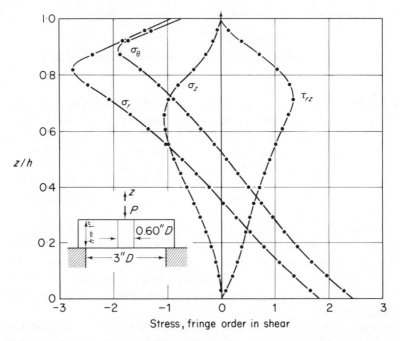

Figure 16.20.   Stress distribution on section *A* (radius 0·30 in.). Courtesy Cheng

16.21 indicate values of σ_θ obtained from an additional scattered light observation in the $r\theta$ plane; this was made to provide a check on the accuracy of the results derived using the photoelastic data of the rz plane only.

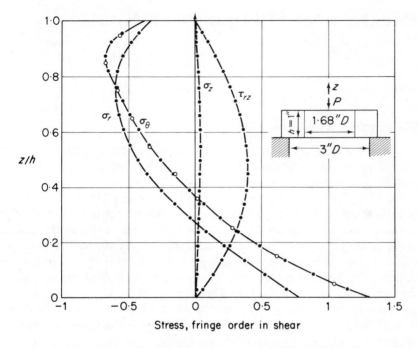

Figure 16.21. Stress distribution on section *B* (radius 0·84 in.). Courtesy Cheng

17

SHELLS AND FOLDED PLATES

17.1 Stresses in shells and folded plates

In shells, folded plates and other thin-walled structures, the stresses in general can be split into membrane stresses and bending stresses while other possible stress components can be neglected.

Along any line normal to the surface of the shell or plate, the magnitudes and directions of the principal membrane stresses are constant. The bending stresses, however, vary linearly along such lines and vanish in the middle plane. At opposite points on the surfaces, the bending stresses are therefore equal in magnitude but opposite in sign. As in transversely bent plates, the directions of the principal bending stresses are approximately constant along lines perpendicular to the surface. The direction of the algebraically greater principal stress will change, however, by 90° at the middle plane. The directions of the principal resultant stresses, i.e. the principal stresses corresponding to the combined stress system, vary continuously along lines normal to the surface and form helical surfaces.

The path curves for such cases can easily be plotted with the aid of the graphical method described in Section 13.4. At points along the normal path of a light beam, the bending stresses are proportional to the distance from the middle plane and have a certain constant direction in one half of the plate or shell; after doubling the angles in accordance with the j-circle method, they have the opposite direction in the other half. They can therefore be represented by two systems of oppositely directed vectors OA_n, OB_n, Figure 17.1, the lengths of which are proportional to the distance from the middle plane. The membrane stress, having constant magnitude and direction throughout the thickness, is represented by a vector CO. This vector is inclined to the vectors OA_n, OB_n at an angle equal to twice the true angle between the membrane and bending stresses.

The path curve obtained by adding the vector CO to each of the vectors OA_n, OB_n in the correct sequence consists of a series of straight line elements. The true path curve, however, is one of continuously varying slope corresponding to elements of infinitesimal length. It can be shown from the

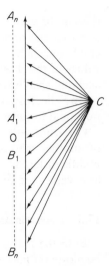

Figure 17.1. Vectorial representation of the superposition of membrane stresses and bending stresses in a shell

geometrical conditions that, in the case of the superposition of bending and membrane stresses, the true path curves are parabolic arcs. The chord connecting the end points of such an arc, Figure 17.2, represents the membrane stress: the direction of the line connecting the mid point of the chord and that point on the arc at which the tangent is parallel to the chord indicates the direction of the bending stress while its length is proportional to one quarter of the maximum bending stress. The latter line is parallel to the axis of the parabola.

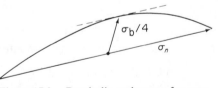

Figure 17.2. Parabolic path curve for combined membrane and bending stresses

17.2 Application of photoelastic methods to shells and folded plates

For the solution of problems on shells, etc., four values have to be determined by optical methods, namely, the magnitudes of the difference of the principal membrane stresses and of the principal bending stresses and the directions of these principal stresses. As explained in Section 13.3, only three optical values, namely m_k, ε, δ, can be determined from one observation using monochromatic light. Hence some additional technique must be applied to determine the fourth value. Different methods by means of which this can be accomplished have been developed. One method is based on the fact that the optical data depend on the wavelength of the light used. By employing light of two different wavelengths in turn, two different sets of values of the three quantities can be determined. Alternatively, this can be accomplished by employing two different values of the load since, in contrast to two-dimensional cases, the optical values are no longer directly proportional to the load.

Certain disadvantages of the above procedure can be avoided by means of another method in which a model containing a reflecting layer at its middle surface is used. By observing a point from opposite sides of the model, four values are obtained (since in this case δ vanishes).

17.3 Determination of optical data

Both the graphical and the analytical methods allow the optical data to be plotted from known values of the stresses but not vice versa. To allow the determination of stresses from measured optical data, the latter were evaluated for a great number of possible combinations of membrane and bending stresses with different angles included between them. From these data, charts were prepared from which the stresses can be read. The values necessary for plotting the charts were determined numerically with the aid of an electronic computer. Complete sets of the charts, applicable with the procedures described in this chapter, are available (see Preface).

The general system of differential equations (13.14) can be transformed into a special system valid for parabolic path curves. The equation of the parabola

$$y = ax^2, \tag{17.1}$$

can be used in preparing the program for the computer. With

$$dm = ds = \sqrt{(dx^2 + dy^2)},$$

and

$$dm \cos 2\varphi = dx, \qquad dm \sin 2\varphi = dy,$$

equations (13.14) read

$$dm_k = dx(\cos 2\varepsilon + 2ax \sin 2\varepsilon), \tag{17.2a}$$

$$d\varepsilon = dx(2ax \cos 2\varepsilon - \sin 2\varepsilon) \cot 2m_k, \tag{17.2b}$$

$$d\delta = dx(2ax \cos 2\varepsilon - \sin 2\varepsilon) \tan m_k. \tag{17.2c}$$

The integration of this system by means of a finite difference method starting at some point on a particular parabola produces an infinite number of values as every point on the parabola is the end point of some parabolic arc. The values of m_k and δ obtained by the integration are obviously independent of the position of the parabola or parabolic arc. If the parabola (i.e. the path curve) is rotated, however, the value of ε will change by an amount equal to the angle of rotation.

The requisite data can be obtained in the above manner by applying the system of equations (17.2) to a number of different parabolas, i.e. for different values of the constant a in equation (17.1), starting the integration at different points. Instead of actually exercising the latter variation, the values obtained starting at one point only on each parabola can be used. These can be transformed into values pertaining to other starting points by means of the graphical method described in Section 13.4 or by means of equation (13.7) as described in Section 13.5.

The values of m_k, ε and δ at the starting point of the integration can easily be determined; m_k and δ are zero while ε should be chosen parallel to the direction of the first element of the path curve. If this value is incorrectly chosen, the resulting inaccuracy of the result is negligible since after the first step of the integration its effect practically vanishes.

From the computed data, two groups of charts were prepared showing the optical values as functions of σ_b and σ_n, which here denote the differences of the principal bending stresses and principal membrane stresses, respectively. Each chart corresponds to a constant angle α (i.e. 5°, 10°, 15°, ..., 45°) between the directions of the stresses. The charts of group A contain lines $\varepsilon = $ constant while group B contain lines $m_k = $ constant and $\delta = $ constant. These charts are generally applicable to any problem involving the superposition of membrane and bending stresses and enable the values required in any particular case to be derived.

17.4 Transmission methods

With models of uniform material (i.e. without a reflecting middle layer) the requisite number of optical data can be obtained either by using light of two different wavelengths in turn or by altering the magnitude of the load acting on the model. The effect of both of these possible variations is theoretically the same. Varying the load, however, has the practical advantage that the ratio of the two different loads can be chosen over a wider range than that of the wavelengths; the range of wavelengths of the visible spectrum is between 4000 and 7700 Å only. Using either procedure, any initial double refraction present in the unloaded model should be taken into account. This can be done by the method described in Section 8.7.

17.4.1 *Method of load variation*

When using the method of load variation, the lower value of the load and hence of the stresses produced should be small enough that the latter can be expected to lie within those regions of the charts near the origin where $\varepsilon = 0$. In the charts, ε is the angle of inclination with respect to a reference direction coinciding with the direction of σ_n; the value of ε measured experimentally thus indicates the angle between σ_n and the chosen reference direction. A more accurate procedure is to apply two or more loads, each small enough that ε will lie within those regions of the charts near the origin where it varies linearly with the stresses. Extrapolation of the values of ε measured with these loads to zero load supplies the exact direction of σ_n.

The angle α between σ_n and σ_b is then determined from values of ε, m_k and δ measured at a load which is great enough that they lie within a region of the charts remote from the origin where they are no longer linearly proportional to the load. Group C of diagrams containing lines $2\alpha = $ constant, from which the value of α can be determined, were prepared from those of groups A and B. The coordinates of these diagrams are δ and m_k and each relates to two values of the parameter ε differing by $90°$, i.e. $0, 90°$; $5°, 95°$; . . . . The diagram of group B having the relevant parameter of α is then used. The required values of σ_b and σ_n are given by the coordinates of the point of intersection of the appropriate curves $m_k = $ constant and $\delta = $ constant.

17.4.2 *Method of wavelength variation*

For the method of wavelength variation, vapour lamps can be used since they produce light of different characteristic wavelengths. With mercury arc lamps, the difference in wavelengths between the blue line (4360 Å) and the yellow line (5780 Å) is sufficiently great to allow reliable measurements. Interference filters for these wavelengths are commercially available.

The first step towards determining the stresses from the two sets of values measured using the different wavelengths, i.e. m_{k1}, ε_1, δ_1 and m_{k2}, ε_2, δ_2 where the index 1 refers to the shorter wavelength, consists of finding the angle α. For this, the values $(\varepsilon_1 - \varepsilon_2)$, m_{k1} and m_{k2} are used. A set of diagrams, group D, containing lines $2\alpha = $ constant with coordinates m_{k1}, m_{k2} was prepared, each of which relates to four values of the parameter $(\varepsilon_1 - \varepsilon_2)$, corresponding to $(\varepsilon_1 - \varepsilon_2)$, $90° - (\varepsilon_1 - \varepsilon_2)$, $90° + (\varepsilon_1 - \varepsilon_2)$ and $180° - (\varepsilon_1 - \varepsilon_2)$. The charts are divided into regions, each of which is relevant to the parameter indicated. The angle α can be read from these diagrams, if the ratio of the wavelengths used is $4360/5780 = 0.755$. The chart of group B having α of the corresponding parameter is then used to determine σ_n and σ_b.

It would have been possible to prepare a further set of charts, each containing the values m_{k1} and m_{k2} or δ_1 and δ_2 pertaining to both wavelengths. This would not be too complicated and could be accomplished by combining with the original diagram a copy of it to a reduced scale, the origins of both diagrams coinciding. Such a diagram would enable the stresses σ_b, σ_n to be read from the point of intersection of the corresponding lines $m_{k1} = $ constant and $m_{k2} = $ constant on the chart of the angle α previously determined. It was found more practical, however, to use merely the original generally applicable charts and to determine the point of intersection of the two lines by means of a simple mechanical device which divides the distance of any point from the origin by the ratio of the wavelengths used. This device, which is represented in Figure 17.3, consists of six links a to f connected at the

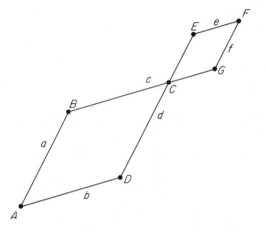

Figure 17.3. Principle of the link device used to read values relating to one wavelength of light from charts relating to another

points A to G. The whole device can be rotated about A which is fixed at the origin of the chart while the point F can be moved around in its field. The points A, C, F are collinear. The ratio AF/AC is constant, depending only on the lengths of the links. To determine the required point of intersection, the point F is moved along the line m_{k1} = constant of parameter equal to that measured. During this movement the point C is observed and the point at which it crosses the line m_{k2} = constant of the parameter measured is the required point.

Theoretically, the magnitudes of the stresses can be determined from the values of m_k only, disregarding those of δ. In some instances, however, the result of the former is less reliable than that of the latter. It is therefore advisable to read the stresses from the points of intersection of the lines m_{k1} = constant and δ_1 = constant as well, thus obtaining a mutual check of the results.

Finally, the directional angle of σ_n is determined from the diagram of group A containing lines ε = constant. As mentioned previously, the values of ε in these diagrams relate to a reference direction coinciding with σ_n. The angle between the chosen reference direction and that of σ_n is thus equal to the difference of the measured value of the angle ε and that read from the diagram. The values of ε_1 and ε_2 should both produce the same result.

The directional angle of σ_b is found by adding α to the directional angle of σ_n.

The value of σ_b cannot be obtained by the above method when the directions of σ_b and σ_n coincide. The same holds true if the stresses or phase differences involved are small or have certain values where all of the lines $\alpha = $ constant intersect in one point.

17.5 Reflection method

In an alternative method, the model is provided with a reflecting layer in its middle surface and observations are made using a reflection type polariscope. By observing a point on the model in the direction of the normal from opposite sides, two pairs of values are obtained, i.e. m_{k1}, ε_1, and m_{k2}, ε_2. With the light reflected back along its original path, the path curve is symmetrical about its mid point. Considering a tip curve, the first half of which terminates at the circumference of the j-circle, it is evident that the projection of the second half, lying on the opposite side of the sphere, will coincide with the first. The starting and end points of the complete curve thus coincide so that the angle δ vanishes.

Three groups of charts have been prepared for use with this method. The determination of the angle α is again the first objective, and is accomplished using the values of $(\varepsilon_1 - \varepsilon_2)$ and m_{k1}, m_{k2}. In each chart of group E, $(\varepsilon_1 - \varepsilon_2)$ has one of two constant values differing by $90°$ ($0°$, $90°$; $5°$, $95°$; ...). These charts contain lines $\alpha = $ constant plotted against m_{k1}, m_{k2}. Here again, the field is divided into zones, the numbering of which indicates the parameter to which each is relevant. Using the chart of the parameter corresponding to the value of $(\varepsilon_1 - \varepsilon_2)$ determined experimentally, the value of α can be read from the values of m_{k1}, m_{k2}.

In each of the charts of groups F and G, lines m_{k1}, $m_{k2} = $ constant and ε_1, $\varepsilon_2 = $ constant are plotted as functions of σ_b, σ_n for a constant value of α between $0°$ and $45°$. These charts are also valid for values of α between $90°$ and $45°$, it being necessary merely to interchange the indices of m_k and ε.

The values of the stresses σ_b, σ_n are then read from the charts of group F using the values of m_{k1}, m_{k2} and α. Here, lines m_{k1}, m_{k2} of given parameters have two points of intersection so that the stresses are not indicated single valued. The correct pair of values σ_b, σ_n may be selected by referring to the charts of group G; the required stresses are those which produce the same value of $\varepsilon_1 - \varepsilon_2$ as obtained experimentally at the point considered. In general, it will suffice to determine only a few single-valued points in this way; the remaining values can be correctly selected observing the condition of continuity of the stresses.

The directions of σ_b, σ_n are derived from the charts of group G. Using the chart having the relevant value of α, the values of ε_1, ε_2 can be read from those of σ_b, σ_n. It is necessary to determine only one of these two angles, it being optional which is chosen. As before, the direction of σ_n is given by the difference of the value of ε measured experimentally with respect to the chosen reference direction and that derived from the charts. Finally, the directional angle of σ_b is found by adding α to the directional angle of σ_n.

17.6 Practical application of charts

Interpolation between different charts of the same group may be avoided by plotting supplementary curves of the parameters involved. For example, curves ε or $(\varepsilon_1 - \varepsilon_2) = 0°, 5°, 10°, \ldots$, may be plotted from the experimental values of ε or $\varepsilon_1 - \varepsilon_2$ respectively. The values of α may then be read from the charts of groups C, D, or E at points on these curves without the need for interpolation. Similarly, from the values of α so obtained, curves of $\alpha = 0°$, $5°, 10°, \ldots$, may be drawn. The values of σ_b, σ_n or ε may then be read directly from the charts of groups A, B, F or G at points on these curves.

Due to the periodicity of the optical data, some groups of charts contain several curves having the same parameter. In cases of doubt, the correct curves to use can easily be identified from the condition of continuity. Starting at a point where no doubt exists, e.g. on a load-free edge or section of symmetry where the directions of σ_b, σ_n coincide, the curves pertaining to neighbouring points follow in succession.

The scales of the charts of groups A, B, F and G correspond to the compensation angle in two-dimensional photoelasticity measured in degrees which, of course, is proportional to the difference of the principal stresses. The values of σ_b, σ_n in N/m^2 can be obtained by multiplying the values at the scales by the factor $f/180d$ where f is the material fringe value in N/m fringe and d is the model thickness in m. In the charts of groups A and B the abscissa yield the values of σ_b while in groups F and G they produce the difference of the maximum bending stresses on opposite sides of the plate, i.e. $2\sigma_b$.

The charts containing curves α (or 2α) = constant are restricted to values of α between $0°$ and $+45°$ while, in practice, α may have any value between $-90°$ and $+90°$. The actual value of α in any given case depends on the ratio m_{k1}/m_{k2} and on the sign of $(\varepsilon_1 - \varepsilon_2)$ and can be obtained by replacing the value read from the chart in accordance with the following tables derived from the geometrical conditions:

(a) Transmission method with load variation:

	$\delta > 0$	$\delta < 0$
$\varepsilon > 0$	α	$90° - \alpha$
$\varepsilon < 0$	$-(90° - \alpha)$	$-\alpha$

(*b*) Transmission method with wavelength variation:

$$\delta > 0 \qquad \delta < 0$$

$$(\varepsilon_1 - \varepsilon_2) > 0 \qquad \alpha \qquad 90° - \alpha$$

$$(\varepsilon_1 - \varepsilon_2) < 0 \qquad -(90° - \alpha) \qquad 180° - \alpha$$

(*c*) Reflection method:

$$m_{k1} > m_{k2} \qquad m_{k1} < m_{k2}$$

$$(\varepsilon_1 - \varepsilon_2) > 0 \qquad \alpha \qquad 90° - \alpha$$

$$(\varepsilon_1 - \varepsilon_2) < 0 \qquad 180° - \alpha \qquad -(90° - \alpha)$$

17.7 Practical examples of shells and folded plates

17.7.1 *Transmission method using light of two different wavelengths: Torsion of a thin-walled beam*

A beam 1000 mm long having the thin-walled cross-section shown in Figure 17.4(*a*) was supported at its ends and loaded by a twisting moment of 20 kp/cm applied through a disc cemented at its centre, Figure 17.4(*b*).

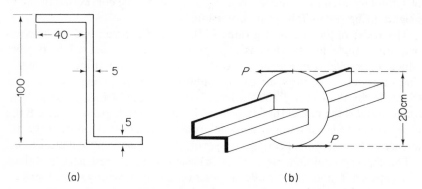

(a)   (b)

Figure 17.4. Thin walled beam investigated using light of two different wavelengths. (*a*) Dimensions in mm. (*b*) Method of loading

The supports were designed in such a manner that the ends of the beam were free to warp while warping of the central section was completely prevented. Under these conditions, the state of stress in the flanges was equivalent to the superposition of bending and membrane stresses.

Point-by-point measurements with the aid of a photomultiplier were made of the values ε, m_k and δ in the flanges using in turn the yellow (5780 Å)

and blue (4360 Å) bands isolated from a mercury vapour source by means of interference filters. The curves of these quantities over the flanges are shown in Figures 17.5 to 17.7 where the subscript 1 refers to the shorter wavelength. The lines $\varepsilon_2 - \varepsilon_1$ obtained by subtraction are given in Figure 17.8. From these families of curves, the values of $\varepsilon_2 - \varepsilon_1, m_{k1}, m_{k2}$ at chosen points were read.

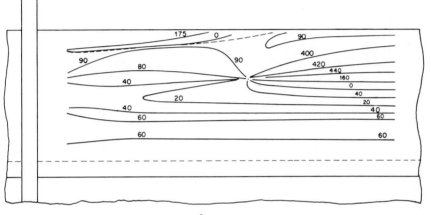

(a) ε_1 = const

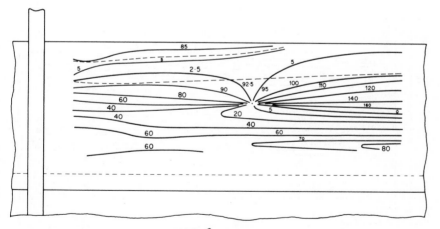

(b) ε_2 = const.

Figure 17.5. Curves of ε = constant obtained for the flanges of the beam shown in Figure 17.4 using (a) mercury blue and (b) mercury yellow wavebands

The angle α between the bending and membrane stresses was determined from the values of m_{k1}, m_{k2} using the charts of group D having the relevant parameters of $\varepsilon_2 - \varepsilon_1$. The curves of α derived in this way are given in Figure 17.9.

Finally, the stresses were determined from the charts of group B by locating the points of intersection of the lines m_{k1}, m_{k2} by means of the device described in Section 17.4.2. The results were checked by the alternative procedure employing the values of m_{k1} and δ and are given in Figures 17.10

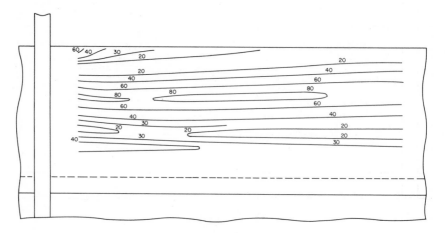

(a) $m_{k1} = $ const.

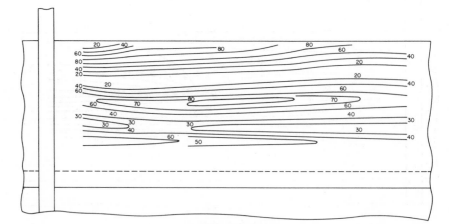

(b) $m_{k2} = $ const.

Figure 17.6.   Curves of $m_k = $ constant

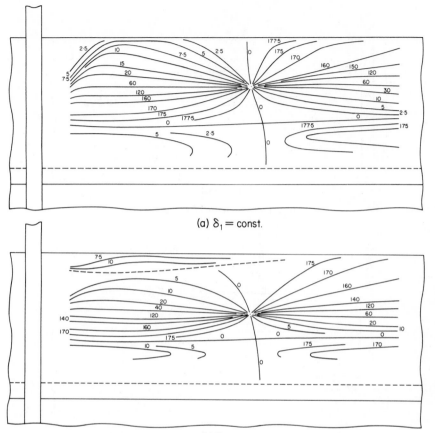

(a) $\delta_1 = $ const.

(b) $\delta_2 = $ const.

Figure 17.7. Curves of δ = constant

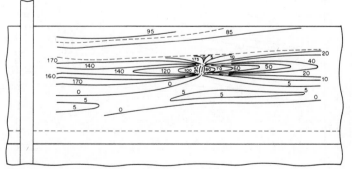

$\epsilon_2 - \epsilon_1 = $ const.

Figure 17.8. Curves of $\varepsilon_2 - \varepsilon_1 = $ constant

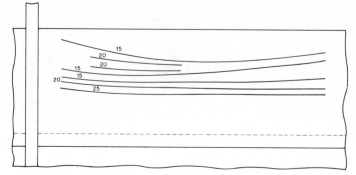

α = const.

Figure 17.9.   Curves of α = constant

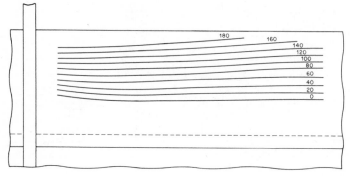

σ_n = const.

Figure 17.10.   Curves of σ_n = constant

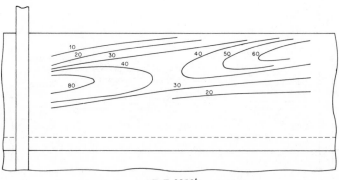

σ_b = const.

Figure 17.11.   Curves of σ_b = constant

and 17.11. On account of the relatively small bending stresses produced, the directions of the stresses could not be determined accurately in this case.

17.7.2 *Reflection method: Semi-cylindrical shell subjected to a central concentrated load*

A semi-cylindrical shell with plane end walls was supported at the four corners and a concentrated load $P = 55$ kp was applied at the centre, Figure 17.12(*a*). The dimensions of the shell are given in Figure 17.12(*b*). The

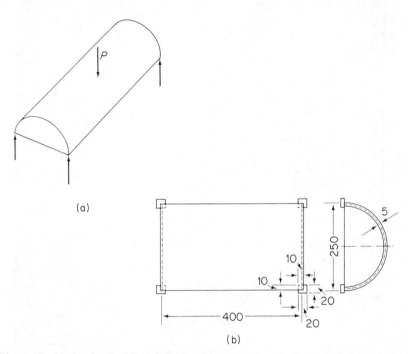

(a)

(b)

Figure 17.12.   Semi-cylindrical shell investigated by the reflection method. (*a*) Method of loading. (*b*) Dimensions in mm

model was manufactured from cold setting epoxy resin and was provided with a reflecting middle surface.

Corresponding points of the shell were observed in turn from each side using a reflection polariscope and the values of m_k and ε measured. From these values the curves given in Figures 17.13 and 17.14 for one quarter of the cylinder were plotted; here, the indices 1 and 2 denote respectively the inner and outer sides of the model. The curves $(\varepsilon_1 - \varepsilon_2) = $ constant, obtained by subtraction, are given in Figure 17.15.

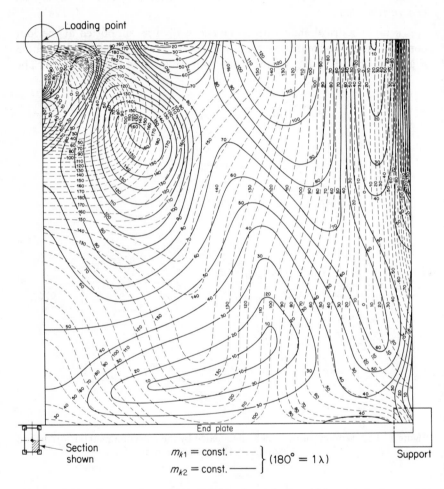

Figure 17.13. Curves of m_{k1} (inner side) and m_{k2} (outer side) = constant over one quarter of the developed surface of the shell

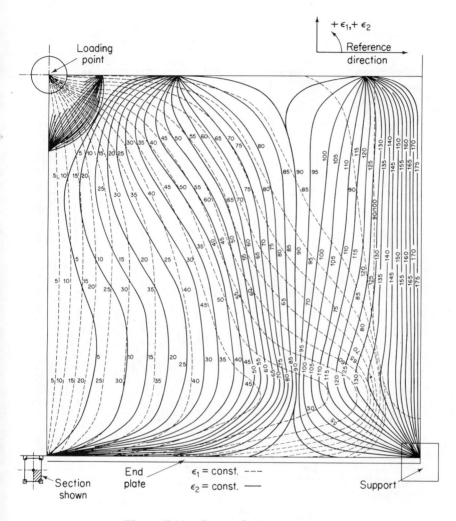

Figure 17.14. Curves of ε_1, $\varepsilon_2 =$ constant

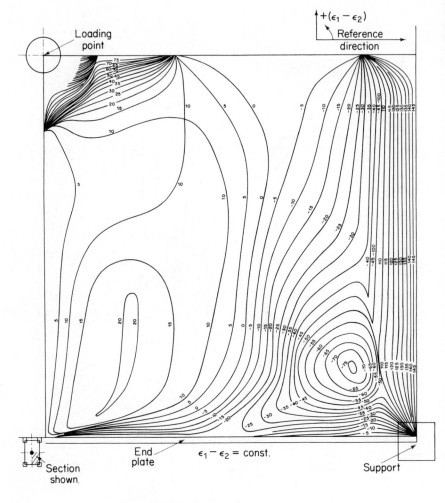

Figure 17.15.   Curves of $\varepsilon_1 - \varepsilon_2 =$ constant

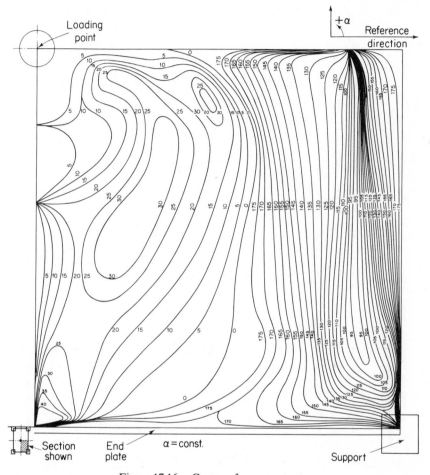

Figure 17.16.   Curves of $\alpha =$ constant

The curves α = constant, derived from values read from the charts of group E, are given in Figure 17.16. Finally, the magnitudes and directions of σ_b, σ_n were determined using the appropriate charts of groups F and G respectively as described in Section 17.5. The results are given in Figure 17.17.

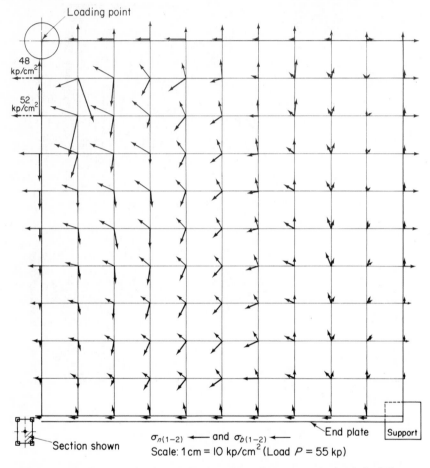

Figure 17.17. Distribution of membrane and bending stresses in the shell of Figure 17.12

18

PHOTOTHERMOELASTICITY

18.1 Thermal stresses

Thermal stresses are induced in a body whenever its thermal expansion or contraction or that of any part of it associated with a change of temperature is restricted. In a body subjected to a uniform change of temperature, such restriction may be produced by external restraints or by the interaction of dissimilar materials. More commonly, thermal stresses result from temperature gradients within a body. In this case, the restriction on expansion (or contraction) results from the interaction of the elements of the body, the thermal expansions of which vary according to the local temperature. A self equilibrating system of stresses is thus set up corresponding to the expansion prevented.

The free thermal expansion of an element of an isotropic body, being equal in all directions, produces no double refraction in a photoelastic model; differences in principal strains and thus of principal stresses occur only with expansions which vary according to the direction. The investigation of problems of thermal stresses by photoelastic methods is therefore simplified since it is unnecessary to separate the thermal expansions from the strains associated with the stresses.

18.2 Application of photoelastic methods

Various photoelastic procedures can be applied depending on the nature of the problem to be investigated. With two-dimensional problems, the classical methods can be applied with slight modification. Similarly, problems of thermal bending stresses in plates can be investigated by the methods described in Chapter 12. For the solution of general three-dimensional problems, some of the usual methods such as birefringent coatings and the scattered light method are applicable. The former method is applicable, however, only when the body to be investigated and the coating have the same coefficient of expansion.

The frozen stress method is not valid for the investigation of thermal stress problems. It has been suggested that thermal stresses caused by temperature differences above the softening point be conserved by gradually lowering the temperature level below the softening point while maintaining the temperature differences constant. This procedure fails, however, because the coefficient of thermal expansion of all model materials changes suddenly at the softening point. As a result, additional effects are introduced so that the frozen in stresses or strains do not represent the correct solution of the problem.

18.3 Model laws

For analogous conditions to exist between a model and its prototype, it is not only necessary that the two are geometrically similar but also that the temperature distribution is analogous; since thermal stresses are independent of the absolute values of the temperatures but are proportional to the temperature differences, this means that the ratio of the temperature difference between any two points of the model and that between the corresponding points of the prototype should be constant. The temperature differences in the model may be greater or smaller than those in the prototype. The stresses also depend upon the model material but not upon the scale.

The stresses σ_p in the prototype and σ_m in the model are related by

$$\frac{\sigma_p}{\sigma_m} = \frac{E_p \alpha_p \Delta T_p}{E_m \alpha_m \Delta T_m},$$

for states of plane stress and

$$\frac{\sigma_p}{\sigma_m} = \frac{1 - v_m}{1 - v_p} \frac{E_p \alpha_p \Delta T_p}{E_m \alpha_m \Delta T_m},$$

for plane strain, where

E_p, E_m are the elastic moduli of the materials,

α_p, α_m are the coefficients of thermal expansion,

$\Delta T_p, \Delta T_m$ are the temperature differences within the two bodies,

v_p, v_m are the values of Poisson's ratio for the materials.

With steady heat supply, no further change of temperature distribution or stress distribution will occur after a certain time has elapsed. Under transient conditions, however, analogous temperature distributions and hence stress distributions exist after the elapse of different times t_m in the model experiment and t_p with the prototype. The ratio of these times depends on the materials and the scale:

$$\frac{t_p}{t_m} = \left(\frac{l_p}{l_m}\right)^2 \frac{a_m}{a_p}, \tag{18.1}$$

where

l_m/l_p is the scale of the model

a_m, a_p are the thermal diffusivities of the materials.

The value of a is given by

$$a = k/\rho c,$$

where

k = thermal conductivity

ρ = density

c = specific heat

Some values of a are:

plastics, model materials	$6 \text{ cm}^2/\text{h}$
steel	$500 \text{ cm}^2/\text{h}$
concrete	$19 \text{ cm}^2/\text{h}$

18.4   Experimental techniques

It can be seen from equation (18.1) and the given values of a that the time relation to be applied in experimental investigations of transient thermal stress problems can vary over a wide range. For example, with a plastic model of a steel part of an internal combustion engine made to a scale of $2:1$, the ratio t_m/t_p is 3333, i.e. one minute in the experiment corresponds to 0·018 sec. with the prototype. This allows the study of rapidly changing thermal stresses in such parts. On the other hand, for a plastic model of a concrete dam constructed to a scale of $1:200$, the ratio t_m/t_p is 1/12631. Hence, 3 minutes in the experiment correspond to 27 days, the time of heat development by hydration of cement.

Since thermal stresses produced in photoelastic models are often fairly small, the temperature difference should be chosen as large as possible in order to obtain sufficient optical effect. The materials best suited for such experiments are hot curing epoxy resin and polycarbonate which allow maximum temperatures of about $80\,^\circ\text{C}$ and $130\,^\circ\text{C}$ respectively. At higher temperatures, considerable creep will occur resulting in an apparent reduction of Young's modulus.

Heat can be supplied to a model in various ways, for example by electrically heated wires, by hot air, or by liquids at controlled temperatures. Some investigators prefer cooling of the model as this allows greater temperature differences and hence higher thermal stresses to be obtained; control of the heat supply is difficult, however, and there are certain other disadvantages if quantitative results are required. The supply of heat will usually require to be controlled; in an electrical circuit this can be read from a wattmeter. Temperatures can be measured by means of thermocouples.

For the investigation of plate problems, the supply of heat may be required at the edges only or through the faces. In the former case, heat transmission at the faces should be prevented by insulating, e.g. with expanded polystyrene. In experiments under both steady state and transient conditions, the insulation should only be removed immediately before photographing.

18.5   Separation of principal stresses

The equations of equilibrium are valid in all cases of thermal stresses and hence all of the methods of separating the principal stresses based on those conditions are applicable. The conditions of compatibility differ, however, from those of static external loading in most cases. The equation of compatibility is identical for both states only in the case of plates to which heat is supplied at the edges only while heat transmission through the faces is prevented. Thus, the graphical, numerical or experimental methods of stress separation based on the condition of compatibility can be applied to problems of this type only.

18.6   Examples of photothermoelasticity

18.6.1   *Thermal stresses in cylinder head of a diesel engine*

Figure 18.1 shows the isochromatics of a simplified model made to a scale of 2:1 of the cylinder head of a diesel engine. The large holes on the left and right represent the inlet and exhaust valve ports respectively; the small lower hole is that for the injector. The edge of the intake valve hole was kept cool while those of the other two holes were heated by means of electrical resistance wires, the leads to which are visible in the photograph. Heat transmission through the faces was prevented by insulating with plastic foam since this transmission would be negligible in the actual engine but would dominate in the relatively thin model. The amount of heat supplied corresponded to measurements made at the engine. The photograph shown was taken after heating for about one minute.

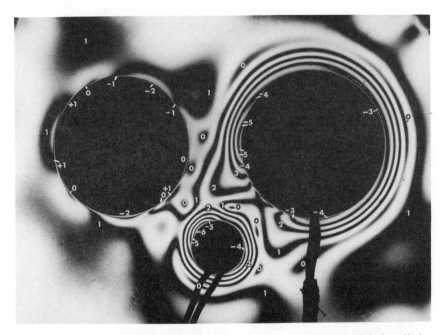

Figure 18.1. Isochromatic pattern produced by thermal stresses in a model of a cylinder head of a diesel engine

It can be seen that the stresses at the edges of the heated holes are everywhere compressive while at that of the intake valve both tensile and compressive stresses are developed in two opposite regions. The signs of these stresses will be reversed when cooling occurs.

18.6.2 *Thermal stresses in a concrete dam during construction*

In fresh concrete, a certain amount of heat is generated by the hydration of cement. In thick-walled structures this heat may produce considerable thermal stresses before it is transmitted to the environs. Figure 18.2 shows the isochromatics with parallel polaroids of a two-dimensional model of a concrete dam. The model was made from hot curing epoxy resin to a scale of 1:200.

Heat developed in the dam is conducted to the environs, i.e. through the edges to the air and through the faces to the rocks and soil according to the thermal capacity of the ground. In order to simulate this process, the model at room temperature was immersed in an oil bath at a temperature of 63 °C, thus reversing the heat transmission.

The modulus of elasticity of the ground below the dam was half that of the dam while the thermal diffusivity of both materials was the same. To simulate

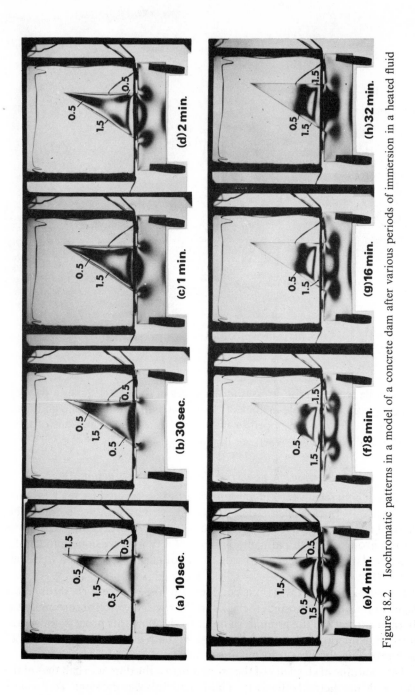

Figure 18.2.   Isochromatic patterns in a model of a concrete dam after various periods of immersion in a heated fluid

these relations, that part of the model corresponding to the ground was made half as thick as the remainder and was covered on both sides by thin plates of plexiglas which conducted the heat without influencing the stresses in the epoxy resin. This part of the model was not immersed in the oil bath.

During test, the model was immersed in the upside down position; the photographs of Figure 18.2 are turned through 180° in order to show the dam in its natural position. The time elapsed between immersing the model and the instant when each individual photograph was taken is registered in Figure 18.2. The temperature of the model slightly below its surface was checked by means of a thermocouple.

18.6.3 *Stresses in gas turbine blades due to thermal shock*

As part of an investigation of the temperature and thermal stress distributions in gas turbine blades under unsteady operating conditions, the transient thermal stresses were determined by the birefringent coating method. To avoid errors arising from the use of materials having different thermal properties, both blade and coating were made from the same material, i.e. Araldite CT200.

The simplified blade profile shown in Figure 18.3 having a plane lower surface was adopted. The coating consisted of a sheet approximately 2 mm

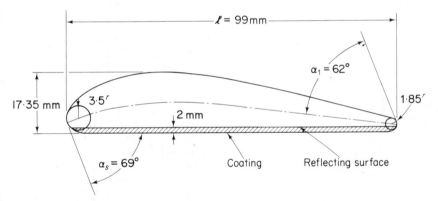

Figure 18.3. Simplified blade profile used for photoelastic investigation of thermal stresses in gas turbine blades

thick sliced from the body of the blade parallel to this surface and cemented back in place after forming a reflective coating of finely powdered aluminium at the interface in the manner described in Section 12.5. Final shaping to the correct profile was then performed.

The temperature distribution was determined by means of thermocouples embedded in a neighbouring blade of the test cascade of four blades.

For the determination of the principal stresses and their directions, the following information was required:

1. Isoclinics of different parameters with initially plane polarized light and the isochromatics with circularly polarized light under normal incidence.

2. The isochromatics with circularly polarized light under oblique incidence.

To improve the accuracy, it was additionally desired that both the whole- and half-order isochromatics should be recorded.

The general optical arrangement is represented diagrammatically in Figure 18.4 while the test layout is illustrated in Figure 18.5. From the approximate point source of a stroboscope, two beams of light were isolated and collimated by means of a specially constructed illuminating head. A system of mirrors directed one of these beams to fall with approximately normal incidence and the other with oblique incidence on the model. After reflection by the model, the light from each beam was directed by further mirrors into the telephoto lens of a cine camera.

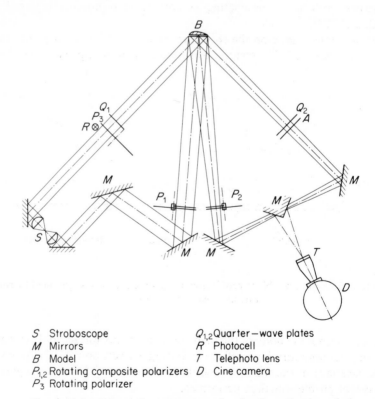

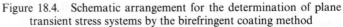

S Stroboscope	$Q_{1,2}$ Quarter–wave plates
M Mirrors	R Photocell
B Model	T Telephoto lens
$P_{1,2}$ Rotating composite polarizers	D Cine camera
P_3 Rotating polarizer	

Figure 18.4. Schematic arrangement for the determination of plane transient stress systems by the birefringent coating method

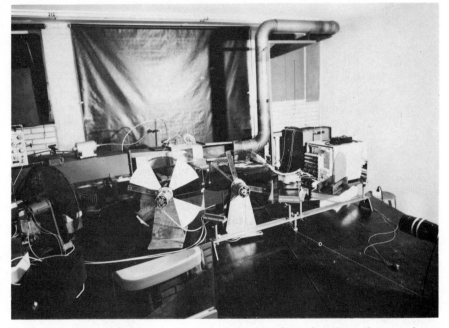

Figure 18.5.   Experimental setup for the investigation of transient thermal stresses in gas turbine blades

Since the stresses varied with time, it was necessary that all of the optical data required for their determination should be recorded in sufficiently rapid succession that they could be regarded as relating to a particular instant during the event. This was accomplished by means of a synchronously rotating optical system. For the multiple measurements required with normal incidence, the two composite polarizers shown in Figure 18.6 were constructed. One half of each polarizer consisted of a simple linear polarizer, the two being maintained in the crossed position for the registration of the isoclinics. The other halves were circular polarizers, formed by the usual combination of linear polarizers and quarter-wave plates, but oriented in such a manner as to produce the isochromatics with both a dark background (upper quadrants) and a bright background (lower quadrants).

The optical system in the path of the oblique incident and reflected beams was that usually employed for circularly polarized light with the exception that the polarizer in the path of the incident beam was rotated by means of an electric motor; in this way, circularly polarized light with a dark and a bright background was obtained in turn whenever the polarizing axis was inclined at 45° to the principal axes of the quarter-wave plates.

In order to avoid the superposition of the optical effects of light reflected from both beams arising from the diffuse reflectivity of the model, it was

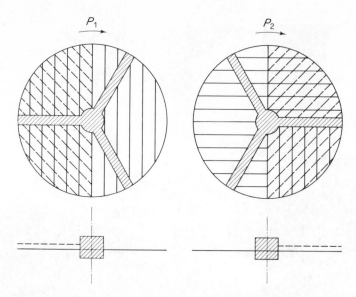

Figure 18.6.   Construction of motor driven composite filters used
in paths of normal light beams

necessary to interrupt the illumination of one while the optical effects of the other were being recorded. This was accomplished by screening off sectors of the polarizers to form rotating shutters.

The synchronism of rotation necessary with the above arrangement was obtained using synchros, by means of which the rotation of the two composite polarizers in the normal path was controlled by that of the polarizer in the oblique path.

The flash rate of the stroboscope was controlled by a photocell responding to thin tin foil markers mounted at the circumference of the rotating polarizer in the path of the oblique beam.

The blades were tested in an air duct which is visible in Figure 18.5. The rate of flow of the air was controlled by a fan mounted within the duct while its temperature could be raised by means of an electric heater or lowered by means of a cooler through which brine was circulated. For the photoelastic investigation, the thermal shock was produced by sudden cooling of the air from 80 °C to 20 °C.

Since isochromatics of low order only were produced, their order was determined by matching the colours with those of a calibration piece which was photographed simultaneously, Figure 18.7. The calibration piece was of the wedge type, similar to those described in Section 6.4, except that a reflecting layer similar to that of the model was provided at the interface between the stressed and unstressed wedges.

(a) $\varphi = 0°$

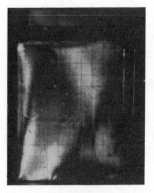
(b) $\varphi = 15°$

(c) $\varphi = 30°$

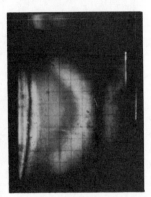
(d) $\varphi = 45°$

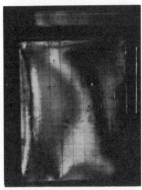
(e) $\varphi = 60°$

(f) $\varphi = 75°$

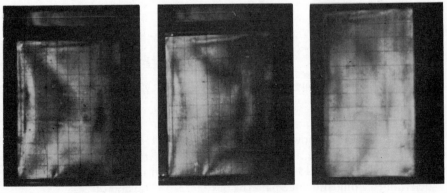
(g)

(h)

(i)

Figure 18.7.   Fringe patterns due to thermal stresses in gas turbine blades obtained by the birefringent coating method. (a) to (f) Isoclinics (and isochromatics), (g) dark field isochromatics, (h) bright field isochromatics, (i) dark field isochromatics with oblique incidence

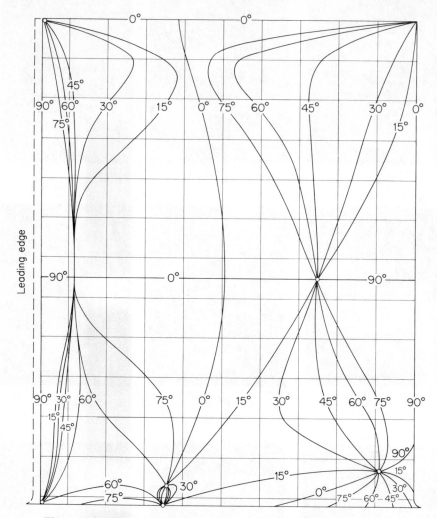

Figure 18.8.   Isoclinic pattern for the front surface of a gas turbine blade

The isoclinics are shown in Figures 18.7(*a*) to (*f*); from these, the isoclinic pattern given in Figure 18.8 and the system of stress trajectories in Figure 18.9 were derived. Figure 18.10 shows the distribution of the isochromatics as determined from the fringe patterns of Figures 18.7(*g*) and (*h*).

The edge stresses obtained directly from the order of the isochromatics along the leading and trailing edges and over the blade tip are plotted non-dimensionally in Figures 18.11 and 18.12 respectively for an instant corresponding to a value of the dimensionless time parameter N_f (Fourier number)

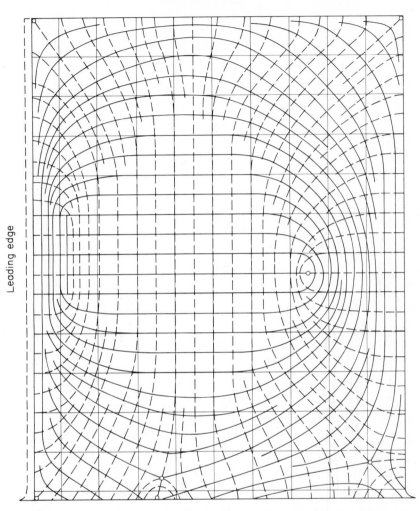

Figure 18.9.   Stress trajectories for the front surface of a gas turbine blade

of 0·27. The stress Σ represents the non dimensional group $\sigma/\bar{E}\bar{\alpha}(T^* - T_0)$, where $\bar{E}$ and $\bar{\alpha}$ are the mean values of the modulus of elasticity and the coefficient of linear expansion while $(T^* - T_0)$ is the temperature drop of the air. The dimensionless co-ordinates ζ/l and ξ/l are defined in Figure 18.12. On account of the curvature of the coating at its edges, it was necessary to determine the stresses at the leading and trailing edges for actual values of ξ/l of 0·02 and 0·99 respectively.

With the additional use of the isochromatics in oblique incidence, Figure 18.7(*i*), the individual principal stresses were determined along the lines

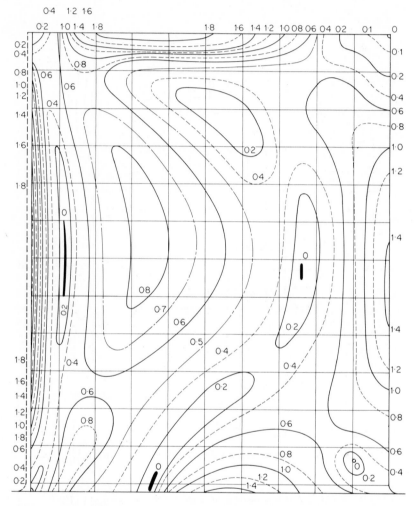

Figure 18.10. Distribution of isochromatics over the front surface of a gas turbine blade

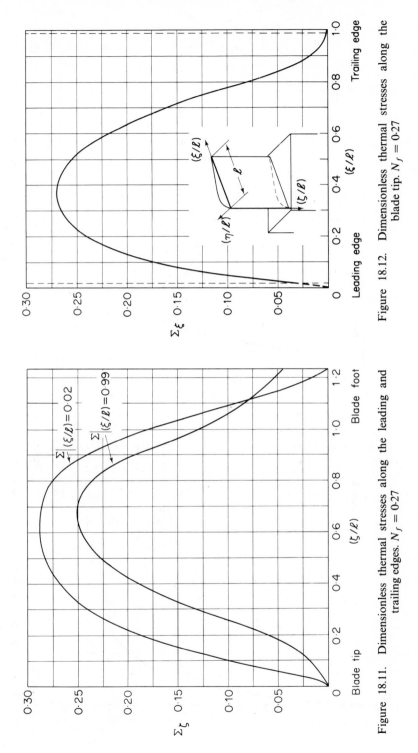

Figure 18.11. Dimensionless thermal stresses along the leading and trailing edges. $N_f = 0.27$

Figure 18.12. Dimensionless thermal stresses along the blade tip. $N_f = 0.27$

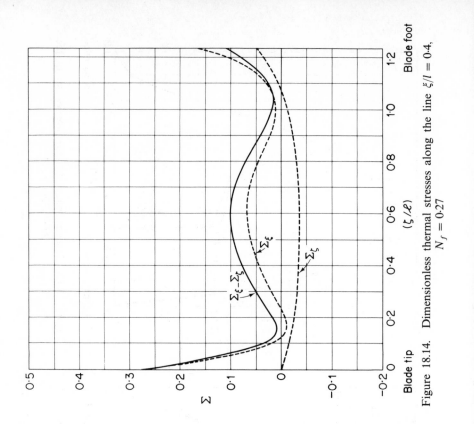

Figure 18.14. Dimensionless thermal stresses along the line $\xi/l = 0.4$, $N_s = 0.27$

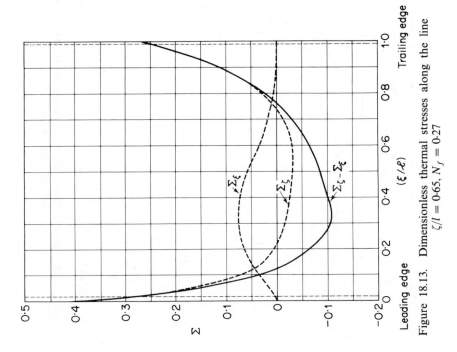

Figure 18.13. Dimensionless thermal stresses along the line $\zeta/l = 0.65$, $N_s = 0.27$

452

$\zeta/l = 0.65$ and $\xi/l = 0.4$ which coincided with the zero isoclinics and also intersected the edges at the points of maximum stress. The results are shown in Figures 18.13 and 18.14 respectively.

The variations with time of the maximum thermal stresses at the blade tip and at the leading and trailing edges are shown in Figure 18.15.

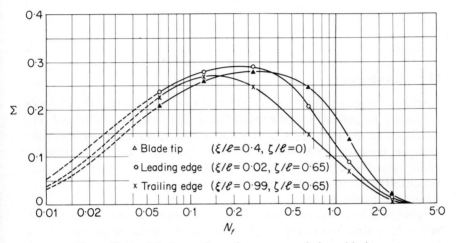

Figure 18.15. Maximum thermal stresses—variation with time

19

DYNAMIC PHOTOELASTICITY

19.1 Introduction

When a structure vibrates or is subjected to impact loading, the whole or part of it is accelerated with the result that inertia forces are introduced. The inertia forces affect the stress distribution in the structure, causing it to differ from that produced under comparable static loading to an extent depending on the rate of loading. Under dynamic loading the stresses vary during and after loading so that a particular state of stress exists only at a corresponding instant during the process.

Additional stresses due to the effects of inertia arise whenever the forces acting on a structure change such as during the initial application of the loads when the forces vary from zero or some other value to their final values, especially if the 'final' value is exceeded in the process. In many cases, however, when the loads are gradually applied or change slowly, the dynamic effect is insignificant and can be neglected. With suddenly applied loads the effect of inertia forces must be taken into account and in extreme cases such as impact or resonance vibration, the dynamic effect predominates.

Although differences in the stresses produced under static and dynamic loading are due mainly to inertia forces, additional effects may arise from differences in the properties of the material under the respective loading conditions. The differences in strength of most materials at low as compared with high rates of loading or in fatigue tests are well known. For some materials, particularly those used in the manufacture of photoelastic models, the relation between stress and strain depends on the rate of increase of stress or on the frequency of vibration. This effect is due to internal friction, which obviously affects dynamic processes only. If the internal friction is low it affects the attenuation of these processes only. With high internal friction, the stress distribution may also be affected.

The mechanics of fracture are also different with gradually and suddenly applied loads.

When the stresses due to dynamic loading exceed the elastic limit, phenomena occur which are quite different from those observed in static cases, e.g. plastic or shock waves may be formed.

454

Dynamic stresses of significance may occur in structures such as bridges (especially railway bridges), craneways, machine foundations, presses and so on. Many machines contain fast-moving parts which are subjected to accelerations causing considerable inertia forces. External forces which change suddenly affect road and railway vehicles, ships and aircraft which must be designed so that the dynamic stresses induced do not cause damage.

The dynamic effect is of decisive influence in events such as earthquakes, explosions, impact of missiles and collisions of all kinds. In some industrial processes, the dynamic effect is utilized to obtain desired results such as in forging, pile driving and explosive forming.

19.2 The propagation of stress waves

The propagation of stress waves in solids can be considered as a part of the field of acoustics or ultrasonics depending on the conditions.

As mentioned above, the stresses produced under dynamic conditions deviate from those in corresponding static cases mainly because of the influence of inertia forces. These forces are proportional to the accelerations. Considering an element of dimensions $dx\,dy\,dz$ the inertia force acting on it is

$$\rho\,dx\,dy\,dz\frac{\partial^2 s}{\partial t^2} = ma, \tag{19.1}$$

where ρ is the density of the material, s the displacement at time t, m the mass of the element and a the acceleration.

This force can be resolved into components parallel to the x, y and z axes. We then have

$$\rho\,dx\,dy\,dz\frac{\partial^2 u}{\partial t^2} = ma_x, \tag{19.2a}$$

$$\rho\,dx\,dy\,dz\frac{\partial^2 v}{\partial t^2} = ma_y, \tag{19.2b}$$

$$\rho\,dx\,dy\,dz\frac{\partial^2 w}{\partial t^2} = ma_z, \tag{19.2c}$$

where u, v, w are the components of the displacement parallel to the x, y and z axes respectively and a_x, a_y, a_z are the corresponding components of acceleration.

The kinetic energy of the element at an instant during the dynamic process is

$$\tfrac{1}{2}mv^2 = \tfrac{1}{2}\rho\,dx\,dy\,dz\left(\frac{\partial s}{\partial t}\right)^2, \tag{19.3}$$

where v = velocity (not to be confused with the displacement v above).

The strain energy δU of the element at the instant considered is

$$\delta U = \tfrac{1}{2}\,dx\,dy\,dz(\sigma_1\epsilon_1 + \sigma_2\epsilon_2 + \sigma_3\epsilon_3)$$

$$= \frac{\delta V}{2E}[\sigma_1^2 + \sigma_2^2 + \sigma_3^2 - 2v(\sigma_1\sigma_2 + \sigma_1\sigma_3 + \sigma_2\sigma_3)]$$

$$= \frac{\delta V K}{2}\left[e^2 + \frac{2v}{1-v}(\epsilon_1\epsilon_2 + \epsilon_1\epsilon_3 + \epsilon_2\epsilon_3)\right], \tag{19.4}$$

where $\sigma_{1,2,3}$ are the principal stresses, $\epsilon_{1,2,3}$ are the principal strains, e is the dilatation, K the bulk modulus and δV the volume of the element.

The momentum of the element is

$$\rho\,dx\,dy\,dz\frac{\partial s}{\partial t} = mv. \tag{19.5}$$

When momentum is transferred from one body to another or from one part of a body to another within the time interval $t_2 - t_1$, the force acting is given by the principle of impulse and momentum:

$$\delta(mv) = \int_{t_1}^{t_2} F\,dt, \tag{19.6}$$

where the force F is a function of time.

The identities expressed by equations (19.1) to (19.5) simplify for uniaxial or biaxial states of stress. For example, the stresses in prismatic bars subjected to longitudinal impact are approximately uniaxial. Equations (19.1) to (19.5) then reduce to

$$\rho A\,dx\frac{d^2s}{dt^2} = ma, \tag{19.7}$$

$$\tfrac{1}{2}\rho A\,dx\left(\frac{ds}{dt}\right)^2 = \tfrac{1}{2}mv^2, \tag{19.8}$$

$$\delta U = \tfrac{1}{2}A\,dx\sigma_x\epsilon_x = \frac{A\,dx}{2E}\sigma_x^2 = \frac{A\,dx}{2}E\epsilon_x^2, \tag{19.9}$$

$$\rho A\frac{ds}{dt} = mv. \tag{19.10}$$

Under dynamic conditions, the body forces acting on an element are the inertia forces. Substituting from equations (19.2) in equations (1.11) for the body forces X, Y, Z per unit volume, we obtain for the equations of equilibrium

$$\frac{\partial\sigma_x}{\partial x} + \frac{\partial\tau_{xy}}{\partial y} + \frac{\partial\tau_{xz}}{\partial z} = \rho\frac{\partial^2 u}{\partial t^2}, \tag{19.11a}$$

$$\frac{\partial \sigma_y}{\partial y} + \frac{\partial \tau_{xy}}{\partial x} + \frac{\partial \tau_{yz}}{\partial z} = \rho \frac{\partial^2 v}{\partial t^2}, \tag{19.11b}$$

$$\frac{\partial \sigma_z}{\partial z} + \frac{\partial \tau_{xz}}{\partial x} + \frac{\partial \tau_{yz}}{\partial y} = \rho \frac{\partial^2 w}{\partial t^2}. \tag{19.11c}$$

The boundary conditions, which are special conditions of equilibrium, are identical under static and dynamic conditions. In both cases, for example, one of the normal stresses and the shear stress vanish at a load-free edge, i.e. a load-free boundary coincides with a stress trajectory.

The relations between stress and strain are also identical in static and dynamic cases. The stresses in equations (19.11) can therefore be replaced by the displacements using equations (2.11), (2.12) and (2.1). In this way we obtain the equations of equilibrium in terms of the displacements:

$$(\lambda + \mu)\frac{\partial e}{\partial x} + \mu \nabla^2 u = \rho \frac{\partial^2 u}{\partial t^2}, \tag{19.12a}$$

$$(\lambda + \mu)\frac{\partial e}{\partial y} + \mu \nabla^2 v = \rho \frac{\partial^2 v}{\partial t^2}, \tag{19.12b}$$

$$(\lambda + \mu)\frac{\partial e}{\partial z} + \mu \nabla^2 w = \rho \frac{\partial^2 w}{\partial t^2}. \tag{19.12c}$$

The equations of compatibility are the same in static and dynamic cases if they involve the displacements or strains only, i.e. equations (2.23) to (2.25) apply in each case. When expressed in terms of stresses, however, these equations differ.

If we differentiate equations (19.12a), (19.12b) and (19.12c) with respect to x, y and z respectively and add we obtain

$$(\lambda + 2\mu)\nabla^2 e = \rho \frac{\partial^2 e}{\partial t^2},$$

or

$$\frac{\partial^2 e}{\partial t^2} = c_1^2 \nabla^2 e. \tag{19.13}$$

Equation (19.13) shows that the dilatation is propagated with the velocity

$$c_1 = \sqrt{[(\lambda + 2\mu)/\rho]}, \tag{19.14a}$$

or, expressing λ and μ in terms of v and E as given by equations (2.10),

$$c_1 = \sqrt{\left(\frac{(1 - v)E}{\rho(1 - v - 2v^2)}\right)}. \tag{19.14b}$$

A wave of the form represented by equation (19.13) is called a dilatation wave.

If we now differentiate equation (19.12b) with respect to z, equation (19.12c) with respect to y and subtract, we obtain

$$\mu\nabla^2\left(\frac{\partial w}{\partial y} - \frac{\partial v}{\partial z}\right) = \rho\frac{\partial^2}{\partial t^2}\left(\frac{\partial w}{\partial y} - \frac{\partial v}{\partial z}\right),$$

or

$$\frac{\partial^2\omega_x}{\partial t^2} = \frac{\mu}{\rho}\nabla^2\omega_x = c_2\nabla^2\omega_x, \tag{19.15}$$

where as given by equation (2.2), ω_x is the rigid body rotation about the x axis. Equation (19.15) shows that this rotation is propagated with the velocity

$$c_2 = \sqrt{\mu/\rho} = \sqrt{[E/2\rho(1 + v)]}. \tag{19.16}$$

It can be shown in a similar manner that the rotations about the y and z axes are propagated with the same velocity. Waves of this type are called distortion waves. From equations (19.14b) and (19.16) we have

$$c_1/c_2 = \sqrt{[(2 - 2v)/(1 - 2v)]},$$

from which it is obvious that the velocity c_1 of the dilatation wave exceeds that of the distortion wave.

In addition to the two basic types of wave described above, a third type known as surface waves or Rayleigh waves can be generated in the material adjacent to the free surfaces of a body. The particle motion of these waves is in planes perpendicular to the free surfaces and parallel to the direction of propagation and is attenuated rapidly with depth. It can be shown that the velocity of Rayleigh waves can be obtained from the solution of the equation

$$\kappa^6 - 8\kappa^4 + (24 - 16\alpha^2)\kappa^2 - (16 - 16\alpha^2) = 0,$$

where κ is the ratio of the velocity of the surface waves to that of the distortion waves. For $v = 0.38$, which is approximately the value of Poisson's ratio for most photoelastic materials, this equation produces $\kappa = 0.941$.

For plates in which a state of plane stress can be assumed, the equations of equilibrium (19.11) reduce to

$$\frac{\partial\sigma_x}{\partial x} + \frac{\partial\tau_{xy}}{\partial y} = \rho\frac{\partial^2 u}{\partial t^2}, \tag{19.17a}$$

$$\frac{\partial\sigma_y}{\partial y} + \frac{\partial\tau_{xy}}{\partial x} = \rho\frac{\partial^2 v}{\partial t^2}. \tag{19.17b}$$

The normal strain ϵ_z perpendicular to the plane of the plate is given by

$$\epsilon_z = -\frac{v}{E}(\sigma_x + \sigma_y).$$

From equations (2.6) we obtain

$$\epsilon_x + \epsilon_y = \frac{1 - v}{E}(\sigma_x + \sigma_y),$$

so that

$$\epsilon_z = -\frac{v}{1 - v}(\epsilon_x + \epsilon_y).$$

The dilatation $e = \epsilon_x + \epsilon_y + \epsilon_z$ can therefore be expressed by

$$e = (\epsilon_x + \epsilon_y)\left(1 - \frac{v}{1 - v}\right) = \frac{1 - 2v}{1 - v}(\epsilon_x + \epsilon_y),$$

from which

$$\epsilon_x + \epsilon_y = \frac{1 - v}{1 - 2v}e = \frac{\partial u}{\partial x} + \frac{\partial v}{\partial y}. \tag{19.18}$$

Replacing the stresses by displacements as before, the equations of equilibrium now become

$$\left[\lambda + \frac{\mu(1 - v)}{1 - 2v}\right]\frac{\partial e}{\partial x} + \mu\nabla^2 u = \rho\frac{\partial^2 u}{\partial t^2}, \tag{19.19a}$$

$$\left[\lambda + \frac{\mu(1 - v)}{1 - 2v}\right]\frac{\partial e}{\partial y} + \mu\nabla^2 v = \rho\frac{\partial^2 v}{\partial t^2}. \tag{19.19b}$$

Differentiating (19.19a) with respect to x, (19.19b) with respect to y and adding we obtain

$$\left[\lambda + \frac{2\mu(1 - v)}{1 - 2v}\right]\nabla^2 e = \rho\frac{\partial^2}{\partial t^2}\left(\frac{\partial u}{\partial x} + \frac{\partial v}{\partial y}\right)$$

$$= \rho\frac{1 - v}{1 - 2v}\frac{\partial^2 e}{\partial t^2},$$

where

$$\nabla^2 = \frac{\partial^2}{\partial x^2} + \frac{\partial^2}{\partial y^2}.$$

Expressing λ and μ in terms of E and v this reduces to

$$\frac{E}{1 + v}\nabla^2 e = \rho(1 - v)\frac{\partial^2 e}{\partial t^2},$$

or

$$\frac{\partial^2 e}{\partial t^2} = c_L^2\nabla^2 e,$$

where c_L is the velocity of the dilatation wave:

$$c_L = \sqrt{\left(\frac{E}{\rho(1 - v^2)}\right)}.$$ (19.20)

If we now differentiate equation (19.19a) with respect to y, equation (19.19b) with respect to x and subtract, we obtain

$$\mu \nabla^2 \left(\frac{\partial \mu}{\partial y} - \frac{\partial v}{\partial x}\right) = \rho \frac{\partial^2}{\partial t^2} \left(\frac{\partial u}{\partial y} - \frac{\partial v}{\partial x}\right),$$

from which

$$\frac{\partial^2 \omega_z}{\partial t^2} = \frac{\mu}{\rho} \nabla^2 \omega_z = c_T^2 \nabla^2 \omega_z$$

where

$$c_T = \sqrt{\frac{\mu}{\rho}} = \sqrt{\left(\frac{E}{2\rho(1 + v)}\right)}$$ (19.21)

is the velocity of a distortion wave. Dilatation and distortion waves in plates are usually referred to as longitudinal and transverse waves respectively.

Finally, we consider the transmission of plane longitudinal stress waves along a rod of uniform thin section. Taking the axis of the rod as the x axis, the equation of motion is

$$\frac{\partial \sigma_x}{\partial x} = \rho \frac{\partial^2 u}{\partial t^2}.$$ (19.22)

For uniaxial stress, $\epsilon_x = \sigma_x/E$, $\epsilon_y = \epsilon_z = -v\epsilon_x$, so that $e = (1 - 2v)\epsilon_x$ from which

$$\epsilon_x = \partial u/\partial x = e/(1 - 2v) \qquad \text{and} \qquad \sigma_x = Ee/(1 - 2v).$$

Differentiating equation (19.22) with respect to x and substituting for $\partial u/\partial x$ and σ_x produces

$$\frac{E}{1 - 2v} \frac{\partial^2 e}{\partial x^2} = \frac{\rho}{1 - 2v} \frac{\partial^2 e}{\partial t^2},$$

or

$$\frac{\partial^2 e}{\partial t^2} = \frac{E}{\rho} \frac{\partial^2 e}{\partial x^2} = c^2 \frac{\partial^2 e}{\partial x^2},$$

where

$$c = \sqrt{(E/\rho)}$$ (19.23)

is the velocity of propagation of longitudinal waves in the rod.

In plates and rods of elastic material, longitudinal waves are propagated with the velocities given by equations (19.20) and (19.23) provided the wavelength is large compared with the thickness of the plate or cross-sectional dimensions of the rod. If the wavelength is comparable with these dimensions, the velocity depends on the wavelength and approaches that of Rayleigh surface waves when the wavelength is very short.

The principle of impulse and momentum expressed by equation (19.6) enables some elementary laws applicable to impact problems to be established. While these do not yield the quantitative values usually required such as the forces acting and the stresses, they are nevertheless useful for checking results. Let us consider, for example, the collision of two bodies having masses m_1 and m_2 respectively, the centres of gravity of which have initial velocities v_1, v_2 respectively. Assuming no external forces act on the bodies, the impulse is zero so that, as can be seen from equation (19.6), the total momentum remains unchanged throughout the event, i.e.

$$m_1 v_1 + m_2 v_2 = \text{constant}.$$

This is the well known principle of conservation of momentum and holds true in every case whether the behaviour of the material during impact is purely elastic, purely plastic or partially elastic. It gives the time integral of the forces acting only. The duration of contact and the variation with time of the forces can be determined only if the ratio between the force acting and the compression of each body is known. Since this ratio is different in static cases from dynamic cases if the results are influenced by oscillations, its application to dynamic problems will produce approximate values in some cases while in others no useful results are possible.

19.3 Hertz theory of impact of spheres

The problem of the collision of two spheres has been solved by Hertz.[70] This solution is useful for predetermining the duration of contact to be expected in dynamic experiments in which the results will not be affected by oscillations. The conditions under which this method can be applied are discussed in Section 19.4.

According to Hertz's theory, the force F acting between the spheres is related to their total compression δ, i.e. the distance through which the spheres approach one another after the first instant of contact, by

$$F = C_1 \delta^{3/2} \tag{19.24}$$

where

$$C_1 = \frac{4}{(1/\rho_1 + 1/\rho_2)^{1/3}(1/E_1' + 1/E_2')}$$

$\rho_{1,2} = $ radii of curvature of the spheres; $E_{1,2}' = \dfrac{E_{1,2}}{1 - v_{1,2}^2}$.

If the initial relative velocity of the spheres is v, the maximum compression is

$$\delta_{max} = \left\{ \frac{5m_1m_2}{4C_1(m_1 + m_2)} \right\}^{2/5} v^{4/5}, \tag{19.25}$$

and the maximum compressive force acting between them is

$$F_{max} = k_1 v^{6/5}, \tag{19.26}$$

where

$$k_1 = \left(\frac{5m_1m_2}{4(m_1 + m_2)} \right)^{3/5} C_1^{2/5}.$$

The time of contact is

$$t_c = 2 \cdot 9432 \delta_{max}/v,$$

or

$$t_c = \frac{k_2}{v^{1/5}}, \tag{19.27}$$

where

$$k_2 = 2 \cdot 9432 \left\{ \frac{5m_1m_2}{4C_1(m_1 + m_2)} \right\}^{2/5}.$$

The radius a of the surface of contact between the spheres is given by

$$a = \left\{ \frac{3F\left(\dfrac{1}{E'_1} + \dfrac{1}{E'_2} \right)}{4\left(\dfrac{1}{\rho_1} + \dfrac{1}{\rho_2} \right)} \right\}^{1/3}, \tag{19.28}$$

and the maximum pressure at the centre of the surface of contact by

$$p_{max} = \left\{ \frac{6F\left(\dfrac{1}{\rho_1} + \dfrac{1}{\rho_2} \right)^2}{\pi^3 \left(\dfrac{1}{E'_1} + \dfrac{1}{E'_2} \right)^2} \right\}^{1/3}. \tag{19.29}$$

These equations can often be applied to obtain approximate values in problems involving the collision of bodies of any shape, the actual bodies being replaced by equivalent spheres having the same weights and elastic moduli as the bodies and with radii equal to the radii of curvature of the bodies at their points of contact. Equation (19.27) is particularly useful since it enables the approximate time of contact between the bodies to be predicted. It is shown later that this time is of decisive influence on the nature of the dynamic response to be expected. Since the time of contact is inversely proportional to the fifth root of the relative velocity of the colliding bodies, the influence of this velocity is small within certain limits. The time of contact thus depends mainly on the masses of the bodies, their dimensions and their elasticity.

19.4   Classification of dynamic problems

As mentioned previously, the dynamic effect, i.e. the influence of inertia forces on the process of stress development in a body, depends on the dynamic loading conditions. Three groups of typical phenomena can be distinguished. These are (1) quasistatic states of stress, (2) vibrations, (3) stress waves. The limits between these groups are not clearly defined, however, and frequently the phenomena associated with more than one group can occur in the same dynamic event.

The dynamic response of a body in any particular case depends not only on the magnitude of the forces acting but also, to a decisive extent, on their rate of change. Thus, while stress waves are produced by the change of forces, the frequency of these waves is determined by their rate of change. If the change of forces is due to the impact of a striking body, this means that the response of the body struck depends on the time of contact between the two bodies.

When the forces acting on a body change slowly so that the frequency is very low, the length of the wave is usually great compared with the dimensions of the body. In terms of impact, this requires that the time of contact between the bodies is long compared with the periodic time of free vibration of the body considered or the time required for a stress wave to pass through it. In such extreme cases, the stress distribution is independent of the rate of change of the forces. Although the stresses vary in magnitude during the process, being proportional at any instant to the corresponding instantaneous values of the forces acting, their distribution remains the same throughout and is identical with that under corresponding static loading. The external forces acting on the body are in equilibrium throughout the event and all stresses vanish when these forces cease to act. Problems in which the behaviour follows this pattern are called quasistatic.

When the frequency of the loading cycle is of the same order as the resonance frequency of the body, the stress waves and their reflections cause vibrations, e.g. longitudinal or flexural vibrations. Due to inertia forces the stress distribution will differ to some extent from that in comparable static or quasistatic cases and the external forces are not in equilibrium throughout the event.

If the rate of change of the forces acting on a body corresponds with a high frequency, i.e. with the generation of waves which are short compared with the dimensions of the body, the effect of stress waves predominates. In such cases the stress distribution differs greatly from that produced under static or quasistatic conditions.

With many problems it can be predicted that the response will conform more or less with that of one or other of the above three groups. In other cases it is not so obvious to which class the problem belongs. Certain problems may involve the behaviour associated with more than one group.

The first step towards the practical solution of a dynamic problem should consist of determining, either theoretically or experimentally, the class to which it belongs. With some previous experience on similar problems the time of contact can be estimated. The subsequent procedure depends on the class to which the problem belongs and on the time of contact to be expected.

In quasistatic problems it is usually sufficient to determine only the maximum values of the forces acting by dynamic methods. When these values have been obtained, the solution can be completed either analytically or experimentally using static methods. A problem frequently encountered is that of determining the stress concentration factor under dynamic conditions involving stress waves or vibrations at a notch or other irregularity in the shape of a body. In such cases the procedure to be adopted depends on the relative dimensions of the wavelength and the notch. If the dimensions of the notch are small compared with the wavelength, the stress distribution in the neighbourhood of the notch will be similar to that under comparable static loading. Such loading applied to a model of the relevant small part of the body will therefore produce the same stress distribution. Further investigations such as comparing different shapes of notch can then be accomplished using static methods.

When the length of a stress wave is of the same order or smaller than the dimensions of a body or, for instance, of a notch in it, dynamic methods must be applied. This is also true in the case of vibrations.

Since a stress concentration factor depends on the length of the stress wave involved, it is obvious that there is no generally applicable dynamic factor of stress concentration.

19.5   Photoelastic investigation of dynamic problems

A considerable number of problems of the dynamic response of beams and plates loaded by coplanar forces have been investigated photoelastically. The models used are similar to those for static investigations and may be made from either hard materials such as epoxy resin or soft ones such as urethane rubber. Dynamic problems of plates or other thin-walled structures subjected to transverse bending can be investigated using models formed from two layers of the same material with a reflecting middle layer or alternatively, from two layers of different materials having different photoelastic constants as described in Chapter 12.

In three-dimensional dynamic problems, surface stresses can be determined using the birefringent coating method. Stresses at internal points can be determined by the scattered light method using, for example, a laser as light source or by using polarizers embedded in the model. Some of these methods are rather delicate, however, and their application is restricted to simple cases.

19.6 Recording of transient optical data

In many dynamic problems the stresses change rapidly. As a result the isochromatics and isoclinics which they produce in a photoelastic model move with considerable velocity. In the case of elastic stress waves, the maximum possible velocity of the isochromatics is equal to the velocity of sound in the material which, for the usual hard model materials, is about 2000 m/s. It is obvious that the effects produced in this type of event must be recorded using different techniques from those employed in static investigations.

Several methods suitable for the photoelastic investigation of dynamic problems have been developed. In most of these, photographs are taken of the isochromatics and isoclinics while in others the variation of stresses is determined from point by point measurements of the corresponding variation of intensity of the transmitted or reflected light using photomultipliers or other photosensitive devices.

19.7 Photographic methods

When photographing dynamic fringe patterns, the exposure time should be short enough to arrest all sensible movement of the image over the surface of the film. The technique and equipment most suitable for any specific problem depends on the velocity of the isochromatics to be expected and on the nature of the record required. In some cases, a single photograph taken at a particular instant during the event may be sufficient while in others a series of pictures taken at successive instants may be required to record the complete history of the event.

For the study of impact problems, very short exposure times and high picture repetition rates are required. When photographing isochromatics moving with a velocity of 2000 m/s the time of exposure should not exceed 1 μsec. since in this interval the isochromatics will travel a distance of 2 mm. In single picture methods, such brief exposures can be obtained with a continuous light source of high intensity using a camera with a very fast working shutter such as a Kerr cell shutter. A more common method is to control the exposure time by the duration of the source itself. This can be done using a flash light source such as a gas-discharge flash tube or an electric spark which produce very short but intense flashes. With this method an ordinary still camera or cine camera capable of taking single shots can be used in conjunction with a delay generator which allows the flash to be triggered at the desired instant during the event.

Single pictures obtained in the above way can be studied individually. Alternatively, if the event is multiply repeatable, a continuous film record can be built up by taking a series of single pictures for successively increasing values of the delay period between the instigation of the event and the instant of the flash. The cine projection of this film will reproduce in slow motion the photoelastic effects produced during the event. Pictures from films produced by this method are shown in Section 19.16.1.

The single picture technique can be extended to obtain a limited number of pictures taken at different instants during the one dynamic occurrence. This is useful when dealing with problems such as those involving plastic stress waves or fracture, each repetition of which would require another model. One method based on a multiple spark system developed by Cranz and Schardin[71] is to use several flash light sources and a corresponding number of cameras in the arrangement shown in Figure 19.1. By means of a suitable

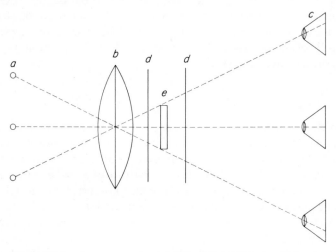

Figure 19.1.   Arrangement producing several pictures during a
single dynamic event

triggering device, the sources are caused to flash in very rapid succession. The light from each source in turn is concentrated by the lens b into the corresponding camera c after transmission through the polaroids d and the model e.

The laser has special advantages as a light source in dynamic investigations. The light emitted is monochromatic and this allows photographs to be taken on black and white film without the necessity of using a colour filter which absorbs most of the energy of flash lights. The types of laser known to the authors, however, are not capable of producing several flashes during a single dynamic event.

Various forms of shutterless high-speed film cameras are available by means of which a limited number of pictures can be taken in rapid succession during a single dynamic event. Because of the limitations in strength of the film, intermittent film movement with its consequent high accelerations is not practicable with very high framing rates and consequently such cameras are characterized by employing either continuously moving or stationary film. In some high-speed cameras, the film is transported continuously past the film gate and the successive images are projected on to its surface by means of a rotating prism. In the Hycam rotating prism camera,* the exposure time is controlled by a rotating segmented shutter. This camera is capable of framing rates up to 40,000 pictures per second.

The problem of obtaining an adequate intensity of illumination for photography at extremely short exposure times is reduced with the image converter camera which reproduces the image with large amplification of the intensity. In this camera, an image of the object is projected by the objective lens on to the photocathode of an image converter tube. The photocathode converts the distribution of light intensity into a corresponding distribution of emitted electrons. The electron image is projected on to the photoanode where the original image is reproduced with greatly increased intensity. This image is finally projected by a lens system on to a photographic film. The Imacon image converter camera* provides $50 \times$ amplification of the light intensity and allows a limited number of pictures to be recorded at exposure times as short as 5 nanoseconds.

Picture repetition rates up to about 8×10^6 pictures per second can be obtained using an ultra high speed drum type camera in which a film of limited length is attached to the cylindrical surface of a drum. In some cameras of this type the drum is rotated at high speed while in others it is stationary and the film is swept by the light from a rotating mirror system. Since only a limited number of pictures can be accommodated during each operation of the camera, the duration of an event which can be recorded is reduced proportionately as the framing rate is increased. Accurate synchronization is therefore required to ensure the projection of the desired portion of the event on to the film.

Very high framing rates are required when recording impact phenomena. For the study of elastic stress waves propagation in the usual hard photoelastic materials by cine projection at 16 frames per second, the framing rate should be of the order of 10^6 pictures per second. This can be obtained

* Manufactured and supplied by John Hadland (Photoelastic Instrumentation) Ltd, Newhouse Laboratories, Bovingdon, Herts., England.

using the Barr and Stroud ultra-high-speed mirror camera* which allows from 28 to 117 pictures at maximum framing rates between 1.6×10^6 and 8×10^6 pictures per second.

19.8   Method of streak photography

A continuous photographic record of the isochromatics during rapid dynamic events can be obtained by the method of streak photography developed by Tuzi and Nisida[72] and later by Frocht and Flynn.[73]

In one form of streak camera, the film is run continuously behind a diaphragm provided with a narrow slit *a* (Figure 19.2). An image of the model

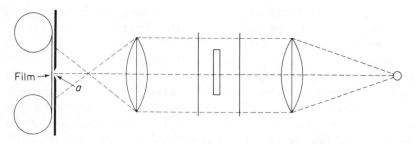

Figure 19.2.   Schematic arrangement for streak photography

is projected in the plane of the diaphragm and is positioned so that a chosen line on the model coincides with the slit. This method does not produce a series of discrete pictures of the isochromatics but provides instead a chronographic record of their movement along the chosen line. In another form of streak camera the image is swept over a stationary film by means of a rapidly rotating central mirror. All three cameras mentioned in Section 19.7 can be adapted for streak operation.

19.9   Use of low modulus materials

As shown in Section 19.2, the velocity with which stress waves are propagated in elastic materials is proportional to the square root of the modulus of elasticity. By using models manufactured from materials having low elastic moduli, the velocity of propagation can be greatly reduced with the result that the requirements for successful photography are less demanding. High-speed cameras such as the Hycam can then be used instead of the more elaborate and expensive ultra-high-speed cameras. Suitable model materials

---

* Manufactured by Barr and Stroud Ltd., Anniesland, Glasgow, Scotland, and supplied by John Hadland (P.I.) Ltd., Newhouse Laboratories, Bovingdon, Herts., England.

are the polyurethane rubbers such as Hysol 4485 supplied in the U.S. by the Hysol Corporation, New York and Photoflex supplied in the U.K. by Sharples Photomechanics Ltd, Preston, England.

Urethane rubbers are available in different grades covering a wide range in elastic moduli. For the more commonly used grades, the modulus of elasticity has an average value of about $3·5$ MN/m^2 giving a wave velocity of about 60 m/sec. This is only about 1/30th of the velocity of stress waves in the hard plastics. Successions of pictures suitable for cine projection can then be taken at framing rates of about 30,000 pictures per second. If cine film projection is not required and the dynamic response is to be studied by viewing the individual pictures of a series in turn, substantially lower framing rates will usually suffice.

19.10 Stroboscopic method

By using a stroboscopic light source, the photoelastic effects at any particular instant during a multiply repeatable dynamic event can be studied visually without the necessity of taking a photograph. In this method, which was developed by Allison,[74] each flash is arranged to occur at the same instant relative to the instigation of the event during each repetition. By gradually increasing the delay period between the instigation of the event and the occurrence of the flash, the dynamic response can be viewed or filmed in slow motion. The method is restricted in its application to events in which the loading cycle can be repeated at a frequency corresponding to the flash rate required of the lamp. In many investigations the loading requirements can be satisfied by using an electromagnetic vibration generator as a loading device. Since the maximum load which can be applied in this way is limited, photoelastic materials of high optical sensitivity such as urethane rubber are normally required to obtain a satisfactory fringe pattern.

19.11 Light sources

The use of flash light sources in single picture techniques has already been discussed. The high intensity of illumination required to produce satisfactory pictures with the very short exposure times associated with high framing rates and writing speeds can be obtained using a variety of sources. The type of source most suitable for the investigation of any specific problem depends on the exposure time and on the duration and nature of the photographic record required.

In certain cases sufficient intensity can be obtained using continuous sources such as tungsten filament or mercury arc lamps. In others, 'Photoflash' bulbs which usually have a duration of about 20 milliseconds above half peak intensity can be used either singly or flashed in sequence.

Stroboscopic flash light sources giving up to 8000 flashes per second suitable for synchronization with high speed cameras are available.

Flash light sources of the square-wave type giving flash durations up to several hundred microseconds are available for use with ultra-high-speed framing cameras.

19.12 Photometric methods

The distribution of transient stresses under dynamic loading can be determined from point by point measurements of the variations of the intensity of light transmitted or reflected by the model. Since these are not 'full field' methods they are not very popular. The choice between photographic and photometric methods, however, is largely one of parameter. In 'full field' photographic methods, pictures are obtained at one or a series of particular instants during the event. On the other hand, photometric methods provide a continuous time record of the effects at the point under observation. Streak photography also produces a continuous record but is not 'full field'. Another advantage of photometric methods is that the equipment used is simple and inexpensive.

Basically, these methods consist of projecting the light from a small area surrounding the chosen point on the model into a photomultiplier or other photometric device which converts the variations of intensity into variations of electric potential. This signal is displayed against a time base on the screen of an oscilloscope or, in the case of events of lower frequency, can be recorded by means of an ultra violet recorder.

One difficulty encountered in photometric methods is the elimination of cyclic and random fluctuations of potential due to mains ripple and 'noise' which are superimposed on the desired signal. For accuracy, it is essential that the light rays entering the photomultiplier are restricted to those emanating from a very small area of the model. High amplification of the signal is therefore required and this results in a low signal-to-noise ratio. Mains ripple can be eliminated and noise reduced to an insignificant level by using d.c. supplies from batteries for the lamp and photomultiplier. A low-voltage tungsten-filament projection lamp supplied by a lead-acid battery is satisfactory as a source. The light from this source can be rendered highly monochromatic by passing it through a narrowband interference filter. For the photomultiplier, an H.T. supply of the order of 1000 V is required. Small dry batteries giving this voltage can readily be obtained or made up and have a long life.

Since the intensity of the light transmitted is a cyclic and not a single valued function of the difference of principal stresses, difficulty may be experienced in some cases in interpreting the results. This can be avoided by arranging that the maximum retardation produced in the model shall not exceed one half of a wavelength. The method then possesses the additional advantage that model materials of low optical sensitivity such as Perspex or Plexiglas can often be used. These materials are cheaper and usually more readily available than the normal photoelastic materials. Oscillograms obtained using Perspex models are illustrated in Section 19.16.2.

In the study of dynamic problems involving uniaxial states of stress such as longitudinal impact loading of thin rods or flexural vibrations of beams, the variation of stress with time at any point can be obtained from the recording of a single signal representing the variation of intensity of the light transmitted using either plane or circularly polarized light. With plane polarized light, crossed polaroids are set at 45° to the known direction of the stress, e.g. to the axis of the rod or beam in the examples mentioned. Equation (5.14) then reduces to

$$I_{45} = I_0 \sin^2 \frac{\alpha}{2} = \frac{I_0}{2}(1 - \cos \alpha),$$

from which

$$\alpha = \cos^{-1}(1 - (2I_{45}/I_0)). \tag{19.30}$$

The vertical scale of the trace displayed on the oscilloscope screen can be determined either before or after the dynamic test by rotating crossed polaroids until the horizontal lines corresponding to zero and maximum intensity are obtained. The distance between these lines represents the total intensity I_0.

If the same arrangement is used but with the axes of the polaroids parallel, the phase difference is given by

$$\alpha = \cos^{-1}\left(\frac{2I_{45}}{I_0} - 1\right). \tag{19.31}$$

Since the equations for the intensities transmitted in the crossed and parallel circular polariscopes are identical with those for the corresponding plane polariscopes when $\theta = 45°$, the phase differences from comparison with equations (19.30) and (19.31) are:

$$\alpha = \cos^{-1}\left(1 - \frac{2I}{I_0}\right), \tag{19.32}$$

and

$$\alpha = \cos^{-1}\left(\frac{2I}{I_0} - 1\right), \tag{19.33}$$

If the setup employs only a single quarter-wave plate inserted between the polarizer and the model with its axis at 45° to that of the polarizer, it can easily be shown that

$$\alpha = \sin^{-1}\left(\frac{2I_{45}}{I_0} - 1\right), \tag{19.34}$$

or

$$\alpha = \sin^{-1}\left(1 - \frac{2I_{45}}{I_0}\right), \tag{19.35}$$

with crossed or parallel polaroids respectively.

With arrangements giving a dark background, the phase difference α is related to the intensity through

$$\frac{I_{45}}{I_0} = \sin^2\frac{\alpha}{2},$$

and to the stress through

$$\frac{nd}{f}\sigma = \frac{\alpha}{2},$$

with the assumption that the stress optic law, equation (5.4), is still valid. It follows that for values of α less than π, oscillograms obtained in this way can be regarded as distorted stress-time diagrams, the degree of distortion of the stresses varying in the same manner as a $\sin^2$ curve deviates from a straight line. Thus, serious distortion will occur only when I_{45}/I_0 is small or approaches unity while over a considerable range between these values the stress-intensity relation is approximately linear. This is demonstrated in Figure 19.3 which

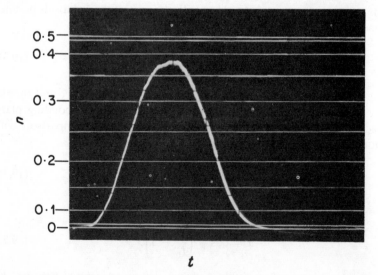

Figure 19.3.   Oscillogram with superimposed scale of isochromatic fringe order

shows an oscillogram produced by the passage of a stress pulse through a plate to a non-linear ordinate scale of fringe order.

In biaxial dynamic problems the variation of the difference of principal stresses and their directions, whether the latter remain constant or vary with time, can be obtained from two photoelectric signals. In one procedure, the intensity of light transmitted in the plane polariscope is recorded for two different settings of the crossed polaroids. It is convenient to choose these positions at 45° to each other. Let θ be the angle between the axis of the polarizer and the principal stress σ_1 in the first setting. The intensity I_θ transmitted is given by

$$I_\theta = I_0 \sin^2 2\theta \sin^2 \frac{\alpha}{2}.$$

Replacing θ by $\theta + 45°$, we obtain for the second setting

$$I_{\theta+45} = I_0 \cos^2 2\theta \sin^2 \frac{\alpha}{2}.$$

Adding we obtain

$$I_\theta + I_{\theta+45} = I_0 \sin^2 \frac{\alpha}{2} = \frac{I_0}{2}(1 - \cos \alpha),$$

from which

$$\alpha = \cos^{-1} \left[1 - \frac{2(I_\theta + I_{\theta+45})}{I_0} \right]. \tag{19.36}$$

Dividing, we obtain

$$\tan^2 2\theta = \frac{I_\theta}{I_{\theta+45}},$$

from which

$$\theta = \tfrac{1}{2} \tan^{-1} \sqrt{\left(\frac{I_\theta}{I_{\theta+45}} \right)},$$

or

$$\theta = \tfrac{1}{4} \cos^{-1} \left(\frac{I_{\theta+45} - I_\theta}{I_{\theta+45} + I_\theta} \right). \tag{19.37}$$

The recordings of I_θ and $I_{\theta+45}$ may be made in turn repeating the event when this is possible or simultaneously using two photomultipliers and a double beam oscilloscope. For simultaneous measurement, two light sources may be employed, the beams of light from each passing through the same point on the model but through separate polaroids and inclined at an angle of a few degrees to the normal. Alternatively, the light from a single source may be split into two beams by means of a half mirror. Since the total intensities of the two beams will not in general be equal, the following equations replacing equations (19.36) and (19.37) should be used

$$\alpha = \cos^{-1}\left[1 - 2\left(\frac{I_\theta}{I_0} + \frac{I_{\theta+45}}{I_0'}\right)\right],$$ (19.38)

$$\theta = \tfrac{1}{4}\cos^{-1}\frac{(I_{\theta+45}/I_0') - (I_\theta/I_0)}{(I_{\theta+45}/I_0') + (I_\theta/I_0)},$$ (19.39)

where I_0 and I_0' are the total intensities of the beams corresponding to I_θ and $I_{\theta+45}$ respectively.

An alternative procedure is to measure the variations of intensity of plane and circularly polarized light transmitted through the model. The intensity I_1 of the plane polarized beam is given by equation (5.14). Denoting the intensity of the circularly polarized beam by I_1^* and the corresponding total intensity by I_0^* we have from equation (5.18)

$$I_1^*/I_0^* = \sin^2\frac{\alpha}{2},$$ (19.40)

which yields

$$\alpha = \cos^{-1}(1 - 2I_1^*/I_0^*).$$ (19.41)

From equations (5.14) and (5.18) we obtain

$$\sin^2 2\theta = \frac{I_1/I_0}{I_1^*/I_0^*},$$

which produces

$$\theta = \tfrac{1}{4}\cos^{-1}\left(1 - 2\frac{I_1/I_0}{I_1^*/I_0^*}\right).$$ (19.42)

If the principal stresses in a two-dimensional system maintain constant known directions, their individual values can be determined by the oblique incidence method (Section 14.7) from measurements of the intensity of two beams, one of which passes normally and the other obliquely through the model. One arrangement allowing simultaneous recording of the signals is shown in Figure 19.4. The beam of light transmitted through the half mirror H falls with normal incidence on the model M. The beam reflected at the half mirror is again reflected at the full mirror F to fall on the model at a suitable angle of incidence i in the plane containing the normal and one of the principal stresses, e.g. σ_2. Before passing through the model, the beams are plane polarized by the polarizers P_1, P_2 set with their axes at 45° to the directions of the principal and secondary principal stresses respectively.

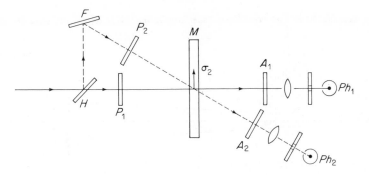

Figure 19.4. Arrangement for simultaneous recording of intensities
of normal and oblique beams of light

After passing through the model and the analysers A_1, A_2 the intensities are measured by means of the photomultipliers Ph_1, Ph_2 respectively.

If α, α_i are the phase differences determined from the intensities by means of equation (19.30), the principal stresses are given by

$$\sigma_1 = \frac{f \cos i}{2\pi d \sin^2 i}(\alpha_i - \alpha \cos i),$$

$$\sigma_2 = \frac{f}{2\pi d \sin^2 i}(\alpha_i \cos i - \alpha).$$

The above arrangement may be varied by using circularly polarized light beams with or without analyser quarter-wave plates as previously described.

Dynamic states of stress may be investigated by the interferometric method of Favre and Schumann described in Section 10.8. Figure 19.5 shows an arrangement used by Haenggi and Schumann[75,76] for the investigation of problems in which the principal stresses maintain constant directions. The system is basically that of Figure 10.7 modified to allow simultaneous recording of the intensities J_1, J_2 of the reflected light vibrating parallel to the principal stresses σ_1, σ_2 respectively. Light from the source S is rendered monochromatic by the filter F. By means of the split polarizer P, one half of the beam is plane polarized in one direction while the other half is plane polarized in the perpendicular direction. The light reflected at the half mirror B passes through the half-wave plate H (which enables the planes of polarization to be rotated into coincidence with the directions of the principal stresses) and the small aperture D on to the model M. After reflection at the faces of the model, the half beams are directed into the photomultipliers Ph_1, Ph_2 and the signals generated are displayed on the screen of an oscilloscope. The specimen N carries electrical resistance strain gauges and is used to calibrate the system.

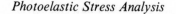

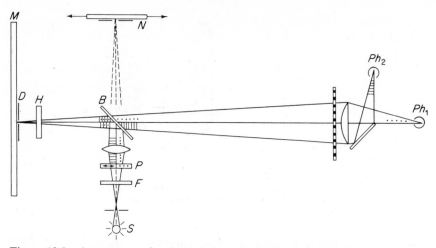

Figure 19.5. Arrangement for determining transient stresses with known constant directions (after Haenggi and Schumann)

In the general case when the directions as well as the magnitudes of the principal stresses are time dependent, the complete state of stress can be determined using an extension of the above method reported by Bohler and Schumann.[77] The setup is shown in Figure 19.6. A continuous gas laser L which produces two monochromatic plane polarized beams forms a convenient source. One of these beams (b_1) is converted into circularly polarized light in the usual way and projected at normal incidence on to the desired point on the model. The additional quarter-wave plate Q_1 serves to eliminate additional interferences produced between the faces of the model and the Fabry–Pérot mirrors in the laser. The light reflected from this beam at the faces of the model is again reflected at the half silvered mirror H_1 and directed into the photomultiplier Ph_1. The other laser beam is split by means of the

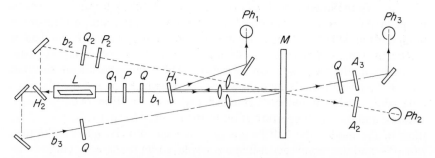

Figure 19.6. Arrangement for the determination of transient stresses with varying directions (after Bohler and Schumann)

half mirror H_2. Each of the two beams thus formed is reflected by a system of mirrors to fall on the desired point of the model at a small angle (3°) to the normal. One of these beams (b_2) passes through crossed polaroids P_2, A_2. The quarter wave plate Q_2 serves merely to obtain uniformity of intensity for all positions of P_2. The other beam b_3 is converted into circularly polarized light, passing through the usual elements of the circular polariscope. The intensities of the light from these beams transmitted through the analysers A_2, A_3 are measured by the photomultipliers Ph_2, Ph_3 respectively.

The relative retardation R $(=\alpha/2\pi)$ and the directions of the principal stresses are determined from the measured intensities of the transmitted beams b_2, b_3 by means of equations (19.41) and (19.42) respectively. Adding equations (10.11), the intensity of the light reflected from the circularly polarized beam b_1 is found to be

$$ J = \frac{J_0}{2}\left[1 - \cos\frac{2\pi}{\lambda}(R_m + R_0)\cos\frac{2\pi}{\lambda}R \right], $$

where J_0 represents four times the intensity of the first reflection. Inserting the values of J, J_0, R and R_0 in this equation, the value of R_m is obtained. Finally, the individual values of the principal stresses are determined by inserting the values of R and R_m in equations (10.16).

19.13 Dynamic behaviour of photoelastic materials

The question of the validity of the stress optic law under dynamic conditions has been the subject of a number of investigations. In this context, the physical origin of birefringence requires consideration; the possibility that this may be due to stress, strain or a combination of both is mainly of academic interest with elastic materials but assumes a practical significance with viscoelastic materials such as the usual model plastics. At the present time, experimental evidence appears to favour the stress hypothesis but is to some extent conflicting and the range of materials and loading rates investigated is inadequate to allow the formulation of general rules.

As indicated by the multiphase theory, the modulus of elasticity under dynamic conditions is higher than the static modulus. It would therefore be expected that its value will depend on the rate of change of stress, which may differ at different parts of the body or at different instants. This was investigated for the materials Araldite and Lekutherm X30 by an improved rotating rod method. A thin-walled cantilever tube of the model material was statically loaded by a pure bending moment and then rotated at different velocities. The vertical deflection of the end of the tube was then inversely proportional to the dynamic modulus of elasticity (while its horizontal displacement was proportional to the internal friction). The results indicated that a limiting value of E was reached asymptotically within the range of speed employed.

This value was found to be less than 2 % greater than that for short-time static loading. This result was confirmed from measurements of the velocity of propagation of stress waves in the material.

The rotating tube method was also employed to determine the dynamic stress-optical coefficient. For this purpose the inner surface of the tube was made reflective by means of aluminium powder on an oil film. Observations of the reflected light showed that the stress-optic coefficient was practically the same as with static loading. Appreciable differences in the values of both the stress-optic coefficient and the elastic modulus for other materials between static and dynamic loading have been reported in the literature, however. It is therefore advisable that the values used in an investigation be determined from calibration tests carried out on specimens subjected to the same loading conditions as the model.

Serious deviations from the correct stress distribution are likely to arise from the high internal friction or viscous damping of viscoelastic model materials. One measure of the influence of internal friction is the distance travelled by a stress wave when its amplitude has been reduced to one half of its original value (corresponding to the 'half-life period' or 'half-value period' in nuclear physics). The half value length was found to be about 3 m for Araldite and Lekutherm X30 and about 1 m for polyurethane rubber. For analogous conditions to exist in the model and prototype, the scale of the model should correspond to the ratio of the half value lengths for the two materials. Since this value for steel is several hundred or thousand metres, it is scarcely possible that this condition can ever be realized in practice. Quantitative dynamic photoelasticity should therefore be restricted to problems where friction has no great effect, such as during the initial phases of response to impact.

19.14 Model laws for dynamic investigations

In addition to the usual requirement of geometrical similarity, in dynamic investigations the shape of the pulse should be similar in model and prototype and it should be to the same scale as the linear dimensions. In problems of impact, this requires that the striking body is also reproduced true to scale and that the densities and elastic moduli of the striking and struck bodies should have the same ratios as for the prototype, i.e.

$$\rho_{sp}/\rho_p = \rho_{sm}/\rho_m \quad \text{and} \quad E_{sp}/E_p = E_{sm}/E_m,$$

where the subscript s refers to the striking body while p and m have the same meaning as before.

For equal inertia forces in rods the equation of motion (19.22) gives

$$\frac{\partial \sigma_p}{\partial x_p} \bigg/ \frac{\partial \sigma_m}{\partial x_m} = \frac{\rho_p a_p}{\rho_m a_m} = 1$$

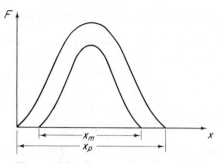

Figure 19.7. Stress pulses for model
and prototype

while similarity of the pulse (see Figure 19.7) requires that

$$\partial F_p/\partial x_p = \partial F_m/\partial x_m \quad \text{and} \quad F_p/x_p = F_m/x_m.$$

The relation between the velocities of the striking bodies follows from equation (19.6):

$$\frac{m_p v_p}{m_m v_m} = \frac{\displaystyle\int F_p \, dt}{\displaystyle\int F_m \, dt} = \frac{\dfrac{A_p}{c_p} \displaystyle\int \sigma_p \, dx}{\dfrac{A_m}{c_m} \displaystyle\int \sigma_m \, dx}.$$

In this equation, $\int \sigma \, dx$ can be replaced by σx or σL; observing also that $m = \rho L^3$ and $A = L^2$, it yields

$$\frac{v_p}{v_m} = \frac{c_m \rho_m \sigma_p}{c_p \rho_p \sigma_m}.$$

Assuming that the requirements for analogous conditions are otherwise satisfied, corresponding stress distributions will exist in the model and prototype at different instants during the event when the stress waves have travelled corresponding distances. This occurs when

$$t_m/t_p = L_m c_p/L_p c_m,$$

where t is the time elapsed since the beginning of the event and c is the velocity of propagation. In rods and in other problems with $v_m = v_p$,

$$\frac{c_p}{c_m} = \sqrt{\left(\frac{E_p}{E_m}\frac{\rho_m}{\rho_p}\right)}.$$

The above equation then becomes

$$\frac{t_m}{t_p} = \frac{L_m}{L_p}\left(\frac{E_p}{E_m}\right)^{1/2}\left(\frac{\rho_m}{\rho_p}\right)^{1/2}.$$

19.15 Separation of principal stresses in dynamic problems

The isochromatics and isoclinics recorded in dynamic problems produce the differences of principal stresses (or strains) and their directions in the same way as in static cases. At load free edges, the isochromatics indicate the stresses directly. At points within the field of a plate, however, separation of the principal stresses is necessary if the state of stress is to be determined completely. Several procedures are available by means of which this can be accomplished.

By the interferometric method described in Section 19.12, the individual principle stresses are determined experimentally so that no separation procedure is required. The application of the oblique incidence method in other photometric procedures has already been described. This method may also be applied to photographic procedures in a manner similar to that for static investigations. Photographs of the isochromatics under normal and oblique incidence may be taken simultaneously or in turn if the event is fully reproducible.

In two-dimensional problems, separation of the principal stresses can be effected using the isochromatic and isoclinic patterns relating to the same instant during the occurrence. Several patterns of the isoclinics of different parameters required can be photographed simultaneously from a single flash using an arrangement similar to that shown in Figure 19.1 or they can be photographed singly if the event is repeatable. In addition, the isochromatic and isoclinic patterns, together with the equations of equilibrium of an infinitesimal element (or of a finite element as shown later) and the equation of compatibility, enable the investigator to detect the existence of a dynamic (inertia) effect and to distinguish between longitudinal and transverse waves. An equation by means of which the presence of a transverse wave may be recognized can be derived from the equations of equilibrium (19.17). Differentiating equation (19.17a) with respect to y, equation (19.17b) with respect to x and subtracting we obtain

$$\frac{\partial^2(\sigma_x - \sigma_y)}{\partial x\,\partial y} + \frac{\partial^2\tau_{xy}}{\partial y^2} - \frac{\partial^2\tau_{xy}}{\partial x^2} = \rho\frac{\partial^2}{\partial t^2}\left(\frac{\partial u}{\partial y} - \frac{\partial v}{\partial x}\right)$$

$$= -2\rho\frac{\partial^2\omega_z}{\partial t^2}, \tag{19.43}$$

where $\omega_z = \frac{1}{2}[(\partial v/\partial x) - (\partial u/\partial y)]$ is the rigid body rotation of an element about the z axis, i.e. about an axis perpendicular to the plane of the plate. Further, from equations (1.6) and (1.7) we have

$$(\sigma_x - \sigma_y) = (\sigma_1 - \sigma_2)\cos 2\varphi,$$

and

$$\tau_{xy} = \frac{1}{2}(\sigma_1 - \sigma_2)\sin 2\varphi.$$

The terms on the left hand side of equation (19.43) can therefore be determined from the isochromatics and the isoclinics alone. Under static conditions, the left-hand side of this equation is zero so that any deviation from zero indicates a dynamic effect. It can be seen from the right-hand side that such a deviation will occur if there is a time-dependent change of the rotation of an element about the z axis, i.e. if there is a transverse wave.

The equation of compatibility in terms of the stresses in a two-dimensional state is

$$\frac{\partial^2}{\partial y^2}(\sigma_x - v\sigma_y) + \frac{\partial^2}{\partial x^2}(\sigma_y - v\sigma_x) = 2(1 + v)\frac{\partial^2 \tau_{xy}}{\partial x \, \partial y}. \tag{2.26}$$

Adding the term

$$\tfrac{1}{2}(1 + v)\left(\frac{\partial^2}{\partial x^2} - \frac{\partial^2}{\partial y^2}\right)(\sigma_x - \sigma_y)$$

to both sides of this equation and dividing by $(1 - v)$ we obtain

$$\nabla^2(\sigma_x + \sigma_y) = \frac{1 + v}{1 - v}\left[4\frac{\partial^2 \tau_{xy}}{\partial x \, \partial y} + \left(\frac{\partial^2}{\partial x^2} - \frac{\partial^2}{\partial y^2}\right)(\sigma_x - \sigma_y)\right]. \tag{19.44}$$

This equation is used for separating the principal stresses in one of the methods described later.

An equation from which it is possible to recognize the presence of a longitudinal wave can be derived in the following way. Differentiating equation (19.17a) with respect to x, equation (19.17b) with respect to y and adding produces

$$\frac{\partial^2 \sigma_x}{\partial x^2} + \frac{\partial^2 \sigma_y}{\partial y^2} + \frac{2\partial^2 \tau_{xy}}{\partial x \, \partial y} = \rho\frac{\partial^2}{\partial t^2}\left(\frac{\partial u}{\partial x} + \frac{\partial v}{\partial y}\right). \tag{19.45}$$

If we now eliminate τ_{xy} from equations (19.45) and (2.26) we obtain

$$\frac{\partial^2 \sigma_x}{\partial x^2} + \frac{\partial^2 \sigma_y}{\partial y^2} + \frac{1}{(1 + v)}\left[\frac{\partial^2 \sigma_x}{\partial y^2} - v\frac{\partial^2 \sigma_y}{\partial y^2} + \frac{\partial^2 \sigma_y}{\partial x^2} - \frac{v\partial^2 \sigma_x}{\partial x^2}\right] = \rho\frac{\partial^2}{\partial t^2}\left(\frac{\partial u}{\partial x} + \frac{\partial v}{\partial y}\right),$$

which reduces to

$$\nabla^2(\sigma_x + \sigma_y) = \rho(1 + v)\frac{\partial^2}{\partial t^2}\left(\frac{\partial u}{\partial x} + \frac{\partial v}{\partial y}\right). \tag{19.46}$$

For plane stress we have from equation (19.18)

$$\frac{\partial u}{\partial x} + \frac{\partial v}{\partial y} = \frac{1 - v}{1 - 2v}e,$$

which, inserted in equation (19.46), produces

$$\nabla^2(\sigma_x + \sigma_y) = \frac{\rho(1 - v^2)}{1 - 2v} \frac{\partial^2 e}{\partial t^2} \tag{19.47}$$

Finally, combining equations (19.44) and (19.47), we obtain

$$\frac{\partial^2 e}{\partial t^2} = \frac{1}{\rho} \frac{1 - 2v}{(1 - v)^2} \left[4 \frac{\partial^2 \tau_{xy}}{\partial x \, \partial y} + \left(\frac{\partial^2}{\partial x^2} - \frac{\partial^2}{\partial y^2} \right) (\sigma_x - \sigma_y) \right]. \tag{19.48}$$

The term on the right-hand side of this equation, which is zero in static cases, indicates the presence of longitudinal waves.

Separation of the principal stresses can be accomplished by the following methods:

The equations of equilibrium (19.17a) and (19.17b) can be integrated only if the acceleration of each point is known. In general, it will be unknown. In quasistatic problems, the acceleration can be neglected. In other problems where it is expected to be of minor influence only, the acceleration can be estimated. With these estimated values an approximate integration can be carried out, the result of which can be checked from the boundary conditions.

Another approximate method based on the conditions of equilibrium can be applied in certain simple cases. From equations (19.17a) and (19.17b) the components of the acceleration parallel to load free edges can be evaluated (Figure 19.8(a)). Moreover, the average values of the components of

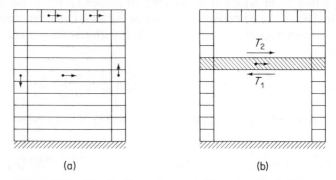

(a)   (b)

Figure 19.8. Sketch illustrating approximate method of separating principal stresses in a dynamically loaded plate, (a) components of acceleration parallel to load free edges, (b) inertia and shear forces on an elementary strip

acceleration in certain directions (e.g. parallel and perpendicular to an edge) can be determined using finite elements (Figure 19.8(b)). If such an element is bounded by load-free edges and parallel sections within the plate, the inertia force acting on the element in the direction parallel to the sections is obviously equal to the difference $T_2 - T_1$ of the shear forces acting on the sections. The shear forces T_1, T_2 are equal to the integrated shear stresses along these sections. By dividing the plate into a series of elements, the average inertia forces can be determined, e.g. in two mutually perpendicular directions. The distribution of these forces can then be estimated. With these estimated values of the forces or accelerations, equation (19.17a) and (19.17b) can be applied in a similar way as in static cases. The accuracy of the values of the normal stresses σ_x, σ_y obtained by this means can be checked by comparing their differences with the values of $(\sigma_x - \sigma_y)$ obtained from the isochromatics and isoclinics.

Separation of the principal stresses can be accomplished after determining their sum $(\sigma_1 + \sigma_2) = (\sigma_x + \sigma_y)$ from equation (19.44) in a similar manner as in static investigations (see Section 9.4.1). While the right-hand side of the corresponding (Laplace) equation (2.31) for static stress distribution is zero however, that of equation (19.44) for dynamic loading varies from point to point in the plate. After determining values of the right-hand side from the isochromatics and the isoclinics, this Poisson type equation can be solved numerically by Liebmann's relaxation method using the modified formula

$$\Sigma_0 = \frac{\Sigma_1 + \Sigma_2 + \Sigma_3 + \Sigma_4}{4} - \frac{d^2}{32} \frac{1 + v}{1 - v}$$

$$\times \left[4 \frac{\partial^2 \tau_{xy}}{\partial x\, \partial y} + \left(\frac{\partial^2}{\partial x^2} - \frac{\partial^2}{\partial y^2} \right) (\sigma_x - \sigma_y) \right] \quad (19.49)$$

where Σ_0 is the sum of the stresses at the centre and Σ_i their sum at the point i of the square and d is the length of the diagonal (Figure 19.9). Since the lines τ_{xy} = constant and $(\sigma_x - \sigma_y)$ = constant depend on the directions chosen for the x, y axes, the diagonals of the squares should be parallel to the same axes.

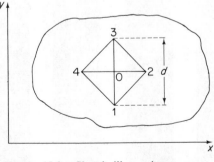

Figure 19.9. Sketch illustrating separation of principal stresses by the iteration method

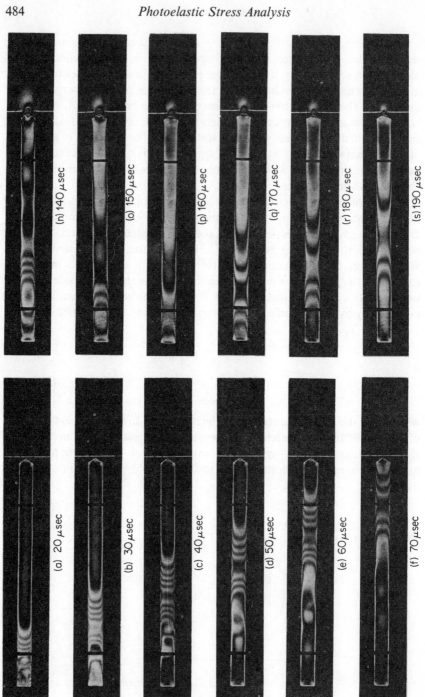

(n) 140 μsec

(o) 150 μsec

(p) 160 μsec

(q) 170 μsec

(r) 180 μsec

(s) 190 μsec

(a) 20 μsec

(b) 30 μsec

(c) 40 μsec

(d) 50 μsec

(e) 60 μsec

(f) 70 μsec

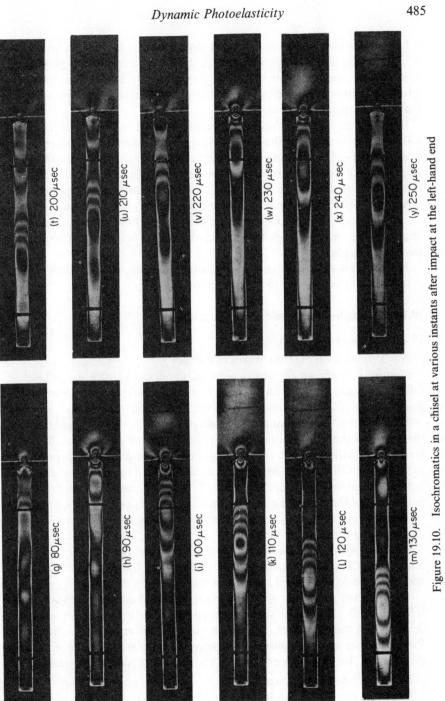

(t) 200 μsec

(u) 210 μsec

(v) 220 μsec

(w) 230 μsec

(x) 240 μsec

(y) 250 μsec

(g) 80 μsec

(h) 90 μsec

(i) 100 μsec

(k) 110 μsec

(l) 120 μsec

(m) 130 μsec

Figure 19.10. Isochromatics in a chisel at various instants after impact at the left-hand end

For separating the principal stresses in transversely bent plates, equations (12.10) can be used for both static and dynamic cases. This method is an approximate one, the accuracy depending upon the influence of the shear forces. In dynamic problems, short stress waves produce relatively higher shear forces than longer ones.

In problems of superposition of membrane and bending stresses in thin-walled structures such as shells and folded plates, separation of the principal stresses can be effected using separate procedures for the two partial stresses. The principal bending stresses can be separated by the method used for transversely bent plates, i.e. equations (12.10) can be applied as in static cases. Separation of the principal membrane stresses by an analytical method would, however, be complicated. The method of oblique incidence would appear to be easier though the procedures of the *j*-circle (see Chapter 13) have to be applied.

The problem of separating principal stresses in general three-dimensional cases does not yet appear to have been solved.

19.16   Examples of dynamic photoelasticity

19.16.1   *Photographic method*

A.   Propagation of stress waves in a chisel

Figure 19.10 shows the stress patterns for a model of a long chisel at different instants after an impact. These photographs were obtained by the single shot method from repetitions of the experiment.

The impact gives rise to a compression pulse which is propagated through the length of the chisel. Although the pointed (right-hand) end of the chisel is in contact with the plate from the beginning, the impact produces no effect in the plate until the pulse arrives at this end. Part of the pulse is then transferred to the plate while the rest is reflected. The curvature of the isochromatics indicates that the reflected pulse is tensile. This change of sign always occurs at a free end of a rod while, at a fully clamped end, pulses are reflected without change of sign.

Since the stiffness of the tapered part of the chisel is less than that of the remainder, the point remains compressed against the plate even after the end of the pulse has moved away from this end.

The reflected pulse is propagated to the left-hand end of the chisel where it is again reflected. As the sign changes again with this reflection, a second compression pulse—but of reduced amplitude—arrives at the pointed end. Part of the energy of this pulse is transferred to the plate while the remainder is reflected. Thus, repeating this cycle, the pulse is gradually attenuated.

B.   Diffraction of stress waves

Huygens' principle postulates that diffraction of stress waves may occur, similar to that experienced by other forms of wave. Thus, an abrupt change of section in a solid produces an effect on a stress wave analogous to that of a slit in a beam of light. Unlike light or sound waves, however, which maintain their original character after diffraction, a longitudinal stress wave is partially transformed by diffraction into a transverse wave and vice versa. The amplitudes of the longitudinal and transverse components produced by diffraction depend on the direction of propagation. Figure 19.11 shows the amplitude

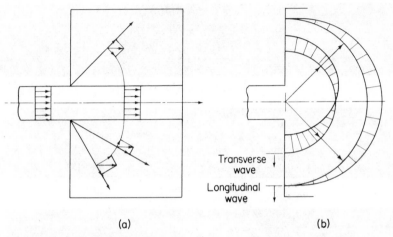

(a)   (b)

Figure 19.11. Diffraction of a longitudinal stress wave at a sudden enlargement of cross-section, (*a*) longitudinal and transverse components produced in different directions, (*b*) distributions of amplitude of longitudinal and transverse component waves

distributions of the longitudinal and transverse component waves produced by the diffraction of a longitudinal wave at a sudden enlargement of section; the amplitude of the longitudinal component varies from a maximum in the original direction of propagation to a minimum in the perpendicular direction while the amplitude of the transverse component varies in the opposite manner.

Figure 19.12 shows the isochromatics in a bar with an enlarged central section at different instants after impact at the left-hand end. The departure from linear propagation of the originally longitudinal wave at the changes of section are clearly demonstrated.

19.16.2 *Photometric method. Stresses in a prismatic bar due to longitudinal impact of a falling weight*

A mass of 107 gm was allowed to fall from a height of 50 mm on to the end of a vertical Perspex bar. The bar was of uniform rectangular cross-section having dimensions 10 mm × 9·54 mm, length 200 mm and mass 21·7 g.

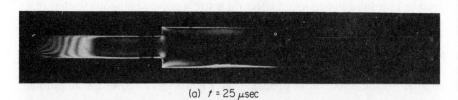

(a) $t = 25\,\mu$sec

(b) $t = 50\,\mu$sec

(c) $t = 75\,\mu$sec

(d) $t = 100\,\mu$sec

(e) $t = 125\,\mu$sec

(f) $t = 150\,\mu$sec

Figure 19.12.   Isochromatics in a bar with an enlarged central section at different instants

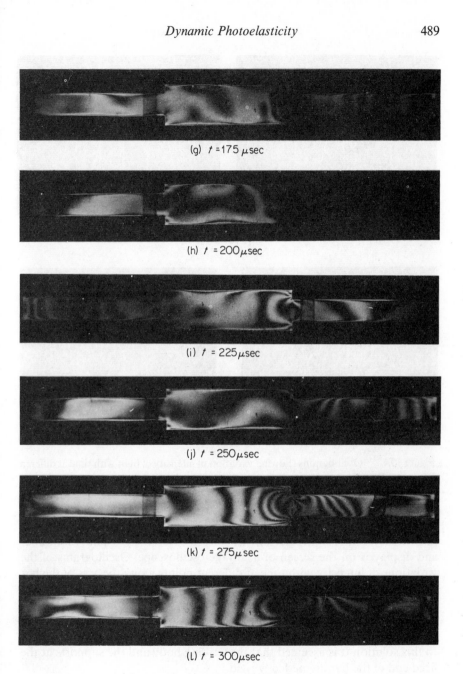

(g) $t = 175 \mu sec$

(h) $t = 200 \mu sec$

(i) $t = 225 \mu sec$

(j) $t = 250 \mu sec$

(k) $t = 275 \mu sec$

(L) $t = 300 \mu sec$

after impact at the left-hand end showing diffraction of an originally longitudinal wave

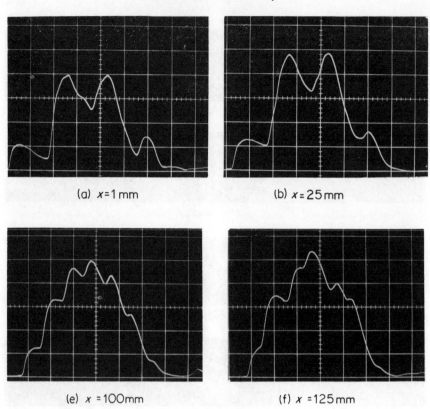

(a) *x* = 1 mm   (b) *x* = 25 mm

(e) *x* = 100 mm   (f) *x* = 125 mm

Figure 19.13. Oscillograms showing variation of intensity of light with time at different

Crossed polaroids were set at 45° to the axis of the bar and the variations of intensity of the light transmitted were measured by a photomultiplier and displayed on the screen of a storage oscilloscope. Oscillograms of the variations of intensity with time at different points along the axis of the bar are shown in Figure 19.13. As explained in Section 19.12, these oscillograms can be regarded as distorted stress diagrams.

A theoretical solution of the problem of the stresses in prismatic bars due to the longitudinal impact of a striking body was obtained by Boussinesq.[78] In this solution it is assumed that the striking body and the supports at the fixed end of the bar are perfectly rigid.

The downward velocity of the particles at the upper end of the bar at the instant of contact is equal to the velocity v_0 of the falling weight at this instant (Figure 19.14). This velocity is transmitted downwards through the particles of the bar with the velocity $c = \sqrt{(E/\rho)}$ of elastic stress waves. The initial

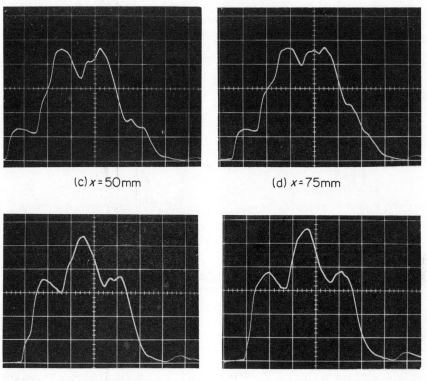

(c) $x = 50$mm   (d) $x = 75$mm

(g) $x = 150$mm   (h) $x = 175$mm.

points along the axis of a bar subjected to impact loading at one end and fixed at the other end

rate of change of momentum is therefore $\rho A c v_0$ where ρ is the density and A is the cross-sectional area of the bar. The equation of motion is

$$\sigma_0 A = \rho A c v_0,$$

where σ_0 is the stress at the upper end of the bar at the instant of contact. We thus obtain

$$\sigma_0 = \rho c v_0. \tag{19.50}$$

Following the initial contact, the velocity of the weight and hence of the particles at the top of the bar diminishes due to the resistance of the bar. If v is the velocity at time t after the instant of contact, the corresponding stress is

$$\sigma = \rho c v. \tag{19.51}$$

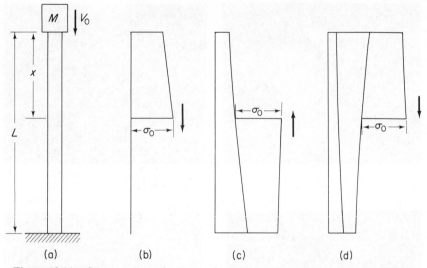

Figure 19.14.   Stresses at a point in a prismatic bar due to the impact of a falling weight, (*a*) instant of impact, (*b*) arrival of initial wave, (*c*) arrival of first reflected wave, (*d*) arrival of second reflected wave

Denoting by M the mass of the falling weight, its equation of motion is

$$-M\frac{dv}{dt} = \sigma A. \tag{19.52}$$

From equation (19.51) we have

$$\frac{dv}{dt} = \frac{1}{\rho c}\frac{d\sigma}{dt},$$

so that equation (19.52) can be written

$$\frac{d\sigma}{dt} + \frac{mc}{M}\sigma = 0,$$

where $m = \rho A$ is the mass of the bar per unit length. The solution of this equation is

$$\sigma = \sigma_0 \, e^{-(mct/M)} = \rho c v_0 \, e^{-(mct/M)}.$$

A compression pulse of diminishing amplitude is thus transmitted down the bar. This pulse travels a distance x in time x/c (Figure 19.14(*b*)). The stress at a depth x at time t is therefore the same as the stress at $x = 0$ at time $t - x/c$, i.e.

$$\sigma = \sigma_0 \, e^{-(mc/M)(t-(x/c))} = \rho c v_0 \, e^{-(mc/M)(t-(x/c))}$$

This equation is valid in the range $x/c < t < (2L - x)/c$. At time $t = L/c$, the wavefront arrives at the lower end of the bar. The pulse is then reflected upwards with its sign unchanged and the reflected wavefront arrives at the point x at time $t = (2L - x)/c$. The stress at this point thus rises suddenly by the amount σ_0 (Figure 19.14(c)).

After time $t = 2L/c$, the reflected pulse arrives at the upper end of the bar. Since the velocity of the mass M cannot change suddenly, this end also behaves as a fixed end and the pulse is again reflected without change of sign. The front of this downward reflected pulse arrives at the point x after time $t = (2L + x)/c$ when the stress again increases suddenly by σ_0 (Figure 19.14(d)). This process is continued with further reflections from opposite ends of the bar arriving at the point considered at time intervals of $2x/c$ and $2(L - x)/c$ respectively. The stress at any instant is due to the combined effects of the direct wave originating at M and the multiple reflections and diminishes to zero at the end of contact.

As can be seen from Figures 19.13, at points near the ends of the bar the reflections from opposite ends approximately coincide producing sudden large increases of stress. At the centre of the bar, reflections from opposite ends arrive at regular time intervals of L/c. In this region the number of fluctuations is twice as great but these are reduced in magnitude. The oscillograms also show that the amplitudes of successive reflections attenuate fairly rapidly. This is due to the transmission of some of the energy to the contacting bodies at the ends of the bar with each reflection and to internal friction.

20

SPECIAL APPLICATIONS
OF PHOTOELASTICITY

20.1 Photoelastic strain gauges

The photoelastic strain gauge, originally patented by Golubovic and later developed by Oppel,[79] consists of a small piece of photoelastic material provided with a reflective coating on one side and covered with a polarizing film on the other. In use, the gauge is bonded to the surface of the part to be tested with its reflective surface towards the part. When the part is loaded, its surface strains are communicated to the gauge and produce an isochromatic fringe pattern or cause a displacement of isochromatics previously frozen into the material from which information regarding the strain in the part can be deduced.

The principal advantage of the photoelastic strain gauge is that it is direct reading and self-contained, no other instrumentation being required. Disadvantages include limited accuracy in regions of high strain gradient as well as those common to birefringent coatings in general, e.g. errors arising from the reinforcing effect of the gauge and its distance from the neutral surface when applied to thin members in bending.

Two basic forms of gauge are commercially available, one of which is applicable to the measurement of uniaxial strain while the other is intended for use in biaxial fields.

20.1.1 *Uniaxial strain gauge*

The birefringent element of the uniaxial gauge, Figure 20.1, consists of a thin rectangular strip of epoxy resin in which two or three uniformly spaced isochromatics have previously been frozen in a direction perpendicular to its length. Such a strip may be obtained by cutting a thin slice from a uniform beam in which the isochromatics produced under pure bending have been frozen as indicated in Figure 20.2.

In use, the gauge is bonded to the test surface using cold setting epoxy adhesive at its ends only to avoid the effects of lateral sensitivity. To ensure no

494

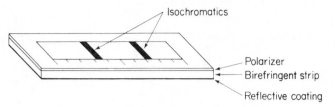

Figure 20.1. Photoelastic uniaxial strain gauge

adhesion of the central part of the gauge, it may be backed by a thin rubber or plastic film.

The strain is determined from the lateral displacement of the isochromatics which occurs when the part is loaded and which is read from a scale incorporated in the gauge. The isochromatic pattern in the unstressed gauge

Figure 20.2. Sketch illustrating method of obtaining a birefringent strip with frozen-in isochromatics for use as a uniaxial strain gauge

corresponds to a stress σ_y acting in the direction perpendicular to its length, the magnitude varying linearly along the gauge. The difference of σ_y corresponding to unit difference of fringe order is $f/2d$. If the spacing between the isochromatics is L, the stress gradient is therefore

$$d\sigma_y/dx = f/2dL.$$

Let the stress corresponding to an isochromatic of arbitrary order n coinciding with a line A, Figure 20.3(a), in the unloaded gauge be σ_{yA}. If a uniform stress σ_x is now superimposed, Figure 20.3(b), the same fringe order will appear on a line B where

$$\sigma_{yB} - \sigma_x = \sigma_{yA}.$$

Hence,

$$\sigma_x = \sigma_{yB} - \sigma_{yA}.$$

If x is the distance between the lines A and B, i.e. the displacement of the isochromatics when the part is loaded, then

$$\frac{(\sigma_{yB} - \sigma_{yA})}{x} = \frac{f}{2dL},$$

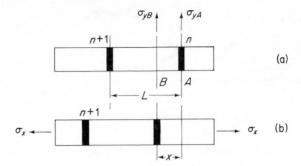

Figure 20.3. Sketch illustrating method of determining stress from the movement of the isochromatics in a
uniaxial strain gauge showing gauge, (*a*) before loading,
(*b*) after loading

i.e.

$$\sigma_x = fx/2dL,$$

or, in terms of the strain,

$$\epsilon_x = fx/2dLE.$$

The sign of the strain is indicated by the direction of movement of the fringes.

Uniaxial strain gauges have a strain sensitivity of between 20 and 80 micro-inches per inch. They may be used to measure the principal strains in a biaxial field provided the directions of the principal axes of strain are known. These directions may be determined using the biaxial type of gauge or by other means.

20.1.2   *Biaxial strain gauge*

In the biaxial gauge, the birefringent element is a thin circular disc with a central hole to act as a stress raiser. Biaxial gauges are manufactured with and without frozen-in concentric circular isochromatics such as would be produced in the element by a hydrostatic stress field. In use, the gauge is cemented to the surface of the part round its periphery only.

A gauge with no initial isochromatic pattern will exhibit one similar to those of Figure 20.5 when the part is loaded. The magnitudes of the principal strains may be obtained by comparing the isochromatic pattern with similar patterns produced under known conditions of strain.

In a gauge with a frozen-in pattern, the circular fringes become distorted under strain, Figure 20.4. The magnitudes of the principal strains are a function of the radial displacement of the fringes on the axes of symmetry

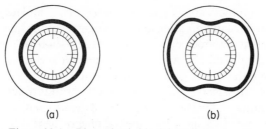

Figure 20.4. Photoelastic biaxial strain gauge with frozen-in isochromatics, (*a*) before loading, (*b*) after loading

and can be read from a graph supplied by the manufacturer. This type of gauge has a strain resolution of approximately 40 micro-inches per inch.

In both forms of gauge, the principal axes of strain coincide with the axes of symmetry of the fringe pattern; their directions can be read to within about 5° from a circular scale attached to the gauge.

20.2 Photoelastic insertion stressmeters

The state of stress in rock masses and in brickwork and concrete structures may be determined from in situ measurements using photoelastic stressmeters. The stressmeter consists of a short solid or thick hollow cylinder of birefringent material which is inserted in a borehole and bonded to the material to be studied. Any change of stress in the surrounding material due to the application of external forces or a redistribution of internal forces causes stresses in the meter. From the optical effects produced in circularly polarized light, it is possible to deduce the state of stress in the surrounding material.

The basic theory of the method—that of a rigid inclusion in an elastic medium—has been developed by several investigators, notably Hiramatsu[80] and Barron.[81] Further developments and applications of the method have been described by Coutinho,[82] Roberts and Hawkes,[83] Dhir[84] and others.

The difference of principal stresses and hence the isochromatic fringe order at any point of the stressmeter depend on its dimensions and physical properties and on those of the surrounding material. Provided that the elastic modulus of the stressmeter is several times greater than that of the surrounding material, however, the stressmeter behaves as a rigid inclusion and the stresses produced within it are practically independent of the elastic modulus of the surrounding material. The sensitivity of the meter is then independent of any non-linearity of the stress-strain relation of the surrounding material. For this reason, and since the level of loading to be measured is usually high, photoelastic stressmeters are normally made of optical glass ($E \simeq 70 \text{ GN/m}^2$).

When it is impracticable to form a borehole completely through the structure, a source of illumination and a circular polarizer may be attached to the stressmeter at its inner end to allow observation of the isochromatic pattern by transmitted light. Alternatively, the rear surface of the meter may be mirrored and the effects observed by reflected light.

The type of stressmeter used depends on the information required. If only the magnitude of the shear stress is to be determined, a meter without a central hole, known as a shear stressmeter, may be used. When inserted in a material in which the state of stress is uniform, the stress and hence the isochromatic fringe order over the whole field of the meter is also uniform. The required shear stress can be determined from the isochromatic order or the characteristic colour produced.

If more detailed information regarding the state of stress is required, a meter with a central circular hole, known as a biaxial stressmeter, is used. In such a meter, a characteristic isochromatic fringe pattern is produced by any change of stress in the surrounding medium. The directions of the principal stresses are indicated by the axes of symmetry of this pattern while the individual stress to which each relates can be distinguished from its shape. The procedure for determining the magnitude of the principal stresses depends on whether the state of stress is uniaxial or biaxial.

20.2.1   *Uniaxial stress*

The magnitude of the stress in a uniaxial field is determined by measuring the change of isochromatic fringe order at a particular point of the stressmeter. The sensitivity of the meter varies, of course, from point to point and the point of observation should be selected such that the sensitivity is as high as possible while its rate of change with radius is small. Optimum points occur on the axes of symmetry and on the bisectors of the angles between these axes at a certain distance from the inner edge depending on the dimensions.

To calibrate the stressmeter, it is bonded within a column or slab of material which is subjected to uniform uniaxial compression in the laboratory. The factor of proportionality relating the isochromatic fringe order at the point of observation and the stress in the material is determined from measurements of these quantities with different values of the load. Provided that the ratio of the elastic moduli of the stressmeter and the material is such that the stressmeter can be regarded as a rigid inclusion, it is unnecessary that it be calibrated in the same material as the structure or even that the elastic modulus of the structure be precisely known. When the elastic moduli are approximately equal, however, the same material must be used.

20.2.2   *Biaxial stress*

It can be shown that, in a biaxial stress field, the factor of proportionality relating the isochromatic fringe order at a point of the stressmeter and, say, the greater principal stress in the surrounding material depends on the ratio of the two principal stresses. It follows that, in order to separate the principal stresses, their ratio must first be determined. This may be accomplished by comparing the fringe pattern with that produced under known conditions of loading. For this purpose, a portable comparator which allows uniform compressive stresses to be applied to a calibration stressmeter in two perpendicular directions may be used, the magnitudes of the loads being adjusted independently until an identical fringe pattern is obtained. When a lesser degree of accuracy will suffice, the fringe pattern may be compared with a series of 'standard' fringe patterns of the type shown in Figure 20.5, that resembling the actual isochromatic pattern most closely providing the required value of σ_2/σ_1.

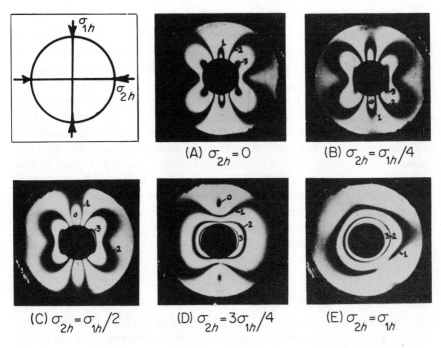

(A) $\sigma_{2h} = 0$ (B) $\sigma_{2h} = \sigma_{1h}/4$

(C) $\sigma_{2h} = \sigma_{1h}/2$ (D) $\sigma_{2h} = 3\sigma_{1h}/4$ (E) $\sigma_{2h} = \sigma_{1h}$

Figure 20.5. Isochromatic patterns for a biaxial stressmeter at third order fringe. Courtesy Dhir

Other methods of determining σ_2/σ_1 have been suggested, for example, from measurements of the birefringence at two points of the stressmeter or from the spacing of the isotropic points which appear within the range $0.2 < \sigma_2/\sigma_1 < 0.9$.

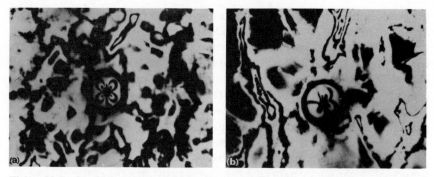

Figure 20.6. Isochromatic patterns for concrete slabs with birefringent coatings and embedded stressmeters, (a) aggregate size less than the diameter of the stressmeter, (b) aggregate size greater than the diameter of the stressmeter. Courtesy Dhir

Figure 20.6, kindly provided by Dr R. K. Dhir, illustrates an application of the biaxial stressmeter to the study of the effect of aggregate size in concrete. The aggregate size determines the area over which the stresses must be integrated in order to measure the average level of stress.

Square slabs were cut from cubes cast with various aggregate sizes. Stressmeters were set in the slabs and birefringent coatings were bonded to their faces. Figure 20.6 shows the isochromatic patterns in slabs under uniaxial compression having aggregate sizes less and greater, respectively, than the diameter of the meter. The unsymmetrical fringe pattern of Figure 20.6(b) indicates that the meter is situated in a non-uniform stress field.

The general conclusions of this study were:

1. If the meter diameter is greater than the maximum aggregate size, its readings are proportional to the average stress level.

2. The stressmeter produces no significant stress concentration and acts as a piece of aggregate in the concrete.

20.3 Application to non-cohesive masses

An interesting application of photoelasticity to the study of load transmission through non-cohesive masses has been described by Dantu.[85] Masses possessing both two- and three-dimensional characteristics were investigated.

(a) Two-dimensional masses. The mass consisted of a pile of small Pyrex glass cylinders laid in rows to form a vertical wall within a rectangular cell having thick glass sides. For part of the investigation, a mixture of cylinders of length 2 cm and diameters varying between 2 and 6 mm was used. This random array—statistically homogeneous and isotropic—could be roughly compared with 'two-dimensional sand'. The pile was subjected to a central vertical load applied at a certain depth below the surface, corresponding to a foundation pile sunk in soil. The upper surface of the 'sand' was constrained by means of an elastically supported cover plate.

The optical effects produced in circularly polarized light when viewed in the direction parallel to the axes of the cylinders are illustrated in Figure 20.7. As can be observed, the load is transmitted along lines of cylinders forming a network resembling the roots of a plant. Cylinders outside these lines are only lightly loaded.

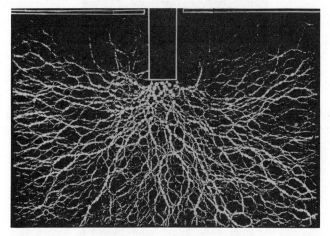

Figure 20.7.   Two-dimensional 'sand' subjected to a vertical load below the surface: Photoelastic observation in circularly polarized light. Courtesy Dantu

It was found that the system of illuminated lines did not change as the load was increased until a state of limiting equilibrium had been passed. Rupture in the form of relative movement of the cylinders, sinking of the central load and disturbance of the lines of load transmission then took place.

(b) Three-dimensional masses. The investigation was extended to the study of a three-dimensional granular medium by using 'sand' consisting of crushed and sieved Pyrex glass. A liquid having the same refractive index as the glass was introduced into the cell to render the contents transparent. The sand was loaded in compression at its upper surface in the manner indicated in Figure 20.8(b).

The optical effects produced in circularly polarized light are shown in Figure 20.8(a). As in the two-dimensional case, it can be observed that the load is transmitted along lines through the sand. Instead of being in disorder, however, the network of lines now conforms closely with a regular system of stress trajectories. Although barely visible in the photograph, the orthogonal network of stress trajectories could also be clearly seen by eye.

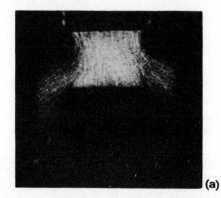

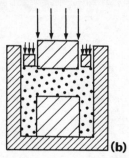

Figure 20.8. Glass 'sand' loaded in compression at its surface, (*a*) photoelastic observation in circularly polarized light, (*b*) diagram of loading. Courtesy Dantu

When viewed in plane polarized light, broad dark bands corresponding to the isoclinics in ordinary two-dimensional photoelasticity could be observed. The isoclinic pattern was traced and from this the network of stress trajectories was drawn in the usual way. The results were found to agree closely with those indicated by Figure 20.8(*a*).

20.4   Plastic and elastoplastic stresses and strains

In both two- and three-dimensional problems, the conventional photoelastic models and procedures suffice to indicate whether or not the yield point of the material of the prototype will be exceeded. By employing suitable techniques and materials, however, photoelastic investigations can be extended to determine the amount of plastic strain and its influence in the elastically deformed region in elastoplastic stress fields. Such investigations may be conducted using either (*a*) birefringent coatings applied to the surfaces of the prototype, or (b) a model similar to that used in conventional photoelastic investigations but in general manufactured from a material having special photoplastic properties.

20.4.1 *Birefringent coatings*

For most photoelastic materials the stress-strain-fringe relation is linear over a range of strain much greater than that at which yielding will occur in metals. The deformation of the prototype may therefore be partly elastic and partly plastic while that of the coating is purely elastic. At any point however, the strain of a properly bonded coating will equal that of the surface of the prototype within the usual limits of accuracy of the method. The principal stresses in the coating can be determined by one of the usual methods. If the stress-strain relation of the prototype material for the actual loading conditions is known, the stresses in the prototype can be read from a combined stress-strain diagram for the two materials in which the scales of strain are the same for each while the stress scales are chosen such that, within the elastic region, both diagrams coincide, Figure 20.9. The maximum stress in the prototype is determined from that in the model in the manner indicated.

In Figure 20.9, the variations of stress and strain at a point within the plastic region of the prototype are represented by the curve *OB*. Unloading is represented by the line *BC*, which indicates that residual compressive stresses are produced in the region where plastic flow occurred. Equilibrating tensile stresses are produced in the surrounding region where the deformation under load was purely elastic. These residual stresses arise from the differential

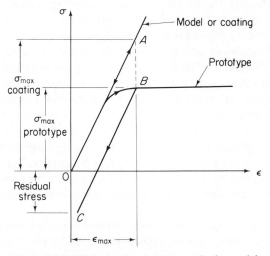

Figure 20.9.   Relation between stresses in the model
and prototype

contractions of the elements on unloading, elements stressed elastically tending to return to their original dimensions while those stressed beyond the elastic limit tend to return to increased dimensions. Such stresses are of special significance in relation to the fatigue strength of the prototype. If the residual stress does not exceed the yield stress (in compression), repeated loading and unloading will simply reproduce the conditions represented by *CB* and *BC*, i.e. no further yielding will take place. This means that fracture by fatigue will not occur provided that the cyclic amplitude of stress is within the fatigue limit. On the other hand, if the residual stress after unloading exceeds the elastic limit, or if the prototype is subjected to cyclic reversed loading, plastic deformation is repeated with each loading cycle and rupture will occur in a few cycles.

20.4.2 *Model methods*

If the region in which yielding occurs in the prototype is small, its influence on the stresses in the elastically deformed region is negligible. In such cases, approximate results can be obtained using a model made of an orthodox photoelastic material. The stresses and strains in the elastic zone can be determined from the isochromatics and isoclinics in the usual way. Approximate values of the stresses in the plastic zone can be determined from the model values using a diagram similar to that of Figure 20.9. The usual rules for the transition of the model results to the prototype must of course be observed.

If the region of plastic deformation is relatively large, it will influence the stress distribution in the elastic region. Problems of this nature may be investigated using models made of materials having suitable photoplastic (creep) characteristics. The stress-strain and stress-fringe curves for such materials are greatly dependent on the time after loading.

Hiltscher[86] employed polystyrene for the investigation of elastoplastic problems. In this material, the elastic and plastic components have stress optic coefficients of opposite sign. The result is that the fringe order increases linearly with stress up to the elastic limit then diminishes with the onset of plastic strain, becoming zero when a state of full plasticity is reached; with further deformation, it increases with reversed sign. These properties allow the limits of the purely elastic and completely plastic zones to be located by direct observation in the polariscope, the first being characterized by maximum and the second by zero fringe order. Polystyrene behaves in the above manner only in compression, and is unsuitable for the investigation of tensile stress systems.

A procedure which yields more detailed information on the conditions in elastoplastic stress fields has been developed by Mönch[87,88] using celluloid (cellulose nitrate) as a material. In common with other photoplastic materials, the stress-strain and fringe-strain diagrams depend greatly on the loading time and hence on the strain rate. To overcome this problem, the model and calibration pieces are loaded at constant strain rates so far as this is possible.

As a calibration test, a number of specimens are loaded in tension at different constant strain rates $\dot{\epsilon}$ and their stress-strain diagrams are drawn, Figure 20.10(a). An 'effective' or 'resulting' curve is then drawn through the

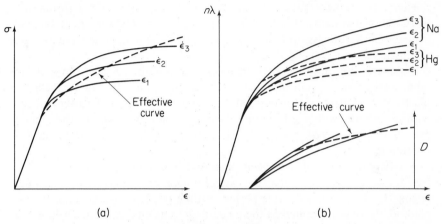

Figure 20.10.   Determination of the effective curves for (a) strain and (b) dispersion at a given time after loading (after Mönch)

points on each diagram which represent a constant loading time t, chosen to agree with the loading time of the model when the isochromatics are recorded. This resulting curve, relating only to the time t and independent of the strain rate, must be similar to the stress-strain diagram of the prototype material.

From the same calibration tests, curves of the retardation $n\lambda$ versus strain are also determined using light of two different wavelengths, e.g. sodium and mercury blue. Photoelastic dispersion, which is negligible up to the yield point, causes curves relating to the same strain rate to diverge with increasing plasticity as shown in Figure 20.10(b), and is thus a measure of plastic deformation. Curves of the dispersion D, arbitrarily defined by

$$D = \frac{(n\lambda)_{\text{Na}} - (n\lambda)_{\text{Hg}}}{(n\lambda)_{\text{Na}}} 100\%,$$

are plotted for the different strain rates using corresponding pairs of values read from the $n\lambda$ curves. An effective curve for D pertaining to the test time t is then drawn in a similar manner as the effective curve for stress.

To approximate to the condition of constant strain rate in the model, the deformation at the loading point, or preferably at the point of maximum strain, is maintained constant. The isochromatics for both wavelengths of light are recorded at the chosen time t after commencement of the loading. The difference of isochromatic order at any point allows the dispersion to be determined. Lines of constant dispersion which, as shown by Jira,[88] are also lines of constant maximum principal strain, may then be drawn. These lines may also be regarded as lines of constant plastic deformation, in particular, the line $D = 0$ marking the boundary between the elastic and plastic regions. With uniaxial systems, the value of the strain ϵ at the test time t can be read from the effective curve for D and the associated stress from the effective stress-strain curve.

The method requires that the strain rate, although varying from point to point in the model, shall be constant at any one point. In a general elasto-plastic state of stress this is impossible to achieve, but is sufficiently realized with the above method of loading if elastic conditions exist in the greater part of the field. With essentially plastic systems, however, the method is unreliable.

A different approach has been proposed by Frocht and Thomson[90,91] based on the observation that the rate of optical creep of celluloid becomes negligible after a considerable period of time under constant load. In this method, the load is applied by increments up to its final value according to a program which is repeated in the calibration tests and is then maintained constant for a considerable period. The isochromatic fringe orders and the isoclinics are recorded when variations in their values due to creep are insignificant within the period of time required for this to be done. The above authors employed an initial loading period of $1\frac{1}{2}$ hr and the data were recorded after a total of 5 hr.

For the purpose of calibration, creep curves of birefringence, Figure 20.11(a) and strain, Figure 20.11(b), are determined for different values of the stress

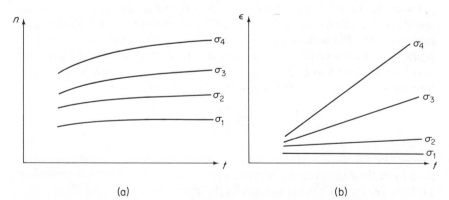

(a)   (b)

Figure 20.11.   Creep curves of (a) birefringence and, (b) strain for celluloid (after Frocht and Thomson)

from a series of tensile tests in each of which the stress is kept constant. From these curves the stress-fringe and stress-strain curves are constructed (Figures 20.12(*a*) and 20.12(*b*) respectively) for the time when the data are recorded in the main investigation.

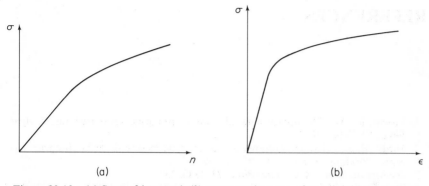

Figure 20.12   (*a*) Stress-fringe and, (*b*) stress-strain curves for celluloid for a given time after loading (after Frocht and Thomson)

For uniaxial states of stress, the stresses can be read from Figure 20.12(*a*) using the measured values of the birefringence. The corresponding strains may then be read from Figure 20.12(*b*).

These investigators also showed that, for the same difference of principal stresses, the stress optic law for biaxial stresses was virtually the same as that for uniaxial stresses. It was also shown that the isoclinics indicate the directions of the principal stresses in elastoplastic as well as in elastic stress fields. It is therefore possible to separate the principal stresses in biaxial elastoplastic fields by the shear difference method.

REFERENCES

1. Frocht, M. M., The optical determination of isopachic stress patterns, *J. Appl. Phys.*, **10**, 1939, 248–257.
2. Tardy, H. L., Méthode pratique d'examen et de mesure de la birefringence des verres d'optique, *Rev. Opt.*, **8**, 1929, 59–69.
3. Senarmont, H. de, *Ann. Chim. Phys.*, **73**, 1840, 337.
4. Post, D., Isochromatic fringe sharpening and fringe multiplication in photoelasticity, *Proc. S.E.S.A.*, **12**(2) 1955, 143–156.
5. Post, D., Fringe multiplication in 3-dimensional photoelasticity, *J. Strain Analysis*, **1**, 1966, 380–388.
6. Post, D., Photoelastic fringe multiplication—for tenfold increase in sensitivity, *Exp. Mech.*, **10**(8), 1970, 305–312.
7. Fessler, H., and Lewin, B. H., A study of large strains and the effect of different values of Poisson's ratio. *B.J.A.Ph.*, **11**, 1960, 273–277.
8. Neuber, H., *Kerbspannungslehre* (Zweite auflage), Springer, Berlin, 1958. English translation: *Theory of notch stresses*, J. W. Edwards, Inc., Ann Arbor, Michigan, 1946.
9. Heywood, R. B., *Designing against fatigue*, Chapman & Hall, London, 1962.
10. Peterson, R. E., *Stress concentration design factors*, John Wiley & Sons, New York, 1953.
11. Siebel, E., Neue wege der festigkeitsberechnung, *Z.V.D.I.*, **90**, 5, 1948, 135–139.
12. Leven, M. M., *Epoxy resins for photoelastic use*. Proc. international symposium on photoelasticity, Frocht, M. M. ed., Pergamon Press, New York, 1962, 145–165.
13. McConnel, L. D., *Practical techniques for photoelastic analysis*. Inst. Phy. & Physical Soc. conference, London, 1965.
14. Durelli, A. J., Lake, R. L., and Phillips, E., Stress concentrations produced by multiple semi-circular notches in infinite plates under uniaxial state of stress. *Proc. S.E.S.A.*, **10**, 1, 1952, 53–64.
15. Durelli, A. J., and Riley, W. F., *Introduction to photomechanics*, Prentice-Hall, Inc., Englewood Cliffs, N.J., 1965, 115–146.
16. Föppl, L., and Mönch, E., *Praktische spannungsoptik*, Springer, Berlin, 1959, 107.
17. Vadovic, F., *Eine neue methode zur trennung der hauptspannungen*. International symposium on photoelasticity, Berlin, 1961.
18. Hiltscher, R., *Development of the lateral extensometer method in two-dimensional photoelasticity*. Proc. international symposium on photoelasticity, Frocht, M. M., ed., Pergamon Press, New York, 1962, 43–56.
19. Post, D., A new photoelastic interferometer suitable for static and dynamic measurements, *Proc. S.E.S.A.*, **12**, 1, 1954, 99–116.

20. Nisida, M., and Saito, H., A new interferometric method for two-dimensional stress analysis, *Exptl. Mech.*, **4**, 12, 1964, 366–376.
21. Nisida, M., and Saito, H., Application of an interferometric method to studies of contact problems, *Sci. Papers Inst. Phys. Chem. Res., Tokyo*, **59**, 3, 1965, 112–123.
22. Nisida, M., and Saito, H., Stress distribution in a semi-infinite plate due to a pin determined by interferometric method, *Exptl. Mech.*, **6**, 1966.
23. Favre, H., and Schumann, W., *A photometric-interferometric method to determine separately the principal stresses in two-dimensional states and possible applications to surface and thermal stresses*, Proc. international symposium on photoelasticity, Frocht, M. M., ed., Pergamon Press, New York, 1962, 3–25.
24. Favre, H., Sur une nouvelle méthode optique de détermination des tensions intérieures, *Revue d'Optique*, **8**, 1929, 193–213, 241–261, 289–307.
25. Gabor, D., Microscopy by reconstructed wave-front, *Proc. R. Soc., A* **197**, 1949, 454.
26. Leith, E., and Upatnieks, J., Wavefront reconstruction with diffused illumination and three-dimensional objects, *J. Opt. Soc. Am.*, **54**, 1964, 1295.
27. Fourney, M. E., Application of holography to photoelasticity, *Exptl. Mech.*, **8**, 1, 1968, 33–38.
28. Hosp, E., and Wutzke, G., Die anwendung der holographie in der ebenen spannungsoptik, *Materialprüfung*, **11**, 12, 1969, 409–415.
29. Hosp, E., and Wutzke, G., Holographische ermittlung der hauptspannungen in ebenen modellen, *Materialprüfung*, **12**, 1, 1970, 13–22.
30. Hovanesian, J. D., Brcic, V., and Powell, R. L., A new experimental stress-optic method, stress-holo-interferometry, *Exptl. Mech.*, **8**, 8, 1968, 362–368.
31. Hovanesian, J. D., and Zobel, E. C., *Application of photo-holoelasticity in the study of crack propagation*, Proc. Seventh Symp. Non-destructive Evaluation of Components and Materials in Aerospace, Weapons Systems and Nuclear Applications, San Antonio, Texas, April, 1969.
32. Hovanesian, J. D., Application of photoholoeslasticity to frozen two-dimensional models with extension to three-dimensional analysis, *Strain*, **5**, 2, 1969, 84–87.
33. Hovanesian, J. D., New application of holography to thermoelastic studies, Proc. Int. Conference on Experimental Stress Analysis and its Influence on Design, Cambridge, England, 1970, 428–435.
34. Mönch, E., Die vollständige bestimmung des dehnungszustandes auf oberflächen durch photoelastische streifenschichten, *Schweizerische Bauzeitung*, **84**, 48, 1966.
35. O'Regan, R., New method for determining strain with photoelastic coatings, *Exptl. Mech.*, **5**, 8, 1965, 241–246.
36. Duffy, J., and Lee, T. C., Measurement of surface strains by means of bonded birefringent strips. *Proc. S.E.S.A.*, **18**, 2, 1961, 109–112.
37. Day, E. E., Kobayashi, A. S., and Larson, C. N., Fringe multiplication and thickness effects in birefringent coatings, *Exptl. Mech.* **2**, 4, 1962, 115–121.
38. Zandman, F., Redman, S., and Riegner, E., Reinforcing effect of birefringent coatings, *Exp. Mech.*, **2**, 2, 1962, 55–64.
39. Holister, G. S., *Experimental stress analysis*, Cambridge University Press, 1967, 220–223.
40. Post, D., and Zandman, F., Accuracy of birefringent coating method for coatings of arbitrary thickness. *Exp. Mech.*, **1**, 1, 1961, 21–32.
41. Lee, T. C., Mylonas, C., and Duffy, J. Thickness effects in birefringent coatings with radial symmetry, *Exp. Mech.*, **1**, 2, 1961, 134–142.
42. Duffy, J., and Mylonas, C., *Experimental study of thickness effects with birefringent coatings*, Proc. international symposium on photoelasticity, Frocht, M. M., ed., Pergamon Press, New York, 1963, 26–41.

References

43. Duffy, J., Effects of the thickness of birefringent coatings, *Exp. Mech.*, **1**, 3, 1961, 74–82.
44. Pih, H., and Knight, C. E., Photoelastic analysis of anisotropic fibre reinforced composites. *J. Compos. Mater.* **3**, 1969, 94–107.
45. Kedward, K. T., and Hindle, G. R. Analysis of strain in fibre-reinforced materials, *J. Strain Analysis*, **5**, 4, 1970, 309–315.
46. Tuzi, Z., and Oosima, H., On the artificial quarter-wave plate for photoelasticity apparatus and its theory, *Sci. Papers Inst. Phys. Chem. Res., Tokyo*, **37**, 1939, 72.
47. Jessop, H. T., and Wells, M. K., The determination of the principal stress differences at a point in a three-dimensional photoelastic model, *B. J. Appl. Phys.*, **1**, 1950, 184–189.
48. Nisida, M., *New photoelastic methods for torsion problems*, Proc. international symposium on photoelasticity, Frocht, M. M., ed., Pergamon Press, New York, 1963, 109–121.
49. Frocht, M. M., Torsional stresses by oblique incidence, *J. App. Mech.* **11**, 1944, A229: *Photoelasticity*, vol. 2, J. Wiley & Sons, New York, 1948, 393–419.
50. Hetenyi, M., Some applications of photoelasticity in turbine generator design. *Trans. A.S.M.E.*, **61**, 1939, A151; **62**, 1940, A80.
51. Newton, R. E., Photoelastic study of stresses in rotating discs, *Trans. A.S.M.E.*, **62**, 1940, A57, A174.
52. Reichner, P., Stress concentration in the multiple notched rim of a disk, *Proc. S.E.S.A.*, **18**, 2, 1961, 160–166.
53. Guernsey, R., Photoelastic study of centrifugal stresses in a single wheel and hub. *Proc. S.E.S.A.*, **18**, 1, 1961, 1–7.
54. Fessler, H., and Thorpe, T. E., Centrifugal stresses in rotationally symmetrical gas turbine discs, *J. Strain Analysis*, **3**, 1968, 135–141.
55. Hendry, A. W., Photoelastic experiments on the stress distribution in a diamond head buttress dam, *Proc. I.C.E.*, **3**, 1, 1954, 370–396.
56. Hendry, A. W., *Photoelastic analysis*, Pergamon Press, 1966, 136–140.
57. Rydzewski, J. R., Experimental method of investigating stresses in buttress dams, *Brit. J. Appl. Phys.*, **10**, 1959, 465–469.
58. Saad, S., and Hendry, A. W., Gravitational stresses in deep beams, *Structural Engr.*, **39**, 6, 1961, 185–194.
59. Weller, R., A new method for photoelasticity in three dimensions, *J. Appl. Phys.*, **10**, 4, 1939, 266.
60. Menges, H. J., Die experimentelle ermittelung räumlicher spannungszustände an durchsichtigen modellen mit Hilfe des Tyndalleffectes, *Z. angew. Math. Mech.*, **20**, 1940, 210–217.
61. Drucker, D. C., and Mindlin, R. D., Stress analysis by three-dimensional photoelastic methods, *J. Appl. Phys.*, **11**, 11, 1940, 724.
62. Frocht, M. M., and Srinath, L. S., *A non-destructive method for three-dimensional photoelasticity*, Proc. 3rd. U.S. Natl. Congr. Appl. Mech., 1958, 329–337.
63. Srinath, L. S., and Frocht, M. M., *Scattered light in photoelasticity—basic equipment and techniques*, Proc. 4th. U.S. Natl. Congr. Appl. Mech., 1962, 775–781.
64. Srinath, L. S. and Frocht, M. M., *The potentialities of the method of scattered light*, Proc. international symposium on photoelasticity, Frocht, M. M., ed., Pergamon Press, New York, 1963, 277–292.
65. Jessop, H. T., The scattered light method of exploration of stresses in two- and three-dimensional models, *Brit. J. Appl. Phys.*, **2**, 1951, 249–260.
66. Robert, A., and Guillemet, E., New scattered light method in three-dimensional photoelasticity, *Brit. J. Appl. Phys.*, **15**, 1964, 567–578.

67. Cheng, Y. F., Some new techniques for scattered light photoelasticity, *Exptl. Mech.*, **3**, 11, 1963, 275–278.
68. Cheng, Y. F., A dual-observation method for determining photoelastic parameters in scattered light, *Exptl. Mech.*, **7**, 3, 1967, 140.
69. Cheng, Y. F., An automatic system for scattered light photoelasticity, *Exptl. Mech.*, **9**, 9, 1969, 407–412.
70. Hertz, H., *Gesammelte Werke*, vol. 1., J. A. Barth, Leipzig 1895, 155.
71. Cranz, C., and Schardin, H., Kinematographie auf ruhendem film und mit extrem hoher bildfrequenz, *Zeits. f. Physik*, **56**, 1929, 147–183.
72. Tuzi, Z., and Nisida, M., Photoelastic study of stresses due to impact, *Phil. Mag.*, **21**, 1936, 448–473.
73. Frocht, M. M., and Flynn, P. D., Dynamic photoelasticity by means of streak photography, *Proc. S.E.S.A.*, **14**, 2, 1957, 81–90.
74. Allison, I. M., Inst. Phy. and Physical Soc. Stress Analysis Group Conference, University College, London, April 1967.
75. Schumann, W., and Haenggi, H., Über einen versuch zur vollständigen bestimmung zeitlich veränderlicher ebener spannungszustände mittels eines photoelektrisch-interferometrischen verfahrens, *Forschung in Ingenieurwesen*, **30**, 3, 1964, 78–85.
76. Haenggi, H., and Schumann, W., Sur une application du laser a la détermination des tensions interieures variables en function du temps; *C.R. Acad. Sc. Paris*, **259**, 1964, 2599–2602.
77. Bohler, P., and Schumann, W., On the complete determination of dynamic states of stress, *Exptl. Mech.*, **8**, 3, 1968, 115–121.
78. Boussinesq, J., Du choc longitudinal d'une barre prismatique fixee a une bout et heurtee a l'autre, *Compt. Rend.*, **97**, 1883, 154.
79. Oppel, G. U., Photoelastic strain gauges, *Exptl. Mech.*, **1**, 1961, 65–73.
80. Hiramatsu, Y., Measurement of variation in stress with a photoelastic stressmeter, *Laboratory of Mining Engineering Rep.*, *Kyotu Univ.*, March, 1964.
81. Barron, K., Glass insert stressmeters, *Report* FMP64/123-MRL, *Mines Branch, Dept. of Mines and Technical Survey*, Ottawa, Oct., 1964.
82. Coutinho, A., *Théorie de la détermination expérimentale des contraintes par une méthode n'exigeant pas la connaissance précise du module d'elasticité*, Proc. Int. Ass. Bridge and Struct. Eng. Congress, Zurich, 1949, 83–103.
83. Roberts, A. and Hawkes, I., Application of photoelastic devices for measuring strata pressures and support loads, *Mine and Quarry Engg.*, July **1963**, 298–308.
84. Dhir, R. K., Measurement of rock pressures in mines using photoelastic techniques, *J. Mines, Metals and Fuels*, July **1965**, 203–209.
85. Dantu, P., *Contribution a l'etude mécanique et géométrique des milieux pulvérulents*, Proc. Fourth Int. Conference on Soil Mechs. and Foundation Engg., London 1957, 144–148.
86. Hiltscher, R., Studies in the theory and application of photoplasticity, *Z.V.D.I.*, **97**, 1, 1955, 49–58.
87. Mönch, E., Die dispersion der doppelbrechung bei Zelluloid als plastizitätsmasz in der spannungsoptik, *Z. angew. Phys.*, **6**, 1954, 371–375.
88. Mönch, E., and Loreck, R., *A study of the accuracy and limits of application of plane photoplasticity experiments*, Proc. international symposium on photoelasticity, Chicago, 1961, Frocht, M. M., ed., Pergamon Press, New York, 1963, 169–184.
89. Jira, R., Das mechanische und spannungsoptische verhalten von Zelluloid bei zweiachsiger beanspruchung und der nachweis seiner eignung für ein photoplastisches verfahren, *Konstruktion*, **9**, 11, 1957, 438–449.

90. Frocht, M. M., and Thomson, R. A., *Studies in photoplasticity*, Proc. Third U.S. Nat. Congr. Appl. Mch., 1958, 533–40.
91. Frocht, M. M., and Thomson, R. A., *Further work on plane elastoplastic stress distributions*, Proc. international symposium on photoelasticity, Chicago, 1962, Frocht, M. M. ed., Pergamon Press, New York, 1963, 185–193.

NAME INDEX

SUBJECT INDEX